Unearthing the Nation

Unearthing the Nation

Modern Geology and Nationalism in Republican China

GRACE YEN SHEN

The University of Chicago Press
Chicago and London

Grace Yen Shen is assistant professor of Chinese history at Fordham University.

The University of Chicago Press, Chicago 60637
The University of Chicago Press, Ltd., London

Printed in the United States of America

23 22 21 20 19 18 17 16 15 14   1 2 3 4 5

ISBN-13: 978-0-226-09040-5 (cloth)
ISBN-13: 978-0-226-09054-2 (e-book)

DOI: 10.7208/chicago/9780226090542.001.0001

Library of Congress Cataloging-in-Publication Data

Shen, Grace Yen, author.
Unearthing the nation : modern geology and nationalism in republican China / Grace Yen Shen.
pages cm
Includes bibliographical references and index.
ISBN 978-0-226-09040-5 (cloth : alkaline paper) — ISBN 978-0-226-09054-2 (e-book) 1. Geology—China—History—20th century. 2. Geologists—China—History—20th century. 3. Nationalism and science. I. Title.
QE13.C6S54 2014
555.1—dc23   2013019832

⊗ This paper meets the requirements of ANSI/NISO Z39.48-1992 (Permanence of Paper).

CONTENTS

INTRODUCTION

Science, Nation, and Chinese Modernity

Knowledge is the beginning of practice; doing is the completion of knowing. . . . That the ancients speak both of knowledge and of doing is only because there are those who foolishly do as they wish without any understanding of thinking and self-reflection.

知是行之始。行是知之成。. . . 古人所以既說一個知，又說一個行者，只為世間有一種人，懵懵懂懂的任意去做，全不解思維省察

—Wang Yangming[1]

When the council of the Geological Society of China announced their Society emblem in the Western-language *Bulletin of the Geological Society of China*, they did so with little fanfare. The new symbol was listed with administrative business and explained quite simply:

> [The emblem] contains essentially four Chinese characters which signify *Water* (on the E), *Earth* (on the N), *Mountain* (on the W), and *Rock* (on the S). These four elements constitute the main object of geology and are well developed each in a definite region of China as indicated on the Symbol.[2]

In the Society's new Chinese-language journal, *Dizhi lunping* (地質論評, or *Geological Review*), however, the Society emblem was headlining news and required two pages of explication.[3] The central character『中』(*zhong*), meaning "middle" or "center," represented China and its Geological Society. Beyond their individual significations, the outer characters were intended to be read in series, with each combination adding a layer of meaning.

Taken in pairs, the characters on the vertical axis,『土石』(*tushi*, or "earth-rock"), referred to the "substance" (質 *zhi*) or "tangible, material object"

1. Emblem of the Geological Society of China, taken from "Dizhijie xiaoxi" 地質界消息 (Geological community news), *Dizhi lunping* 地質論評 (Geological review) 2, no. 1 (1937): 89.

(實物對象 *shiwu duixiang*) of geological investigation. Reading right to left,『水山』(*shuishan*, or "water-mountain") symbolized the forces at work in geological change.[4] Alternatively, rock represented the basement rock of the north China craton; water represented the massive layers of marine sediment formed during China's Paleozoic; mountain represented the major orogenic movements of China's Mesozoic; and earth represented the surface processes of the Cenozoic era. The horizontal and vertical axes of the Society emblem could also be read as poetic renderings of basic geological concepts:『土中石』(earth amid, rock),『石中土』(rock amid earth),『水中山』(water amid mountain),『山中水』(mountain amid water). Finally, the symbol could be understood as an image of China in miniature, with each character representing the dominant physiographic feature of the region that corresponded to its compass position around the center.[5]

From the historian's point of view, the design and significance of the Society emblem can be unpacked even further. For example, the placement of the character『中』(*zhong*) echoes traditional depictions of China as the center of civilization. Yet it is not mentioned at all in the Society *Bulletin*'s English-language description, which instead highlights the emblem's geological symbolism. This absence points to broader issues of representation and communication within Chinese geology and raises questions about the transparency of China's early participation in the international scientific community.[6] The timing of the emblem is even more telling. Within months after the *Bulletin*'s publication, Japanese forces attacked Lugouqiao (Marco Polo Bridge) and launched a full-scale invasion that the Chinese fought for eight years in the aptly named War of Resistance (1937–45). Interested Japanese geologists would easily have read the emblem as a claim for Chinese territorial integrity.

In this way, the Society emblem addressed both modern geological understandings and current political circumstances, but it was also reminiscent of much older symbols of unification. Upon establishing his capital at Beijing in 1420, the Ming emperor Yongle constructed a platform honoring the Gods of Land and Grain (社稷壇 *Shejitan*) that used five different colors of soil to symbolize the five elements (metal, wood, water, fire, and earth) and thereby represent the imperial territory: black soil to the north, red to the south, white to the west, green to east, and yellow in the center.[7] The design of the Geological Society emblem reminds us that Chinese modernization always remained in dialogue with voices from the past. Whatever other meanings lay hidden within this image and its dissemination, the small sampling above should suggest the tangle of issues involved in modern Chinese geology, and demonstrate that, for Chinese geologists, nation and science were inextricably linked.

From the late Qing dynasty (1644–1911) to the early years of the Communist era (1949–present), questions of nation and national identity dominated China's political, social, and cultural horizons. By the late 1800s, Chinese assumptions about civilization and empire appeared increasingly incompatible with full and equal participation in the world community, and the form and basis of the polity came under fire from thinkers across the political spectrum. With the imperial order no longer inviolable, both the content and value of Chinese culture were reassessed. In fact, the more the idea of nationhood took hold, the more definitions of "China" and "Chineseness" became contentious.

The role of science and technology in catalyzing this ferment was complex. During the late Qing, gunboats and textile processes shifted the military and economic balance of power. Later, sanitation and transportation measures in treaty ports challenged ideas about living standards and urban space.[8] Explorations of the countryside and its natural history that seemed innocuous at first soon felt offensive and threatening to Chinese sovereignty.[9] Though they were often only experienced in limited forms, modern science and technology did more than any other aspect of Western civilization to stimulate Chinese self-examination and change.

As science emerged as the key to new technology and modernization in the early 1900s, many Chinese began to conceptualize what was happening to them in scientific terms. For many, ideas like social Darwinism or Marxist materialism not only illuminated the mechanisms of modern imperialism, but they often also recommended themselves, with just a few modifications, as potent antidotes.[10] Thinking through the idea of science helped the Chinese devise a coherent picture of Western dynamism and a standard for calibrating their own progress. Theories as disparate as democracy, socialism, capitalism, and feminism were often interpreted through the "scientific" lenses of method, objectivity, empiricism, and natural law.[11] Science was at once a lure and a threat, an explanatory framework and a practical solution. Throughout the first half of the twentieth century, ambitious young Chinese were convinced that science would help them multiply their nation's wealth and power and catapult China into the modern world.

To China's geological pioneers, the connection between nation and science was even more basic. Whether they were collecting rocks and fossils or elucidating earth processes, they were in a sense studying China directly and fitting it into a global narrative.[12] The centrality of the land to both their research and the nation allowed Chinese geologists to toggle between China as an object of loyalty and China as an object of study. By considering the multiple valences of the land itself—as territory, resource, physical environment, and native place—this book analyzes the ways that models of science and nation converged in geological activity. Choices about one frequently affected understandings of the other, and Chinese geologists' national and scientific projects were mutually reinforcing in ways that were unimaginable to scientists removed from the particularities of place.

The Rocky Road to Science

The standard story of the development of modern geology in China starts with foreign imperialism. A select few had the insight to see that science fueled Western (and later Japanese) successes and went abroad to acquire this new knowledge. Once initiated, these "returned students" (留學生 *liuxuesheng*) then carried the techniques and theories of geology back to China. They taught these to their compatriots and dedicated themselves to both strengthening the nation and improving scientific knowledge by overcoming all obstacles to study the geology of their homeland. Some narratives stop here, while others continue on to the institutional achievements of the 1920s and 1930s, the expansion of regional geology, or notable technical accomplishments in stratigraphy, paleontology, structural geology, etc. These versions often end with the disruption of the War of Resistance. Others use the entire period from the late 1800s to 1949 to establish the solid scientific foundations of the greatly enlarged geological programs that followed under the Communist regime.[13]

Though this story is obviously a very crude collapsing of much more detailed and nuanced research, it allows us to ponder some of the inherent limitations of framing Chinese geology in terms of modernization and using it as an exemplary case of science transfer.[14] By treating science as an objective, portable "thing," the acquisition of which is an inevitability of progress, this historical approach may have room for varying degrees of success, different forms of agency, and complex hypotheses about mechanism; but the contingency is still only in the details. Moreover, this approach completely ignores questions of meaning and significance that are crucial to understanding the many parts science played in imagining (and reimagining) Chinese modernity. How did science emerge as a critical factor at all? Why did historical actors choose geology? What happened when personal, scientific, or national objectives clashed? Who had the right to speak for Chinese territory, and how did nonscientific interests make sense of the geological program?

My root contention is that, as faith in the old order was shaken and questions of nationhood came to the fore, diverse groups in China began to covet foreign science in the name of modernity long before they fully knew what science was, what it entailed, or what its impact on the broader culture might be. The "Self-Strengthening" reformers of the late Qing (fl. 1860s–1890s), who advocated the utility (用 *yong*) of scientific knowledge as a complement to Chinese essence (體 *ti*), assumed that foreign "techniques" could be abstracted from their philosophical and cultural underpinnings without any compromise or danger. In the 1920s, post–May Fourth Movement intellectuals anthropomorphized "Mr. Science" (賽先生 *Sai xiansheng*) as a savior of the young Chinese republic, and claimed that scientific attitudes and the scientific method would not only bring wealth and power, they would also transform the people and their stale, ineffectual culture.

The scientific and technical disciplines that grew up alongside these political and cultural movements were quieter and less controversial, but no less invested in the quest for a new model of Chinese modernity. Unlike

Self-Strengthening and May Fourth activists, who rarely had any technical training, China's early scientists and engineers had their initial views on science constantly reshaped by natural phenomena, international norms, the logistics of organizing research, and the experiences of daily practice. While, from a distance, science might have seemed like a neat and coherent unity, it regularly resolved itself into a series of everyday problems and unspectacular activities with as many negotiable elements as reliable facts. This was especially true for geologists, whose science defied the strict controls of the laboratory and required direct contact with often inaccessible and inconsistent physical environments.[15]

What is interesting about the development of modern geology in China, then, is not how (or how successfully) "real" science was reproduced in China, but how Chinese scientists squared the desires and realities of their local context with the practices and imperatives of active research in the field. Chinese geologists learned science by doing, and this book examines the ways that vague ideas of science in the national discourse translated into a geological culture that was meaningfully Chinese *and* global. The central concern of this work is not how science travels, but how it is refracted in a new medium, and how this process illuminates both the workings of science itself and the distinctive features of its novel surroundings.

Near the end of his study of nineteenth-century geology, Mott Greene writes that "geology in 1912 was as fragmented, from a theoretical point of view, as at any time since its eighteenth-century beginnings," and it "functioned without a universally accepted framework for decades thereafter."[16] It is an important point for Greene because it has been so tempting to see the theory of continental drift proposed by Alfred Wegener in 1912 as the beginning of the plate-tectonic synthesis that has unified so much of contemporary geology (and its related disciplines). For our purposes, it reminds us that as the Republican period (1911–49) began in China, international geology was a highly contested space.

The great debates of neptunism versus plutonism and uniformitarianism versus catastrophism that defined eighteenth-century geology had produced a general acknowledgment of the complexity of earth processes, the great age of the earth, and the importance of present-day phenomena for puzzling out the geological past. But there was little agreement on key issues, such as the nature of crustal movements, their relationship to the internal structure of the earth, the mechanisms of mountain building, and the role of physics in geological theory. Divisions were further underscored by what Homer LeGrand calls the "localism" of place and training, and major geological communities in Britain, the Continent, and America each relied on their own mix of methodological commitments to develop and weigh new theories. Despite well-established practices of fieldwork and data collection, there was no one "Western" theory of geology to master or debunk, and "Chinese geologists . . . like their Western colleagues, were free to pick and choose theories from rather overstocked shelves."[17] It would clearly be foolhardy to think of the development of modern geology in China in merely kinematic terms when the object in motion was so unstable, but even more remiss not to consider geology's own growing pains from the unique perspective of the Chinese case.

Time and again, studies of geological controversies have given us windows into how geology developed into a distinct scientific enterprise; how social networks and class shaped scientific authority; and how major theoretical shifts were integrated into geological practice.[18] Not surprisingly similar insights can be gained from the contests and stresses involved in the emergence of new geological communities. As outsiders developed an interest in geology and tried to access the mainstream of the science to satisfy their own objectives and circumstances, they began to bump into previously invisible assumptions, and their activities often stretched the boundaries of the discipline. Coinciding as it did with the disunity and flux of early twentieth-century geology, the Chinese experience allows us to look at critical issues like the geographical contingency of geological theory, the cultural specificity of scientific sociability, and the geopolitics of expertise without constant reference to the future synthesis of plate tectonics.[19] Instead, the fragmentation of geology played out in the variety of influences available to the Chinese, and we can follow native geologists as they explored a range of credible paths to scientific independence.

The emergence of modern Chinese geology was inarguably shaped by the context of imperialism, but it did not follow the classic model of colonial science.[20] Foreign powers found it more expedient to carve out spheres of influence than to openly colonize Qing territory, and they did not establish full-fledged scientific outposts.[21] Thus, despite the persistent activity of foreign expeditions in the late Qing era and the early Republican period, Chinese understandings of geology were never dominated by any one scientific tradition. In fact, the four foundational figures of Chinese geology, Zhang Hongzhao (章鴻釗), Ding Wenjiang (丁文江), Weng Wenhao (翁文灝), and Li Siguang (李四光), studied geology in Japan, Scotland, Belgium, and England. Their most intimate foreign collaborator, Amadeus Grabau, further contributed an idiosyncratic blend of American and German ideas to the mix, and this influenced generations of younger Chinese geologists.

Contrary to the standard colonial model, Chinese researchers established the country's first geological institutions themselves, and China's pioneering geologists began training their own technical personnel as soon as the new republic was founded. By 1920 the Chinese were inviting well-respected figures like Grabau, Johan Gunnar Andersson (Sweden), and Teilhard de Chardin (France) to work as guests within flourishing native organizations. However, the Geological Survey of China (est. 1913), the Research Institute of Geology (est. 1928), and other important academic centers remained overwhelmingly Chinese, and they all struggled with a common set of problems: the scarcity of qualified native researchers, the fickle interest and weakness of political authorities, and the unreliability of access to key territories. Fortunately, Chinese exposure to a wide array of geological traditions allowed them to respond flexibly to shifting domestic realities, and they were generally able to move between alternative and equally fruitful lines of research whenever the situation demanded. To borrow (and adapt) a term used by Naomi Oreskes in her analysis of the American geological community, geology's lack of a unified theoretical framework was "enabling" for Chinese geologists, and it gave them theoretical and methodological options to "do their science" in the face of political and economic uncertainties.[22]

On the other hand, the fragmentation of geological theory made it difficult for Chinese scientists to find a secure footing in the international community. Established geological communities could bicker about facts, interpretations, and meaning without compromising their legitimacy, but newcomers had to prove themselves and, in the absence of a recognized norm, they could never be sure of their standing. Chinese research was always welcome, but Chinese researchers had to fight to be noticed, and China's newly minted geologists spent much of the Republican period deciding whether it served the nation better to assert themselves or to assert that they were exactly the same as everyone else.[23] Geology was global in scope, but like so many other sciences it was still largely national in practice, and Chinese geologists struggled with the possibility that the "scientific universality" they longed for might simply be "international science."[24]

A Nation Never Set in Stone

Theories of nationalism (like many influential theories of science) generally presuppose that the phenomenon is radically modern, and, because they trace its origins to Western European intellectuals, the Western European model becomes something of a standard. Non-Western cases are commonly analyzed in the context of colonialism (and postcolonialism), and most influential theories have either neglected the Chinese case completely or made sweeping statements about it in order to emphasize a more proximate point.[25] Few, if any, major theories of nationalism have been able to account for the forms of self-conscious, politically bound, and often totalizing community that have existed in premodern East Asia and China in particular.[26]

Like Ernest Gellner's nation-states struggling for congruence between culture and polity,[27] the imperial Chinese state based its legitimacy on a universal ordering and civilizing project that promoted Confucian cultural norms.[28] Chinese elites participated in a literary and cultural world that transcended the boundaries of village or province and, as inscriptions on stelae frequently attest, they shared something akin to Benedict Anderson's idea of "anonymous solidarity" from a fairly early period.[29] Arguments can even be made for "print-capitalism" in the Song dynasty (960–1279), and it is well-known that the triennial ritual of civil-service examinations brought scholarly "pilgrims" to the capital before dispersing them across the empire as state officials.[30] However, these unifying and identity-forming practices oriented participants toward emperor, class, and worldview, rather than "nation" (國家 *guojia*), and, historically, the Chinese referred to their territorially extended community by dynasty (朝代 *chaodai*) or as simply "all under heaven" (天下 *tianxia*).

Clearly "nation" in the sense that interests us—as a dominant, automatic, and territorially restricted self-identifier—was as new to Chinese civilization as it was elsewhere. But what do we make of the vast history that comes before and its relationship to (and meaning for) the Chinese of the modern period? Were historical forms of collective identity viable alternatives to Western European nationalism? Did those indigenous identities prefigure the future nation? Or did all Chinese of the late nineteenth and early twentieth centuries who aspired to a modern nation have to break with their incorrigible past? As scholarly titles such as "How China's Nationalism Was Shanghaied," "The Incoherent Nation," and *Rescuing History from the Nation* suggest, the only consensus among contemporary scholars is that Chinese nationalism has been problematic.

So what can we say about—let alone learn from—the national and scientific projects of Republican Chinese geologists? In the absence of a universally satisfactory theory of nationalism, are our hands really tied?

To make some headway, it is useful to step back and begin again in medias res. Republican Chinese geologists were openly preoccupied with defending Chinese interests and found a common purpose in accepting responsibility for "saving the nation through science" (科學救國 *kexue jiuguo*). Many of them experimented with other forms of patriotic service (medicine, politics, journalism, etc.), but ultimately they coalesced around the physical, empirical, and social imperatives of modern geology as established by early pioneers who returned from abroad.

From this, three key features stand out. First, Republican geologists and other Chinese scientists were part of a broader movement of elite urban nationalists (including overseas Chinese and rural gentry with ties to treaty ports) who were shaped by contact with foreign ideas and peoples.[31] Rather than thinking of this as a dent created by some Eurocentric notion of "impact," we should see this as a "China-centered" outgrowth of our actors' personal experiences and engagement.[32] More importantly, understanding Chinese geologists (including those late Qing individuals who first turned to geology as a critical problem) as part of this outward-looking moiety allows us to explore the terms "China" and "nation" as actors' categories.[33]

William Kirby points out that "linguistically, we know that China was 'China' abroad earlier than at home," but Chinese geologists were quick to appropriate this terminology as a suitable way to communicate the object of their own attachments, for this object was distinguished by its silence anyway.[34] As we shall see, China as a nation was first perceived by would-be geologists as a site of local inaction; it was an entity whose significance was felt when outsiders studied it firsthand and local residents suffered from their own shameful ignorance. For those Chinese who first championed modern geology, the very act of feeling China's loss as one's own assumed both the existence of the nation and the membership of the self in that unit, and this provided a starting point for constructive action.

Second, as part of the larger backdrop of urban nationalism, Chinese geologists rejected the Confucian values of the political and social order and associated them with parochialism and complacency. However, they not only accepted the deeply Confucian role of the intellectual as servitor-cum-guide to state and society, but they also managed to identify this role with progressivism and morality by taking it as a call to self-criticism and renewal. Though Chinese geologists (especially soil scientists) were extremely attentive to agricultural needs, Chinese geology can be seen, in part, as a program for claiming the land from quiescent elements using active, empirical methods. If the idea of China as a nation emerged from the site of injury (i.e., the unresisted imperialist gaze and its practical consequences), then geologists' shared sense of Chineseness grew out of their admission of guilt and their dedication to self-transformation. Geology was a discipline that would reshape its practitioners and resuscitate the nation on the verge of extinction.

This third and perhaps most revealing feature linked Chinese geologists to other urban elites, who identified with China's frailty and couched their own national missions in terms of survival and salvation, rather than simply improvement or renewal. The formula "X to save the nation" was a constant refrain from the 1920s to the end of the War of Resistance, and it recalled a litany of humiliations and threats extending back to the Opium War. It also warned of future dangers in the "struggle for existence" and proposed an endless roster of exotic cures and prophylactics, such as "women's education," "art," and "physical fitness." However, as John Fitzgerald has emphasized, it never specified "who or what was the nation to which [they] . . . all referred when they resolved to 'save the country' (救國 *jiuguo*)."[35] Was it one of the tenuous postimperial regimes that supported foreign treaty claims in China's name? The Han ethnicity? The Chinese "spirit"? Or some unspecified portion of former Qing territory? Could it be all of these, and was it necessary to decide?[36]

My intuition in this study is that our desire to find concrete definitions of "nation" for the sake of logical coherence misconstrues the climate of the times in late Qing and early Republican China, and stymies productive investigation so much that many scholars are choosing to eschew the problem of modern Chinese nationalism for other less monolithic and more "natural" collectivities (e.g., region, ethnicity, voluntary organizations, diasporas, etc.). As Prasenjit Duara pointed out at the beginning of this turn, historians must give up theories of nationalism that accept the nation as not only the *subject* (self-conscious, organic agent) of history, but as *the* subject (hegemonic, durable, and inevitable). Instead we must try to uncover what other meanings and narratives have been suppressed in the assertion of specific designations of Chineseness, and our work should emphasize the processes of contestation that have accomplished this.[37]

I echo this commitment to plurality and its implicit rejection of determinism, but not the emphasis on contention and exclusion. If we take the refrain of "national salvation" seriously, we find an emphasis on acting rather than defining, and while many groups fought to limit the meaning of the "China" they wanted to save, just as many others made use of the concept's open-endedness to rally in greater numbers For them, the idea of the Chinese nation served as a placeholder for future designation when the feeling of peril no longer constrained their options, and the content of "China" was expected to emerge from their timely exertions.

Rather than decentering China and straining to hear the cacophony, I focus on what historical actors were able to agree upon, and listen for improvised riffs in the resulting calm. In this sense, I am taking inspiration from Emily Honig's work on Subei people in Shanghai.[38] Though the category "Subei ren" (or, actually, "Jiangbei ren") was pejorative and not exalted like that of the nation, it functioned as if it were real, well-defined, and objective precisely because its multiple meanings allowed for slippages from one referent to another. Its effective force derived from Shanghai users' determination not to be pinned down, and Honig allows the elements of class and ethnicity built into the Subei identity to emerge from colloquial practice.

In my own research the relevant practices were not linguistic but scientific, and it seems that, if by "identity" we mean a thing which is adopted because its content matches our self-image, then Republican geologists did not look to the nation for identity. The nation, as they often lamented, was few of the things that they took it upon themselves to be: active, modern, progressive, independent, effective. Instead, the nation was an object of loyalty, an entity bracketed out by a sense of crisis as the ultimate prize to be protected.

Like other urban elites who framed modernization and nationalism as problems of existence, Chinese geologists acknowledged "China" as a special category demanding action on its behalf. But unlike ideologues who fought over definitions of the nation to sanction their undertakings, geologists had a ready-made mission and did not have to suppress their chaotic attachments and aspirations to function as a community. Chinese geologists were aware of their different affinities—Zhang Hongzhao was nostalgic for classical learning, Weng Wenhao longed for a stable political form, Li Siguang sought ideological conviction, and Ding Wenjiang detoured into eugenics and anthropology to understand race—but they recognized each other as committed patriots, and, joined by a common scientific practice, they were comfortable filling "China" with different dreams.

In *The Spirit of Chinese Politics*, Lucian Pye writes that "in the modern era the Chinese have had little doubt about their identities as Chinese, and the more they have been exposed to the outside world the more self-consciously Chinese they have become."[39] Couched in terms of "the Chinese" as a whole (or, worse, as a fate-bound collective) this is a controversial point that is difficult to support in the face of twentieth-century paroxysms.[40] Considered in terms of Republican geologists, however, there is more than a little that rings true, for they did not worry about whether they were Chinese; they worried about what to do for China and why.

To continue with Pye's assessment:

> The basic problem in development for the Chinese has been that of achieving within their social and political life new forms of authority which can both satisfy their need to reassert a historic self-confidence and also provide the basis for reordering their society in modern terms.[41]

Here, Pye himself is concerned with the crisis of political authority, but his words suggest other possibilities. Once Chinese geologists linked the survival of the nation to science, science became a new and hopeful form of authority that opened the door to both reassertion and reordering. As a group, geologists' nationalism was defined by China as an object of loyalty, and geology as a framework for expressing it. Any viable understanding of the nation had to suit the twin criteria of protecting Chinese existence and promoting geological activity, and often the boundaries of one effort would shift to accommodate the other.

This does not mean that Chinese geologists tried to construct a unique identity around a scientifically sanctioned theory or phenomenon, nor does it indicate an attempt on their part to naturalize the nation through geohistory or geography.[42] Such narratives look for construction of a unified vision and do not fit the Republican Chinese case, which permitted multiple viewpoints under the guise of an uncontested patriotism. Geologists like Ding, Weng, Li, and Yang Zhongjian (楊鍾健) were all influential and outspoken public figures with their own styles of political activism.[43] But, as Peter Perdue suggests, the practices of nationalism can be more eloquent than "the pontification of intellectuals," and what is important is not what these men said as social critics, but what they did in the context of everyday science.[44] Geology had rules and demands of its own, and as they negotiated these within the constraints of their existing resources. Chinese geologists often expressed preferences and adaptability regarding the nation in unpremeditated ways.

What I am interested in, and the focus of this book, is the site where voices and potentialities implicit in the devotion to "China" were activated and expressed through the immediacy of their contact with nature and the practical choices this entailed. Though China as an object of loyalty is front and center in my approach, China as a self-conscious nation is not the subject of my study. In trying to understand the ways that single-minded dedication to science and patriotism interacted to project multiple images of Chinese modernity, my primary concern is how Chinese geologists respond to the changing international and domestic circumstances of their work.

With this goal in mind, the Republican period is the perfect time to observe the diversity and mutability of attachments bound up in the idea of

"China." The Republican period was dominated by the search for a new national form that would legitimize its political authority over a unified China, but none of the central regimes that nominally ruled the country before 1949 accomplished this convincingly. The Guomindang (Nationalist Party) government (est. 1928) came close, but in real terms, the country was divided by regional warlords and political factions from the end of dynastic rule until the founding of the People's Republic.

Though nationalism preoccupied Chinese intellectuals, the absence of a compelling central power made room for a wide range of relational claims, including race, native place, and historical lineage, to coexist. The weakness of the Republican government provided a unique combination of latitude and urgency that channeled geological efforts toward the nation without binding them irrevocably to any ideology of national identity. Once a unified, penetrative state that could monopolize allegiance in China's name was established, the diversity of opinions and affections that gave the troubled Republican period its resilience and energy looked disturbingly like shades of disloyalty.

Map of the Chapters

The key link between loyalty to China and the authority of geology in the Republican period was the land. As Chinese geologists reconfigured their working relationships with the land, they foregrounded different meanings of "Chinese geology" and entertained multiple possibilities for the shape of Chinese modernity. Though the chapters here follow a roughly chronological order, each one is organized around a problem of geological development tied to the land but conceived and addressed through the imperative of saving the nation.

Chapter 1 sets the stage by asking why a small number of Chinese became interested in modern geology at all. I begin by tracing Qing-era modes of thinking about the empire, its spatial diversity, and territorial knowledge. I then link early foreign geological exploration in China to subsequent territorial and economic encroachment, and explain how the mechanical attempts of Self-Strengthening scholar-officials to introduce the methods of geology and mining failed to generate domestic interest. I argue that the desire to learn modern geology only arose when Chinese students with overseas experience began to view their territory as a vast unknown and reconceptualize their homeland as an object of study. As such, the immediate political and economic challenges posed by foreign domination demanded attention to the materiality of the land—its structure, composition, history, and properties—rather than the simple location of mineral resources.

The materiality of the land required a new, physical way of interacting with it, and chapter 2 examines the role of fieldwork in grounding the Chinese geological enterprise. I follow Zhang Hongzhao, Ding Wenjiang, Weng Wenhao, and Li Siguang from their study-abroad experiences through their establishment of training and research institutions under the new republican government. Fieldwork was not only important as a critical skill for conducting geological research, but it was also the primary means for transmitting geological concepts to uninitiated students and assimilating Western learning to the native environment. I argue that the strenuous physical nature of fieldwork challenged traditional Chinese notions of proper behavior and helped reinvent the modern Chinese intellectual as a vigorous, active individual capable of productive labor. Better geologists equaled better men for a new China.

Fieldwork might have given geologists a new relationship to the land rooted in direct experience, but geologists still could not reach an international audience or stem the tide of foreign exploration. Chapter 3 analyzes Chinese efforts to forge community and participate in international science, from early Chinese proposals for geological organization to the Geological Society of China's high-profile activities in the 1920s. Rather than seeking to displace foreign researchers, the Society framed Chinese territory as a shared concern and made themselves indispensable as hosts to otherwise competitive research interests from around the world. This move gave Chinese geologists exposure to foreign research networks and access to foreign investigations of remote frontier regions. I argue that by projecting a unified image of China and a cosmopolitan vision of Chinese geology, the Chinese actually strengthened their claim to represent their home territory to the world.

Having established itself as an active research community and carved out a niche in international scientific circles, the Chinese geological community then struggled to preserve its autonomy from increasingly central political institutions. Chapter 4 looks at how Chinese geologists tried to balance their own ideas of geology to save the nation and the growing demands of the Guomindang state in the "Nanjing Decade" (1927–37). The Geological Survey of China used its access to nongovernmental funding to set its own agenda while giving lip service to the central administration, but as war with Japan seemed imminent, Weng Wenhao entered into high-level economic planning and blurred the lines between the Survey and the state. In contrast, the Research Institute of Geology led by Li Siguang was founded under Guomindang auspices but quickly distanced itself from the regime's instrumentalist approach to science. I argue that though their strategies were radically different, both the Research Institute and the Survey extended the territorial scope of Chinese geology and pursued the kind of research they perceived to be in China's best interests.

Chapter 5 examines how Chinese geologists dealt with the Japanese occupation of China's coastal regions during the War of Resistance. Here I describe the great inland migration of Chinese scientific institutions and the problems of wartime scarcity. Under these conditions, Chinese geology expanded to include new members without the same intellectual pedigrees as the prewar community, and existing institutions had to develop new mechanisms for initiating them into an active research community. The move to the Interior also opened up previously inaccessible frontier territories to Chinese geological investigation, and the urgency of China's fuel shortage brought the Geological Survey to these outlying areas in search of petroleum. I argue that far from being a setback to the development of Chinese geology, the wartime experience galvanized the community and gave it the confidence to privilege the physical evidence of the earth over naysaying foreign authorities. However, the isolation of war also narrowed Chinese focus, and Chinese geology showed signs of becoming strictly the geology of China by the Chinese, rather than an open program of scientific exchange.

I conclude with the Chinese geological community's postwar efforts to find a new sense of mission as Guomindang and Chinese Communist Party forces vied for control of China. In contrast to the wartime era, when research continued despite all obstacles, the domestic conflict ground fieldwork to a halt, and faced with the choice between two Chinese states, Chinese geologists chose the Chinese land.

ONE

The Lay of the Land: Territoriality, Foreign Exploration, and Modern Geology in Qing China

> Our Celestial dynasty possesses vast territories, and tribute missions from the dependencies are provided for by the Department for Tributary States, which ministers to their wants and exercises strict control over their movements. . . . If you assert that your reverence for Our Celestial dynasty fills you with a desire to acquire our civilization, our ceremonies and code of laws differ so completely from your own that, even if your Envoy were able to acquire the rudiments of our civilization, you could not possibly transplant our manners and customs to your alien soil.
>
> —Emperor Qianlong to King George III, 1793[1]

By the late Qing period, it was clear to the ruling elites that natural resources such as coal and iron not only fed foreign military and economic power, but they also drove imperialists' appetite for territorial control. It is therefore easy to take for granted that geology was introduced into China as an instrument of domination and appropriated by the Chinese themselves as a form of resistance. Foreign powers used their political and military clout to gain control of China's mineral resources and carve out spheres of influence in strategic areas of Chinese territory. The Chinese, for their part, learned to fight fire with fire and ultimately adopted Western methods to stave off further encroachment.

The rough outlines of this familiar plot are not *wrong*, and yet the story's air of inevitability obscures as much as it reveals. It is unclear, for instance, how the image and practice of geology were actually connected to the aforementioned forms of economic and political aggression in late Qing China. Did foreign powers deploy geologists systematically? Was the contest for resources a cause or an effect of Western (and later Japanese) imperialism?

Why did geology come to dominate Republican Chinese discourse instead of geography or mining, which were as—if not more—implicated in the economics of imperialism? And if geology had something special to offer, why did this only become clear in the last years of the dynasty?

The backdrop of imperialism turns out to be a bit of a red herring, and falling for it exposes our own tendency to see local adoption of modern science as an obvious, natural, even necessary response to foreign power. Instead of retracing this narrative and trying to plug geology into a ready-made progression, I want to take a completely different tack and consider the shifting views of Chinese territory that made geological investigation of China meaningful in the first place. What made a particular group of Chinese suddenly *want* geological knowledge in the late Qing? Though inextricably linked to the legacies of "gunboat diplomacy" and "unequal treaties," the Chinese thirst for wealth and power did not automatically translate into geological activity; if anything, it directed attention toward fields like law, economics, and industrial management, which were both more practical and more familiar.[2] Ultimately, the advent of modern geology in China was an unexpected outcome of a new way of thinking about territory and belonging, rather than a foregone conclusion of modernization. To understand this shift, we must dig deeper into the ways that different parties' access to and familiarity with the land continually reshaped notions of Chinese authority and sovereignty.

Until the beginning of the eighteenth century, Westerners who were curious about the nature of China's landscape willingly deferred to the authority of Chinese geographical sources. After the Opium Wars, foreigners gained privileged access to the interior, and Chinese territory was suddenly opened to Western experience. As eager travelers began to explore the previously remote terrain, they exploded the belief that it could be satisfactorily understood (or managed) on Chinese terms alone. Soon Westerners were not only studying Qing China for themselves, but they were also using their findings to wrest control of the mineral wealth within its borders.

Even so, the rush of new researches brought on by the expansion of foreign influence was frequently more opportunistic than strategic, and it represented quite a mixed bag of motives, disciplinary approaches, and expertise. The semicolonial, multilateral nature of foreign interests in China presented a haphazard, often inscrutable face to Chinese modernizers, and did not offer a coherent model (let alone a transformative scientific program) for them to appropriate. Whether they aroused admiration or alarm, those aspects of Western field studies (including geography, geology, mining, natural history, anthropology, etc.) that caught the attention of Chinese elites were filtered through existing frameworks of both natural knowledge and territorial dominion until the very end of the Qing dynasty. When geology finally made a lasting impression as a distinct discipline, it did so with young political outsiders who no longer took either of these frameworks for granted and who demanded new forms of knowledge to suit their newfound sense of territorial responsibility.

Chinese Territoriality under the Qing

Throughout Chinese history, the imperial houses of the east feared invasion from the north and northwest, and Chinese dynasties were repeatedly conquered by groups from the Central Asian steppe or Manchurian plain. Despite these lingering hostilities, China traditionally "faced west," with its "front door" at Yumen Gate in Gansu Province rather than at any of its coastal cities, and westward trade flourished along the Silk Route through Chinese Turkestan and onward toward the Middle East and Europe.

The Qing ruling house, which originated outside of the Great Wall, among the Jurchen tribes of Manchuria, was uniquely positioned to straddle these contrasting worlds of steppe and sown. Through long contact with the Chinese to the south, the Manchus turned to agriculture in the fertile inner plains between their bordering mountains to supplement their traditional livelihood as forest nomads. They established a written language and developed both a Mongolian-inspired banner system and a protobureaucracy along Chinese lines. By the time they declared their imperial intentions in 1636, the Qing ruled the Manchus, Eastern Mongols, and frontier Chinese, and they already held Korea as a vassal state.

When the rebel Li Zicheng captured the Ming dynasty capital at Beijing in 1644, Qing forces seized the opportunity to drive out Li and quickly installed themselves as the new rulers of agrarian China. Over the next forty years the Qing fought Ming loyalists and former collaborators until they finally quelled Chinese resistance in 1681. Meanwhile, during the reign of emperor Kangxi (康熙, 1661–1722), Mongolia, Chinese Turkestan, and Tibet were incrementally incorporated, and the completion of this conquest in the 1750s under the emperor Qianlong (乾隆, 1736–96) marked the golden age of the Qing.

We can see the complexity of Qing rulership in its administrative machinery. Top positions of the Qing imperial bureaucracy were filled by dual Manchu and Chinese appointments, and Mongols, Manchus, and Chinese collaborators from the north were organized into hereditary banners that constituted a standing military loyal to the emperor. To handle relations with tribute states—including Korea, Vietnam, Thailand, Burma, Nepal, and Afghanistan—the Qing took over the "Son of Heaven" mantle and preserved the Ming dynasty's Board of Rites; but for the governance of the northern and western frontiers of Inner Asia, the Qing retained its own Lifan Yuan (理藩院, Ministry of Outer Dependencies), a homegrown institution put in place well before the conquest of the Ming.

Despite this territorial diversity, possession—or the pretense to possession—of eastern, agrarian China (or "China proper"), has been the common thread running through standard renditions of Chinese history, and the Middle Kingdom concept is rooted in this region's historical sense of cultural hegemony and exceptionalism.[3] Not surprisingly, then, the process by which the Qing dynasty brought Han Chinese and groups the Han considered "barbarians" under one durable, multiethnic empire has often been understood in terms of sinicization. Qing Empire building, according to this view, depended as much on its speedy adoption of long-standing Chinese traditions of legitimation and governance as on its military prowess, and the territories conquered between the mid-seventeenth and mid-eighteenth centuries were cultural and political frontiers revolving around the enduring Han Chinese center of civilization.

More recently, this trope of sinicization has given way to the "new Qing history," which frames agrarian China as one domain among many in a multilateral imperial enterprise centered on the Qing ruling house.[4] One key to the success of this balancing act seems to have been the Qing emperor's ability to "do one thing while appearing to do another." In particular, the Qing often "drew on Inner Asian rituals that superficially resembled Chinese practices" or legitimated Inner Asian rituals for Chinese audiences by teasing textual support out of Chinese ritual texts.[5] Thus, in one and the same act of imperial power, the Qing court could feed the Chinese belief in its "sinicization" while showing Inner Asian constituencies that it had indeed domesticated the Middle Kingdom.

Interestingly, by highlighting the agility of Qing rulership, this new Qing history forces us to think about the ways in which the Chinese were consuming ritual representations, and it reminds us that Chinese territoriality under the Qing was not only an amalgam of internal differences and history, but it was also a balance between the practices of empire and what Michel de Certeau has called the "practices of everyday life."[6] While the Qing were tailoring different depictions of their empire to specific audiences, Chinese farmers and literati alike were busily filtering the imperial message through countless daily operations, from farming to speech acts to local administration. Thus, even though Chinese attitudes toward the land took into account the Qing dynasty's spatial ambitions, they were largely anchored to the familiar and contingent on context.

When most Chinese in the provinces imagined the extent of their country, or *guo* 國, it was less a matter of political or physiographic units than a mosaic of conceptual associations between meaning and place. Thus, northern "noodle-eaters" traveled in carts on dry roads and slept on heated brick beds known as *kang*, while southern "rice-eaters" often moved by boat on small local waterways and lived in houses designed to maximize shade. Both in terms of climate and trade, northern China was dominated by Inner Asia and southern China was dominated by the sea. In classical times, northern China was the center of cultural activity and southern China was home to several non-Han peoples. As nomadic incursions pushed Han southward, these tribes were incorporated or displaced, and by the Qing dynasty southern China was by far the wealthier, more dynamic region of the country, famed for both its scholarly and commercial achievements.

Whether north, south, or elsewhere, faraway locales were often identified with scenic spots, famous temples, sacred mountains, and the lore surrounding them. Even provinces were identified by historical nicknames rather than their official designations, like Wu-Yue (吳越, two of the Warring States of antiquity) for Zhejiang Province. Since most Chinese did not travel far from home, places were frequently defined by the people or products that originated there and circulated across provincial boundaries.[7] Thus certain regions were associated with specific professions or practices, such as Shanxi and banking or Cangzhou and martial arts, and other places were known through their local varieties of tea or through handicrafts like Yixing teapots.

More generally, Chinese frequently emphasized that their society was built upon agriculture (以農立國 *yinong liguo*), and foreign observers in the Qing were indeed struck by the close relationship between Chinese peasants and the land. Over the course of the dynasty population pressures tightened this bond and many farmers struggled for subsistence. In such a system, few Chinese outside of the educated and privileged classes had the leisure to think about Chinese territory in abstraction. For most, land was an engine of production and a measure of social relations. Materials such as coal and iron were exploited on a small scale for local needs, and people were careful not to disturb the soil unnecessarily, for land was either needed for farming or invested with geomantic significance.

Concepts of territorial space were frequently ambiguous in concrete terms and yet very clear as designations of self and other. Chinese, for instance, often separated themselves from outside groups by distinguishing between *saiwai* (塞外, lit. "beyond the strategic [Great Wall] passes") and *guannei* (關内, lit. "inside the gates") even though by the Qing dynasty the Great Wall had long ceased to demarcate Han from non-Han.[8] Within China proper, terms like *jiangbei* (江北, meaning "north of the Yangzi River") became derogatory identifiers that referred to place without being bound to any specific or stable location (so-called Jiangbei people lived both south and north of the river and not all those to the north were included in this group).[9] For the majority of Chinese who never left their native areas, China was defined by these associations of places and attributes rather than by maps or political boundaries.[10]

Those few who did travel across the country, whether by privilege or necessity, produced an extensive body of travel literature that influenced both the cultural imagination of Chinese territory and its geographic mapping. Information on trade and military routes could be found in merchant manuals and military histories, but the genre of travel writing offered personal observations of scenery, historical detail, maps, and topographical descriptions, as well as musings on the causes of and relations between observed physical features.[11]

Many of these works were written to commemorate foreign missions, religious pilgrimages, the assumption of distant bureaucratic posts, or periods of exile. Others were the product of literati aesthetes engaged in leisure travel to famous ruins and scenic sites such as West Lake and Mt. Tai. A few travelers, such as Xu Xiake (徐霞客), were known for their trailblazing, and in these writings descriptions of scenery often mix with observations of geographical and geological significance. Xu, who explored much of China proper but devoted the greatest attention to the southwestern province of Yunnan, distinguished the channels of the Salween and Mekong Rivers and was the first to discover the true source of the Yangzi River. Though in subsequent generations his writing was primarily read for its literary quality, Xu also identified several notable ore-bearing bodies and provided accurate-enough information for the geologist Ding Wenjiang to verify many of his observations in the early Republic.[12]

For the Qing court and its territorial administrators, travel literature was informative, but maps and atlases were necessary to manage the empire. Western visitors from the seventeenth century onward often marveled at the volume of geographical material produced in China, and a great deal of the information transmitted to Europe by Jesuits and other visitors was based on existing literature. Local gazetteers, called *fangzhi* (方誌), were an almost ubiquitous geographical genre, and official versions were sponsored at every administrative level from the province down to the county. Many more were privately commissioned by wealthy gentry, and there were also gazetteers for famous mountains, shrines, academies, and market towns.[13]

Besides maps and plans of the region in question, gazetteers typically contained detailed descriptions of topographical features, famous places (such as ruins or temples), transportation routes, waterworks, and official buildings, as well as records of natural history.[14] Gazetteers also catalogued statistics such as population (by household), taxation, market prices for important commodities, meteorological information, and agricultural production. Matters of regional interest such as customs, festivals, linguistic variations, successful examination candidates, and biographies of local worthies were also included, as were details of recent events such as uprisings, floods, or droughts.

Gazetteers came into their own as a recognizable genre in the Southern Song dynasty and became increasingly popular in the Yuan and Ming, but more of them were produced in the Qing than in all other dynasties put together.[15] In part this stemmed from the Qing dynasty's attempt to consolidate their vast multiethnic state and to inventory their Chinese and Inner Asian conquests. However, gazetteers also flourished as a result of early anti-Qing resistance on the part of Ming loyalists like Gu Yanwu (顧炎武), whose *Tianxia junguo libing shu* (天下郡國利病書 Treatise on the advantages and disadvantages of the commandaries and states of the empire, 1662) used local gazetteers to catalogue the effects of topography on the political and economic development of each province in an attempt to assess the causes of Ming defeat. Gu Yanwu spearheaded the school of evidential research, which dominated Qing dynasty scholarship, and despite the uneven quality of privately sponsored gazetteers, his emphasis on both careful philological criticism and firsthand observation pushed the genre to new heights.[16]

Of course the Qing court was well aware of the advantages of territorial reconnaissance, and it commissioned several empire-wide geographical studies, the results of which it disseminated to specific audiences in characteristically selective ways. The first of these was initiated only two years after conquest of the Ming in order to administer land use and taxes.[17] The greatest of the surveys, begun under the Kangxi emperor in 1672 and finished in 1743, involved a revision of all provincial gazetteers as well as the production of the *Da Qing yitongzhi* (大清一統志, or *Unified Gazetteer of the Great Qing*).[18] This process was repeated twice under the Qianlong and Jiaqing emperors in 1784 and 1820. Even more notable were the Jesuit-led mapping projects of the Kangxi and Qianlong reigns.[19] From 1708 to 1718 and then again from 1756 to 1759, Jesuits under imperial command conducted trigonometric surveys of the expanding Qing Empire and the resulting maps formed the basis for all domestic and foreign maps of China until the Republican period.[20] The distribution of these Jesuit maps communicated Qing territorial claims throughout Europe in much the same way that Peter the Great's mapmaking communicated his vision of a west-facing Russia.[21]

From Imitation to Imperialism

Before Jesuit contact with the Ming dynasty in the late seventeenth century, European knowledge of Chinese territory was largely based on occasional reports from merchants with trade contacts there. Most were not firsthand accounts, and it was not until the mid-1600s that European scholars firmly associated Cathay and China.[22] Jorge de Barbuda's influential 1584 map of China in the *Speculum Orbis Terrarum* of 1593, depicted an almost random array of inland mountains and waterways in China because the Portuguese navigators who supplied Barbuda with data were only familiar with coastal areas.[23] When Father Martino Martini published his *Novus Atlas Sinensis* in Leiden in 1655, its marked improvement in accuracy derived mainly from careful study of Chinese gazetteers and conveyed a detailed picture of China's heavily populated inland provinces.[24]

From the outset of their mission in China, Jesuits maintained an active correspondence with European scholars and introduced Westerners to elements of Confucianism, Chinese customs, and Chinese arts. Preoccupation with philosophical and spiritual matters surrounding the Rites Controversy of 1715 generally eclipsed geographical material in Jesuit writings until the great Qing surveys of the early eighteenth century. A notable exception was Athanasius Kircher's 1667 *China Monumentis Illustrata,* which included reports of Johann Grueber and Albert D'Orville's travels from Beijing to Tibet in 1661, in addition to the well-known geographical descriptions of Martini and others.[25] After the Kangxi and Qianlong surveys, Father Jean Baptiste du Halde published cartographer J. B. B. d'Anville's rendering of Jesuit survey results in his four-volume *Description géographique, historique chronologique, politique, et physique de l'empire de la Chine et de la Tartarie Chinoise* (1735). Neither du Halde nor d'Anville ever visited China, but the synthesis of available knowledge in *Description* dominated Western understanding of China for several decades, greatly impressing Voltaire and other influential thinkers.

European interest in Chinese territory during the seventeenth and eighteenth centuries was part of a growing desire to picture and catalog the world. Fueled by technical innovations in determining latitude and longitude as well as developments in cartographic representation, mapping was an important part of both absolutist nation building and Enlightenment rationalism. In this schema the unique role China played in the European imagination was reflected in Western attitudes toward Chinese territory.[26] China represented Europe's idealized Other, an idyllically rational society governed since antiquity according to ethical, meritocratic, and efficient systems.[27] Accordingly, lacunae in Western geographical knowledge about China did not yet constitute a terra incognita in the same way that gaps in knowledge of the New World and Africa did. Mapmakers and philosophes alike acknowledged that even if China was unknown to *them*, it was not "unknown" for it was filled with an ancient civilization of profound sophistication and a teeming population of unimaginable size and productivity.

By the late 1700s, territorial acquisitions and the expansion of maritime trade bolstered Europe's cultural confidence, and as knowledge of China increased, criticisms of its decay and despotism multiplied. Diderot and Condorcet contended that China's social systems limited its ability to adapt and grow, and Adam Smith and Thomas Malthus called the robustness of China's economy seriously into question. This new picture of stagnation and decline inverted earlier praise for China's enduring and timeless society, and made liberal thinkers less willing to engage China exclusively on its own terms.[28] Once the Jesuits were suppressed in 1773, commercial interests dominated Europe's contacts with China, and differences in economic philosophy demonstrated by Britain's failed Macartney Mission (1793) and Amherst Embassy (1816) reinforced popular views of Chinese decadence.[29]

The period of the Canton trade (1759–1839) did not significantly change Western access to Chinese territory. Though British diplomatic efforts to expand trade beyond the single port of Canton (Guangzhou) offered brief peeks into China's countryside, none of these missions was successful, and foreign movement remained severely restricted. Even in the city of Canton, Westerners were only allowed in the Factory District, women were barred entry, and the entire foreign population had to vacate each winter.[30]

With the Treaty of Nanjing, which concluded the Opium War in 1842, the first steps toward state-to-state relations were formally established between the Qing Empire and Great Britain. France and the United States rushed to establish matching rights in China as most favored nations, and the treaty system that emerged forced China to conform to unfamiliar Western standards of foreign engagement and trade on terms heavily favorable to the West. According to the Treaty of Nanjing, four new ports were opened to trade (with foreign administration of concession areas), and Hong Kong was ceded to the British. A supplementary treaty negotiated in 1843 established the right of extraterritoriality, or independent consular jurisdiction over foreign nationals, and this remained a sore point between China and the Great Powers for almost a century.

The Second Opium War (or Arrow War, 1856–60), which was fought while the Qing court was embroiled in the devastating Taiping Rebellion (1850–64), produced the Treaties of Tianjin in 1858. Five more treaty ports were opened and foreigners gained access to the Yangzi River valley. These treaties also authorized foreign travel and protection for Christian missionary efforts throughout the interior of China. Western gunboats could anchor at all treaty ports and patrol the Yangzi to protect foreign interests. Britain, France, Russia, and the United States all gained the right of diplomatic residence in Peking through the most favored nation clause. After a resumption of fighting and Lord Elgin's burning of the Summer Palace in 1860, Britain also obtained Kowloon (Jiulong) Peninsula and Russia negotiated control of the territory north of the Amur River and east of the Ussuri River, as well as rights to enter Kashgar. Other treaties followed, in response to manufactured incidents and through attempts to gain competitive advantage over other powers.[31]

Exploring Chinese Territory, 1842–1866

Once the Treaties of Tianjin opened Chinese territory to outsiders, foreign travelers and explorers began to write themselves into the Chinese landscape. Abbé Huc was one of the first to take advantage of this new freedom, though few of the local people he encountered on his trip from Beijing to Tibet were aware of the treaties or their stipulations, and he was expelled from Lhasa in 1846 for proselytizing. Huc's popular account of his journey first appeared in 1850 and inspired several generations of China exploration.[32]

In the wake of the Opium Wars, Western travelers seeking adventure raced to be the first foreign civilians to go up the Yangzi River by steam or to trace the length of the Ming Great Wall.[33] Others wanted to follow in Marco Polo's footsteps, or add China to the list of exotic places they had already "collected."[34] In contrast, missionaries in China often visited the bleakest, least developed areas of the country in the service of Christianity. Whether leisure travelers or servants of God, however, most were careful to record economic conditions and details of trade as well as landscapes and stories of human interest, recoding the exotic into the familiar.[35] Many travelers were also amateur naturalists who collected plant, animal, and rock specimens for research and display. Those trained in geology—as many well-educated gentlemen were—observed exposures in the famous Yangzi Gorges, unconformities in the stratigraphical record, and patterns of erosion amid the scenic landscapes.[36]

Geology was a recently developed science in the West, but in the nineteenth century it was at the height of its popularity and geologists were eager for new data from around the world. Great areas of disagreement existed concerning problems such as the age of the earth, the nature of forces acting on the earth's surface through time, and the mechanism of mountain building. However, by the 1840s many fundamental concepts of historical geology were well established. Local successions of rock strata had been worked out across Europe and North America and correlated through examination of both mineralogy and characteristic fossils.[37] This program emphasized fieldwork and comparative analysis, and with careful description and collection it was easily extensible to unstudied areas like China. Chinese material would not only help complete the stratigraphic record, but it could also test the applicability of existing theories and provide important clues for the resolution of persistent debates in structural geology.

Nevertheless, the first major geological study of China was more a matter of happenstance than scientific planning. Raphael Pumpelly, who trained at Freiberg Mining Academy and would later become Harvard University's first professor of mining, arrived in Shanghai in the spring of 1863. He and another American geologist had been engaged by the Japanese government to study coalfields in Hokkaido, but a tide of antiforeign feeling left them unexpectedly adrift, and on a whim Pumpelly "concluded to leave for China, where foreigners had lately acquired the right of penetration to the interior."[38]

In China, Pumpelly made three geological excursions, only one of which was planned in advance. At first, Pumpelly thought of Shanghai as just a recreational stop on his homeward journey to the United States, but after seeing boats from Sichuan province filled with high-grade anthracite, he determined to sail up the Yangzi to investigate. This journey took about a month and a half, with several delays and interruptions along the way, but he made it past the famous gorges to the coalfields of Sichuan at Badong by mid-June 1863.[39] During the journey, Pumpelly had little opportunity to examine strata at close range or collect rock or fossil specimens, but he sketched sections wherever possible and frequently referred to observations made by previous travelers on the Yangzi.[40] According to Pumpelly, the gorges cut through a great anticlinal region with granitic and metamorphic "basement" rocks surrounded by a massive Devonian limestone formation that he believed to be associated with overlying coal measures throughout China.[41]

Pumpelly's next trip, to the coalfields in the Western Hills of Beijing, again interrupted his plans to return to the United States, this time via Siberia. The Qing government had ordered eight British gunboats to strengthen its navy and hoped to use local rather than Cardiff coal to power the ships. Through his friendship with Anson Burlingame, the American minister to China, Pumpelly was recommended as a mining advisor to the Qing court, and by mid-November he was examining native mines in the suburbs of Beijing. For three weeks, Pumpelly toured the countryside with three Chinese officials and an interpreter. There he collected coal samples and observed mining conditions before returning to Beijing to prepare for a more extensive survey of the entire region. However, he was immediately told that his services were no longer needed, for the Qing court had sold its entire flotilla of gunboats after the British commanding officer refused to take orders from local officials.[42] Pumpelly filed a report to both the Chinese and the United States governments anyway, describing the need for steam pumps to remove water from native mines and recommending the construction of modern transportation facilities to bring local coal to Beijing.[43]

During the winter of 1863–64, Pumpelly contracted smallpox and while he was recuperating he "found natives to glean the scanty data from the huge volumes of ancient and recent writings and missionaries to translate them."[44] This also gave him time to plan an excursion through the loess valleys along the Great Wall and into Mongolia with a Russian acquaintance, Dr. Pogojeff. The two companions set out in early April 1864, with a Chinese servant, a Cossack guard and a mule train. Upon reaching Lake Dai, just inside the Mongolian plateau, Pumpelly learned that the Muslim Rebellion (1863–77) had spread from Kashgar eastward and decided to turn back, but he was able to gather data for a rough geological map. Though he ultimately took an overland route through the Gobi and Siberia on his departure from China, his first Mongolian trip was his last real geological study of Qing territory before returning home.[45]

Raphael Pumpelly reached America just as the Civil War was ending and he found geological circles eager for news of distant places untouched by their domestic catastrophe. James Dwight Dana asked him to publish an account of his travels in the *American Journal of Science*, and Joseph Henry of the Smithsonian Institute immediately accepted Pumpelly's full memoir in 1866. *Geological Researches in China, Mongolia, and Japan* did not take sides in the battle between isostasy and contraction theories that divided Anglo-

American geologists from their European counterparts, but it did cover a far wider range of topics than Pumpelly's fieldwork alone might suggest. His use of Chinese sources allowed him to speculate on historical changes in the course of the Yellow River and study the growth of its delta as well. Pumpelly further surmised that the Gobi was once a vast freshwater lake drained by the gradual uplift of the Mongolian plateau and movements in the course of the Yellow River. Based on this idea, he believed the thick loess deposits of northern China to be lacustrine in origin.

Pumpelly's outline of China's geological history and his "Hypothetical Map of the Structures of China" combined aspects of several competing theories and probably reflected his isolation from well-defined schools of thought while in Asia. In Lyellian fashion, his historical sequence for China alternated uplift and subsidence. But in the spirit of Élie de Beaumont's contraction theory (and his own Continental training), Pumpelly felt that his greatest contribution was the identification of the "Sinian System," which associated a characteristic Chinese epoch of mountain building with the NE-SW trend of the country's major eastern mountains. In fact, he saw the Sinian System as a "link in the chain of evidence toward proving the subordination to harmonious laws" of Chinese—and world—geology.[46]

Ultimately, few of Pumpelly's major conclusions stood the test of time. However, they did show an ingenious use of correlation and historical data to bring China into existing geological discourse, and Pumpelly was very gracious about recognizing his mistakes.[47] His work also laid the foundation for Ferdinand von Richthofen, a German geologist who befriended Pumpelly's mentor Josiah Dwight Whitney. With his appetite for information on the geology of China whet, Whitney soon helped Richthofen set up both funding and publication agreements for a new series of explorations in China.[48]

Assessing and Possessing the Land, 1867–1899

Ferdinand von Richthofen, whose research "laid the foundation for all future geologic work in China,"[49] was born in Karlsruhe in 1833 and trained in geology and geography in Berlin. In 1859 he accompanied Count Friedrich von Eulenburg as secretary of legation on Prussia's diplomatic and commercial mission to East Asia on behalf of the Zollverein states. The mission was motivated by Britain and France's gains in China following the two Opium Wars, but was equally notable for its symbolic unification of pan-German interests under Prussian leadership.[50] Qing authorities had an understandably vague notion of "Germany" and showed little interest in negotiating. The mission proceeded to Japan, the Philippines, and Java before returning to China, where von Eulenburg managed to negotiate an agreement granting Prussia most favored nation status. During the initial breakdown of political negotiations Richthofen explored the various Asian countries where he was stationed, and after the China treaty was signed in September 1861, he examined the prospect of establishing a naval base on Formosa (now Taiwan) before rejoining the mission in Thailand. After the Prussian fleet departed for home in February, Richthofen took an overland route to Calcutta with the idea of exploring China and Siberia on his way back to Europe.[51] However, the rapid spread of the Taiping Rebellion made this idea impracticable, and in June 1862 he sailed for California instead.

In California, Richthofen spent several years reporting on mining for German newspapers and studying the relationship between igneous rock series and ore bodies, particularly gold.[52] He became friends with California state geologist Josiah Dwight Whitney and, after studying Raphael Pumpelly's research in Asia, revived his earlier ambition of studying the geology of China.[53]

Richthofen arrived in Shanghai on September 5, 1868, and in early November he began the first of four field trips funded by the Bank of California. Traveling by cart and by boat with Chinese laborers and an interpreter companion in tow, Richthofen started with a two-month study of the of outcrops in the area around Shanghai including the Yangzi delta, and then expanded his exploration to cover the entire lower Yangzi Valley from Shanghai to Hankou.[54] On his third trip (March–July 1869) he took advantage of the warmer weather to make a great circuit from Shanghai to Liaoning Province in Manchuria, and back through the Shandong peninsula. In Shandong, his stated objectives were "the geological study of mountains and the investigation of various coal deposits," but he also scouted sites for a potential German naval base.[55] After Shandong, he focused on the coal mines around Boyang Lake in Jiangxi and studied kaolin clay deposits that supported the local porcelain industry.[56]

By the end of 1869 Richthofen had completed his study of the lower Yangzi valley and China's northeastern coast, and made the Shanghai international settlement his base between field trips. Contacts with foreign businessmen had already guided some of his route planning, and in 1870 he agreed to conduct a full-scale study of the mining districts of north and west China for the British- and American-led Chamber of Commerce. The Shanghai Chamber promised to cover all travel expenses and let him make his geological findings public. This arrangement suited Richthofen perfectly, and he did not even request a salary. After his "wasted" California years, he simply wanted to accumulate enough scientific "capital" (in data and experience) to secure something better than "an indefinite future and perhaps some claim on a badly paid Professorship" when he returned home.[57]

Setting out on January 1, 1870, Richthofen began by taking a sea route to Guangdong Province in the far south. From there he worked his way up through Hunan, Hubei, Henan, Shanxi, and Zhili (Hebei) provinces, reaching Beijing by the end of May. He reported his findings to the Shanghai Chamber of Commerce in a series of occasional letters from the field. These described the mineral resources, transportation networks, agricultural output, and trading centers of the areas he traversed, with geological detail only as needed. Famously, Richthofen's letters characterized "the whole of southeastern Hunan . . . [as] one great coal-field" and proclaimed that, despite the lack of necessary transportation to the coast, it was "patent that the world, at the present rate of consumption of coal, could be supplied for thousands of years from Shansi [Shanxi] alone."[58]

In June 1871, Richthofen returned to the Lower Yangzi valley to address persistent inquiries about the mining prospects of Zhejiang, Jiangsu, and Jiangxi. These provinces were in easy reach of major treaty ports, and though Richthofen did not expect to find much coal or iron, local demand warranted a careful inspection. As anticipated, Richthofen found the region to be mineral-poor and, where useful ores existed, conditions for mining were largely unfavorable.

Richthofen's last field expedition in China was also his most ambitious. Starting from Beijing, he entered Mongolia and crossed an area similar to that explored by Pumpelly, but by a different route. From Mongolia he passed south through Shanxi and southwest to Shaanxi and Sichuan, before sailing through the Yangzi gorges back to Shanghai. This trip lasted from October 1871 to May 1872 and Richthofen would have extended it to include Gansu and Yunnan, but an attack on his interpreter convinced him that conditions during the ongoing Muslim Rebellion (1862–78 in Gansu) were too lawless for safe passage through such remote regions.[59]

Having concluded this last trip, Richthofen sailed from Shanghai to a newly unified Germany and immediately received the support of Kaiser Wilhelm I for a comprehensive work on the geology and geography of China.[60] He was named president of the Berlin Geographical Society in 1873, and appointed professor of geography at Bonn in 1875, though he deferred teaching duties until after he completed the first volume of his monumental series *China: Ergebnisse eigener Reisen und darauf gegründeter Studien* in 1877.[61]

In its original conception, *China* was to be one massive volume on the structure of Central Asia and China proper and an analysis of its relationship to the development of the Chinese people, incorporating both Richthofen's geological and geographical expertise. Though this work made reference to geographical information in certain classical Chinese texts, Richthofen made explicit the degree to which foreign experiences had rendered China terra incognita after all. For Richthofen, no amount of detail could make up for the lack of exactitude in Chinese descriptions of the land, and *China* was to be the first comprehensive scientific work on what was hitherto an "incomprehensible and unknown" land.[62]

As the project expanded beyond the cultural and physical geography of the first volume, further details of the geology and topography of north and south China were assigned to two further volumes, with a separate atlas covering all of China. Richthofen completed the north China volume and the atlas in 1882, and he began work on south China but died before accomplishing his goal. This volume was finally completed in 1912 by Ernst Tiessen, and other collaborators also produced two other volumes on China's paleontology (1883) and stratigraphy (1911) from Richthofen's notes and specimens.

In addition to establishing a basic historical sequence in several parts of China, Richthofen's most famous geological contributions were his explanation of China's vast loess plateaus and his application of the term "Sinian" in a strictly stratigraphic sense.[63] Loess, a very porous, largely unstratified loam, with characteristic vertical cleavage and a tendency to crumble, was generally associated with glacial margins in Europe and America. In China loess formations in the north and northwest are typically from 500 to 1,000 feet deep and cover over 250,000 square miles.[64] These massive deposits define the landscape of the region, and in the Republican period, research on loess constituted one of the most important activities of the Geological Survey of China.[65] Pumpelly and others assumed loess to be alluvial or lacustrine, but based on his observations that Chinese loess accumulated at both high and low elevations, displayed no bedding, contained only fossils of land animals, and was absent from the lee side of major mountains, Richthofen established the wind-borne origins of most loess formations.[66]

Of the Sinian, Richthofen claimed that "nothing is more characteristic for the geology of China, than the long succession of sediments which we have previously designated by this name."[67] However, Richthofen's Sinian did not refer to the NE-SW structural axis (or orogenic epoch) of Pumpelly's usage, but to a distinct formation of unmetamorphosed sediments at the heart of China's eastern mountain structure.[68] Though debate has continued about the exact definition and periodization of the Sinian, subsequent usage has universally followed Richthofen in using the term as a stratigraphic unit.

Besides presenting these innovations and laying the groundwork for future studies of China's geological structure,[69] Richthofen's research also had a profound economic and political impact. His ties to foreign business interests in Shanghai were explicit, and Richthofen was painstaking in his analysis of not just mineral resources, but their relationship to local transportation networks, regional trade, labor conditions, and complementary industries. Richthofen was the first person to recognize the magnitude of the Kaiping coal measures in Zhili Province. Though Qing officials had persistently rejected all foreign mining proposals on the grounds that they would disturb neighboring populations and upset local *fengshui*, Richthofen's findings were so favorable that they aroused the interest of progressive official, Li Hongzhang (李鸿章). In 1877 Li used his influence to set up China's first fully modernized mine at Kaiping,[70] and over the next decade several Chinese mining ventures tried operating under a similar "government-sponsored, merchant-managed" (官督商辦 *guandu shangban*) joint-stock scheme, though none were as successful or well-placed.

Foreign firms were still barred from directly exploiting China's mineral resources, and many of the military powers represented in China felt the situation to be unsatisfactory, but few were willing to threaten their existing positions by agitating for mining rights without new provocation. In a certain sense, then, "the most influential German contributor to this explicitly colonial approach was the geographer Baron Ferdinand von Richthofen."[71] As early as 1869 and 1870, Richthofen sent a series of memoranda to Bismarck about the desirability of establishing a naval base in China. His earliest suggestion was the Zhoushan Archipelago at the entrance of Hangzhou Bay, which he felt could easily become a "German Hong Kong."[72] However, Britain had already secured a secret promise from China that no other power would be granted rights over Zhoushan, and the matter was dropped. But Richthofen kept the idea of a base firmly in mind, and soon settled on Jiaozhou Bay as an ideal location.

Rather than identifying a naval base with easy access to an already booming commercial area, Richthofen sensed that Germany needed to control an area that was ripe for development but as yet untouched. Jiaozhou Bay offered a deep but protected harbor and was located on the south side of the agriculturally productive, mineral rich Shandong peninsula, but it had no facilities for inland transport and was therefore ignored by foreign traders. In his second volume of *China*, Richthofen focused a great deal of attention on the geology and geography of Shandong. He pointed out that its only existing port, to the north, was hemmed in by mountains and therefore difficult to reach overland, but that a railroad could easily be built across the

Shandong plain to Jiaozhou Bay, and it would cross most of the peninsula's coal producing regions en route.[73] Given China's reluctance to open mining to foreign interests or approve foreign railway activity, this integrated strategy of developing mining, infrastructure, and shipping from scratch indicated Richthofen's desire to see Germany take a more active stance toward its Chinese interests.

Once again, Richthofen's published findings were so persuasive that they caught the attention of modernizing Chinese officials. Just after the Sino-French War (1883–85) the Chinese minister to Germany cited Richthofen and the Western interest he was generating in a memorial to the emperor on the naval value of Jiaozhou Bay, and soon afterward an imperial censor reiterated the bay's strategic advantages. This time, however, Li Hongzhang was more concerned with developing his existing bases on the Gulf of Zhili (Bo Hai), and this cooled Qing interest in fortifying Jiaozhou until the early 1890s, when a few half-hearted measures were finally taken.[74]

The entire situation changed abruptly in 1895, when Japan defeated the numerically superior but poorly organized Chinese navy in the Sino-Japanese War (1894–95) and exposed China to the world as a "paper tiger."[75] As a result, China lost its influence in Korea, and formally ceded Taiwan, the Pescadores, and the Liaodong Peninsula. Alarmed by Japan's expansion into Manchuria, Russia organized a tripartite intervention with France and Germany and succeeded in pressuring Japan to return Liaodong in favor of a cash indemnity.[76] Germany's participation in the intervention was fueled by Kaiser Wilhelm II's growing ambition to challenge British naval supremacy and establish a base in Asia, but despite China's indebtedness, Germany received concessions at Hankou and Tianjin instead of Jiaozhou Bay.[77] Two years later, when German authorities seized on the murder of two priests in Shandong as an excuse to occupy Jiaozhou, they forced China to lease the bay in "grateful appreciation of the friendship shown them by Germany" in 1895![78]

While it is clear that many other individuals, such as Admiral Tirpitz and Bishop Anzer of Shandong, were prime movers in Germany's bid for a commercial and military base in China, their unified focus on Jiaozhou Bay was explicitly rooted in Richthofen's pointed geological studies.[79] The terms of the leasehold not only established German control over Jiaozhou for ninety-nine years, but they also gave Germany the right to build three railways lines across Shandong and operate mines within ten miles of the tracks.[80] Though France had nominally beaten Germany to the first foreign mining and railway concessions in China (near French Indochina, 1895),

Germany established these rights within a comprehensive system that effectively made all of Shandong its sphere of influence.

Space for Geology, 1897–1911

This precedent set off a "scramble for concessions" that marked a new era of foreign imperialism in China, and in the wake of the Jiaozhou leasehold, Russia acquired rights in Manchuria, Britain leased Kowloon and Weihaiwei (on the Gulf of Zhili), and France procured Guangzhou Bay. Britain also secured a promise that China would not give concessions in the Yangzi valley to any other power, and Japan obtained a similar agreement for Fujian Province. Worried about its own interests but too caught up in the Spanish American War to stake a claim, the United States responded by circulating an "Open Door" memorandum in 1899 and trying to assert equal commercial opportunity for all foreign nationals in China.

As railway and mining concessions became an explicit and accepted aspect of China's domestic reality, foreign powers could be much more direct about investigating the land, but this process became increasingly disengaged from geological considerations.[81] While spheres of influence accelerated survey activity around potential mines and rail lines, these were largely practical studies on the engineering problems of mechanical mining or railway construction. Commercial groups were no longer interested in the geology of China or the distribution of minerals across the country; they were simply concerned with the feasibility of extraction in the concession areas they had already been granted.[82] Much of their research was private, and existing reports in international mining journals had little impact on Chinese understandings of geology.[83]

In contrast, the results of the 1903–4 Carnegie Institution–sponsored explorations by US Geological Survey geologist Bailey Willis and his assistant Eliot Blackwelder were so distinct from mining applications that years later rumors still circulated that "Mr. Carnegie must have kept back all the information of an economic value for his own personal use."[84] Though the Carnegie Expedition built directly on Richthofen's research, the three-volume Carnegie report, *Research in China* (1907), focused exclusively on the structural, stratigraphical, and paleontological aspects of its predecessor's work and did not deal with mining, transportation networks, commerce, land use, regional customs, or historical materials. In fact, the project was initially proposed in hopes that ancestral forms of trilobite might be located in the Sinian limestones documented by Richthofen in north China.[85]

In August 1903, Willis and Blackwelder visited Richthofen in Berlin and consulted with him on their route through Shandong, where they hoped to make comparisons with Richthofen's work without duplicating too much of his actual path.[86] They reached Beijing via Moscow in late September and then spent two months unsuccessfully searching for "Adam Trilobite" in Shandong before returning to Beijing to wait for US Geological Survey topographer R. Harvey Sargeant.[87] Blackwelder explored the older strata of the Liaodong Peninsula independently in December, and the whole team set out to examine Sinian formations in Zhili and Shanxi in January 1904. Though vigilant, Willis doubted that they would find the Precambrian fauna they were looking for, and in choosing to make his way west across the Taihang Mountains into Shanxi (February) and then south over the Qinling in Shaanxi (April), Willis claimed, "the mountains call me, challenging me to read their history: what of the earth forces that have raised them? what of the rivers that have sculptured them?"[88] This was clearly a rhetorical flourish for a popular audience, but the scientific reports of his expedition betrayed no interest in mining and railroads, even in the vaunted coal regions of Shanxi. After crossing into eastern Sichuan, the Carnegie team rounded out their exploration by sailing through the Yangzi Gorges, and they departed for America from Shanghai in late June 1904.

Willis and Blackwelder made great headway in correlating formations in different parts of China and North America, and they established an important type section of the Cambrian in Shandong, but the expedition did not discover a precursor trilobite in China.[89] As his statement on mountains suggests, Willis himself seemed much more interested in broader questions of diastrophism. In volume two of *Research in China* he presented a tectonic theory based on an unusual isostatic model in which the thrusting and folding of China's eastern mountain ranges were produced by basaltic ocean crust pressing laterally on lighter continental material.[90] He also speculated on the age of China's mountains and came to surprising conclusions such as the Quaternary uplift of the Wutai of the Taihang range. Much of Willis's theorizing was rapidly abandoned by the international geological community, but the expedition laid out the first narrowly geological study of Chinese territory and deeply colored later interpretations of the structure of the Qinling range. He also established Richthofen's Sinian as largely Cambro-Ordovician, prompting the later Geological Survey of China to redefine the term again in the 1920s.[91]

While Richthofen's research helped ignite a shift from commercial to territorial imperialism in China, the system of spheres of influence and concessions actually brought mining interests into the open and created a space for more theoretical geological research in China. Dissociation from economic objectives produced a very different kind of research project recoded in terms of global rather than national signs. The Carnegie Expedition approached its fieldwork with a well-defined scientific goal, designed its research strategy to both augment and verify Richthofen's earlier findings, and incorporated its findings into wider debates on orogeny and crustal movement. Willis even made an attempt to engage Chinese readers by including Chinese text with several maps.[92] Though this effort was not "of material aid in stimulating an interest in the earth sciences" in China until after the fall of the Qing, it displayed an awareness that geological research was re-presenting China to its own people.[93]

"A Vast Territory with Abundant Resources"

In the immediate aftermath of the first Opium War, it seemed more pressing for several progressive Chinese intellectuals to understand the outside world than to rethink their own territory,[94] and two very influential world geographies were published in the 1840s. Early drafts of Wei Yuan's *Haiguo tuzhi* (海國圖志 Illustrated gazetteer of the maritime states) were circulated within four months of the Treaty of Nanjing in 1842, and parts of Xu Jiyu's *Yinghuan zhilue* (瀛環志略 A brief account of the maritime circuit) made the rounds of concerned intellectuals as early as 1844.[95]

Wei Yuan's gazetteer was partially based on *Sizhou zhi* (四洲志 Gazetteer of the four continents), a collection of translations from foreign newspapers commissioned by Opium War firebrand, Lin Zexu, in 1839, but it also relied heavily on material from foreign missionary organizations like the Society for the Diffusion of Useful Knowledge in China and the Chinese and American Holy Writ Bookstore.[96] According to Wei, his gazetteer was different from earlier maritime geographies because "they are all based on discussions of foreigners by Chinese, while this work is based on discussions of foreigners by foreigners themselves."[97] Xu's *Yinghuan zhilue* was even more firsthand, and combined snippets from foreign newspapers, books, and atlases with accounts from interviews with Western diplomatic and missionary personnel in Fujian.[98] Xu cataloged information about nations from five continents and presented the modern world as a system of states, but pointedly avoided treating China itself in any detail. Nevertheless, his implication that China was no longer the preeminent world civilization soon lost him favor at court.

Despite the popularity of these geographies, most readers ignored the warning that China must understand its recent adversaries as a new breed of outsiders, and few took these works as more than intriguing curiosities.[99] More significantly, Chinese elites were just beginning to be curious about the West and had little reason to borrow its mercantilist lens to view their own country. If "discussions of foreigners by foreigners" vouchsafed knowledge of the West, surely Chinese knowledge of China was equally authoritative. Chinese territory was familiar and the shock of military defeat did not suddenly make it exotic or worthy of new exploration.

By the 1860s, however, Chinese interest in foreign knowledge had noticeably expanded at court, and even more among the provincial governors facing the Taiping and Nian Rebellions directly. Leaders of the Self-Strengthening movement such as Zeng Guofan, Li Hongzhang, and Zhang Zhidong advocated a policy of adopting Western technology to increase China's "wealth and power" and tried to convince conservative detractors that foreign methods would not compromise native principles.[100] Between the Arrow War and the ill-fated Sino-Japanese War, Self-Strengtheners initiated a wide variety of programs both to produce Western goods and propagate Western knowledge. The most famous of these were the Jiangnan Arsenal (1865), the Hanyang Ironworks (1890), the Fuzhou Shipyard (1866), and the imperial interpreters' college, the Tongwen Guan (1862).

Soon after it was established, a translation bureau was added to the munitions factory at the Jiangnan Arsenal in Shanghai, and by 1879 over ninety-eight works were in print, with forty-five more ready for publication and another thirteen in process. Translations were produced through painstaking collaborations between technically adroit Chinese scholars, such as the mathematician and engineer Xu Shou, his son Xu Jianyin, and a group of foreign missionaries under John Fryer.[101] The translation of geological works was particularly difficult, and illustrates the general process used by the bureau. A foreign member with a background in the relevant subject would study a book in its original language and then explain the gist of each sentence to his Chinese partner as best he could in spoken Mandarin. The Chinese writer would then try to decipher the sentence and translate it into accurate literary Chinese for publication.[102] In the case of James Dwight Dana's *System of Mineralogy* (金石識別, 1872) and Charles Lyell's *Elements of Geology* (地學淺釋, 1871), the foreign interpreter, Daniel Macgowan had only a limited knowledge of Chinese, and the Chinese translator, Hua Hengfang (who knew neither English nor geology) had to rely on facial expressions and hand gestures to try to puzzle out the meaning of the text.[103] The process was so trying that for months the names of ores and stones kept Hua awake, and whenever he did sleep he had feverish nightmares of "scaly creatures" and the "bones of fantastic [prehistoric] beasts."[104]

It is no surprise, then, that readers found these Chinese translations cumbersome and confusing. Those hoping for continuity across translations fared even worse, for the awkward translation procedure resulted in completely idiosyncratic transliterations of technical terms. In the translation of Lyell's *Elements*, for instance, the Carboniferous period was rendered as *ka-pu-ye-fei-la-si* (卡蒲业非拉斯) and the Permian as the *po-er-mi-an* (泼而弥安), but in the 1899 translation of J.W. Anderson's *The Prospector's Handbook* (求礦指南), the Carboniferous was referred to as the *chanmeiceng* (产煤层, lit. the coal-producing layer) and the Permian was transliterated as the *pou-mi-an* (剖密安).[105] Even determined readers could rarely tell if they were reading incorrectly or if the texts themselves were faulty.[106]

As the Jiangnan Arsenal began to translate more utilitarian works like *The Prospector's Handbook*, the problem of accuracy and clarity was overshadowed by the ineffectiveness of mere words in transmitting practical techniques and methods to the uninitiated without any physical context. If the translations were intended to foster the development of native mining personnel, they were a failure. The mindset of mining was even harder to convey. Compare the title of Herbert Cox's *Prospecting for Minerals: A Practical Handbook for Prospectors, Explorers, Settlers, and All Interested in the Opening-Up and Development of New Lands* (1899) with its translated Chinese title, *Xiang di tan jinshi fa* (相地探金石法, 1903), which literally means "methods of looking at the land and seeking minerals." The disembodied contents of the Chinese version could not teach the real-world skills of prospecting, and its title did not begin to convey the trailblazing, speculative attitude of the English original either.

With all of the changes wrought by the Opium Wars and the resulting treaty system, the Chinese still had little reason to think of the eighteen provinces as wild or unknown, and they never imagined mining as a "rough and ready" pursuit. Even for Self-Strengtheners like Li Hongzhang and Zeng Guofan, who began to see coal and iron as the secrets of Britain's global power, "earth treasures" (地寶 *dibao*) were primarily needed to support industrial competition, and their value as commodities was still not fully appreciated.[107] Though modernizers challenged the conservative view that China's mineral wealth should be slowly extracted for local needs with as little disturbance to the land as possible, both groups shared the belief that China was "vast in territory and abundant in all things" (地大物博, *dida wubo*).[108] On one hand, faith in the bounty of the land inspired confidence and generated interest in mining, but it sometimes dulled the urgency of checking foreign exploitation and strengthened the case of individuals objecting to projects on the basis of *fengshui*. If resources were inexhaustible, there would always be more to find and other sites to develop, and this attitude allowed bureaucrats to be cavalier about efficiency and planning.

Official Qing interactions with Western mining advisors often reflected both the complacency engendered by China's supposedly limitless resources and a general lack of understanding about realistic mining practices. These factors were further complicated by traditional ideas of class and erratic commitment to resource extraction.[109] Raphael Pumpelly's experiences in 1863 were typical. An opportunity arose that suggested the advantages of improving coal mining near Beijing, and the Qing court procured the first Western "expert" available. Though Pumpelly happened to be formally trained at Freiberg, his appointment was largely an accident of circumstance, and in many other cases poor quality was overlooked for good timing. In the field, Pumpelly was escorted by two Chinese civil officials who were carried in sedan chairs, and one military official who rode horseback, as well as an interpreter. The civil officials bullied local Chinese with whom Pumpelly interacted and tried to prevent him from taking a geologically important route because it was less comfortable to travel over. Everyone Pumpelly met, including miners, found it unseemly that—as a person acting on imperial authority—he would crawl into mine shafts himself to examine operating conditions.[110] Moreover, as soon as the immediate incentive for local coal extraction disappeared, Pumpelly's commission was rescinded and his field report ignored.

Replacing Foreigners, Rethinking Territory

Between 1860 and 1895 over fifty Western geological advisors were hired by the central government or provincial industrial bureaus, but they were used inconsistently and without long-term commitment or financial support.[111] Even at Kaiping, where Richthofen had generated so much interest in coal resources, foreign mining engineers were hired to survey the area three separate times, but there was no government action until Li Hongzhang intervened and brought in private capital.[112] Qing authorities were so unresponsive that hired advisors frequently provided more details of their results in foreign journals than in their official reports. Moveover, many foreign advisors were simultaneously financed by Chinese and foreign authorities, and this created conflicts of interest in which findings were often steered toward outside advantage.[113]

After the Sino-Japanese War, mining and railway ventures multiplied but Chinese interests saw little benefit. Mining regulations drawn up in 1898, in the ambitious climate of the Hundred Days' Reform, attempted to protect

Chinese stakes in new industrial projects by requiring Chinese partnership and personnel.[114] Realistically, however, almost no qualified Chinese technical staff existed, even when token mining schools were set up to train them, and in practice, Chinese-foreign partnerships were often run by foreign mining engineers, staffed by foreign business managers and machinists, and financed by loans from foreign banks.

As Tim Wright describes it, "[Chinese] mine owners were in effect, if not necessarily intention, mere front men for foreign interests, and several important foreign mines originated this way."[115] The Kaiping Mines, for instance became foreign-owned after secret dealings in the wake of the failed Boxer Uprising of 1900. After the uprising was put down by an international military coalition that occupied Beijing and imposed severe terms on the Qing court, Russian troops threatened to interfere with Kaiping operations in Zhili. Fearing confiscation of the mines, Kaiping's Chinese general director empowered a foreign representative to negotiate a transfer of Kaiping property to a newly created Chinese Engineering and Mining Company (CEMC) under British law. This move successfully deterred Russian advances, but the financial details of the transfer watered down Chinese shares and put outright control of the mines in the hands of foreign stockholders.[116] After Japan's surprise victory over Russia in the Russo-Japanese War of 1905, Japan dispensed with both pretense and legal intrigue by simply taking Sino-Russian mining and railway operations in Manchuria over by force.

While provincial gentry tried to resist foreign domination by launching a Rights Recovery movement (1905–11) using boycotts and other pressure tactics, Chinese youth were beginning to strike a different path.[117] For them, the move to take back control of mines and railways was important, but political and economic resistance to foreign encroachment was still too superficial a response to what they now saw as a larger, more fundamental competition between nations and races. Under the influence of modernizing writers such as Yan Fu 嚴父 (1854–1921), who introduced social Darwinism to China, and Liang Qichao (梁啟超 1873–1929), who called for a new Chinese citizenry, a handful of young Chinese began to pursue scientific and technical knowledge in earnest, and many of them employed a language of responsibility rather than rights.

Among this new breed were Gu Lang (顧琅 1880–?) and Zhou Shuren (周樹人 1881–1938), who later became famous as the short story writer and social critic, Lu Xun (魯迅).[118] Gu and Zhou were classmates at the School of Mines and Railways of the Jiangnan Military Academy in Nanjing, and though they were products of existing training programs in geology and mining, they were exceptions that demonstrated the limitations of these early Chinese educational efforts. The School of Mines was added to the Jiangnan Military Academy in hopes of developing a local coal measure, but mining prospects soon faded, and the school only accepted one incoming class before losing funding. Students at the school studied *suanxue* (算學 mathematics), *gewu* (格物 topics in physics and chemistry), *dixue* (地學 earth studies), and *jinshi xue* (金石學 mineralogy) under German instructors who lectured through translators and taught from the Jiangnan Arsenal's impenetrable Lyell and Dana translations.

Both men graduated from the School of Mines with high honors in 1902 and immediately headed to Japan on government scholarships. In Tokyo the two friends roomed together for two years while studying at the Hongwen Academy (弘文学院, *Kobun Gakuin*), a language and preparatory school for Chinese students. Having gone through the frustrating experience of working with translations, Gu and Zhou felt the need to learn science more directly, so in addition to learning Japanese, they improved their German and English on their own. Eventually they learned enough to study foreign geological and mining materials, and they set to work on the original reports available to them for the first time in Japan[119] Their exposure to foreign studies of China alerted Gu and Zhou to the link between geology and mining that they emphasized in their writings, and shocked the two young men into the realization that outsiders knew vastly more about Chinese territory than the Chinese did themselves.

By 1904 Gu was accepted into the Geology Department at Imperial Tokyo University and Zhou was attending Sendai Medical School (which he famously quit in 1906), but the two kept up a correspondence on geological matters, and soon began publishing their efforts. The first of these was the "Brief outline of Chinese geology" (中國地質略論 "*Zhongguo dizhi luelun*"), which Zhou Shuren published in the expatriate journal *Zhejiang chao* (浙江潮 Zhejiang tide) under the pen name "Seeker" (索子 *suozi*) in 1903. This was the first work on Chinese geology written by a Chinese, and the first Chinese use of the Japanese-coined neologism *dizhixue* (地質學, lit. "earth substance studies") for geology as distinct from *dixue* (地學, lit. "earth studies").[120] In the article, Zhou defined "geology" as the "history of the evolution of the earth," encompassing the "history of rock formation and the structure of the earth's crust," as well as the study of fossils. As such, Zhou tried to dispel geomantic notions by emphasizing that the earth was not composed of "deeply mystical and unimaginable elements" and could be studied logically and empirically. Though he clearly promoted the importance of geology as a way to understand China's mineral wealth, he also stressed that a catalog of mineral distribution was insufficient for utilization of the earth's "precious contents," which required genuine understanding of the mechanisms and historical stages leading to the earth's present state.[121]

To illustrate the advantages of this approach, Gu Lang pulled together information about known mineral deposits in China proper from Richthofen's works, local gazetteers, and two confidential Japanese surveys; and Zhou added a lengthy theoretical preface to put the barebones list in geological context. The combined result was published in Shanghai in 1906 as the *Record of Chinese Mineral Resources* (中國礦產志 *Zhongguo kuangchan zhi*), and the authors explained to readers that the two parts of their book "were connected like form and shadow. Without knowledge of geology, there could be no knowledge of mineral resources, and so a discussion of the former must serve as introduction to the latter."[122]

Though the basic claim here displays a naive positivism, neither "A Brief Outline of Chinese Geology" nor the introduction to *Record of Chinese Mineral Resources* aimed to actually guide readers to new mineral deposits through the application of geological principles. Instead, both pieces used geological concepts such as sedimentation, folding, uplift, and metamorphism to highlight the historical materiality of Chinese territory. China was not simply a bounded political space with natural resources below, but the part of the earth's crust belonging to the Chinese people, with resources as an integral part. Physically, "those who [want to] investigate China's development, must first understand the transformations of the eastern part of the Asian land mass" and then mineral deposits, like landforms and the quality of the soil, could be understood as constitutive products of these changes over time.[123] Geology explained the familiar by making it the endpoint of long-term physical processes, and Zhou's approach was less geared toward teaching the principles of geology than using a geological perspective to reintroduce Chinese territory to his complacent readers.

What inflamed Zhou and Gu the most was that the Chinese, who had lived so long on the land, gave so little thought to its substantive nature beyond the practical concerns of soil and planting. As Zhou put it, whenever he wrote about geological phenomena, he "could not help feeling love and admiration, and a deep fear and apprehension as well."[124] He loved and admired geology's ability to connect Chinese territory to the outside world through global history and processes, but was deeply alarmed by how completely this could occur without Chinese input or awareness.

For Gu and Zhou, the idea of China as "a vast territory with abundant resources" was not an excuse for complacency or indulgence, but a reason to be incensed that China was not making use of its own assets. "China belongs to the Chinese," Zhou insisted. "Foreigners are allowed to study

China, but not to explore it. They are allowed to admire it but not to covet it."[125] However, the blame for this was squarely on the Chinese themselves, for "if the owner is negligent, then brigands will have their way. Today we yield and offer, tomorrow we grant special privileges. Anyone with means at all can steal candy from an orphan."[126]

Geological knowledge of the land was therefore a responsibility of ownership, and since the Qing court had not put geology to use to protect the "homeland of the Han,"[127] this concept of ownership set the Chinese apart not only from Western explorers with imperialist ambitions, but from their Manchu rulers as well. While not overtly anti-Qing, Zhou and Gu were very clear that governments, like the geological regimes recorded in the earth's strata, could undergo revolutionary change, and they enjoined Chinese readers to act independently to defend China's natural wealth.[128]

In the spirit of taking action and redressing these wrongs, Gu and Zhou advertised their use of "secret books from East and West," including Richthofen and Willis, which were hard to find in China, and they framed their methods in heroic terms.[129] According to the promotional material for the first edition of *Record of Chinese Mineral Resources*, Gu Lang was shocked to find a confidential map on the desk of one of his teachers at Tokyo University. It was based on Japanese military reconnaissance and data from other foreigners covertly surveying Chinese territory, and Gu immediately made a tracing of the map so he could make it public knowledge at home.

While these claims were sensationalized for marketing purposes, this combination of indignation, self-recrimination, and patriotic initiative struck a chord with both general readers and frustrated officials eager to promote change. The Ministry of Agriculture and Commerce and the Ministry of Education both endorsed *Record of Chinese Mineral Resources* as "necessary reading for national subjects" (國民必讀 *guomin bidu*), and the book was reprinted by the end of its first year, with a public call for readers to send additional books or firsthand information to the authors for inclusion in the expanded 1907 edition.[130] Though Zhou and Gu were willing to rely on Richthofen and other foreign sources to rouse the interest of others in China's geology and mineral deposits, they did not want Chinese to remain permanently bound to outsiders for information about "their own estate."[131]

Conclusion

When Wei Yuan and Xu Jiyu considered Western geography in the 1840s, they were startled by the unfamiliarity of the outside world. When Gu Lang and Zhou Shuren investigated foreign geology a half century later, they were startled by the unfamiliarity of China itself. Though Wei and Xu produced their world gazetteers in the aftermath of the Opium War, they retained the confident territorial sensibility of the eighteenth-century "Pax Sinica." During the decades that followed, this equanimity gave way as treaties granted foreign powers administrative rights at a growing number of ports and leaseholds, allowed foreigners privileged access to the once-sheltered interior, exempted them from Chinese law, and disrupted existing patterns of trade and production. As treaty concessions weakened central Qing control and estranged Chinese from foreign-held areas of their own country, they also recast traditional spatial boundaries of self and other, slowly reminding Han Chinese that their Manchu rulers were outsiders too.

Having come of age amid these changes, Gu and Zhou typified the growing number of disaffected youth who were passionately committed to China but acutely dissatisfied with the Qing establishment's efforts to protect it. Many sought Western-style training through "wealth and power" programs, and hoped to find both an explanation of the national calamity and a way to make a difference. However, subjects like shipbuilding, which had clear practical applications, still only skirted these root concerns, and the available sciences were taught in such confusing, disembodied ways that even motivated students found it hard to relate to the material.

Fortunately for Gu and Zhou, they found new possibilities abroad, and through the lens of foreign research on Chinese territory, they spied an image of geology that was no longer just a mass of indecipherable terms and unfamiliar places, but a coherent system that located their homeland and its material resources within a broader pattern of global history and causation. Suddenly the land they had taken for granted as reliable and familiar became a vast unknown, and this blankness reflected their increasing alienation from the dynastic worldview. Despite all of the Self-Strengtheners' worries about what was in the ground, where it was, and how it could be extracted, no one had paid attention to the materiality of the land itself. Patriotic youth had been struggling to learn "mining" and "geology" when they should have been trying to understand China instead.

Neither "A Brief Outline of Chinese Geology" nor *Record of Chinese Mineral Resources* were scientifically original; in fact, they gained much of their rhetorical urgency from the fact that they were based on research widely available in foreign—and not Chinese—circles. These works, like so much of Zhou Shuren's later fiction, were a "call to arms" that articulated a new way of thinking about the land. By predicating control over Chinese resources on a physical understanding of the land itself, Gu and Zhou suggested to their compatriots that living on the land did not automatically mean possessing it in the fullest sense. Foreigners had traveled to parts of China and experienced it in ways that local Chinese had never even dreamed of. Gu and Zhou described the situation in emotional terms, expressing outrage at the intimacy of this contact and humiliation that the Chinese had neither the awareness nor the power to stop it.

But if geology signified danger, it also promised deliverance, and by offering geological knowledge as a way to defend Chinese interests, Gu and Zhou staked out a novel way for individual Chinese (soon to be "national citizens") to take responsibility for their homeland and contribute to the collective fate.[132] Like other young political outsiders, Gu and Zhou were not invested in the Qing dynasty the way early Self-Strengtheners were. The popularity of their indictment of both Manchu negligence and native bureaucratic complacency marked a budding ethno-nationalist consciousness that took Chinese land to be the birthright of the Han people rather than the demesne of the Qing. To be good stewards of the national territory, the Chinese people had to know what it was made of, whence it came, and how it was connected to the rest of the world.

Gu and Zhou did not have a sophisticated understanding of scientific theories or geological methods, and they did not pretend to have conducted research themselves, but the desire to know the geology of China that they articulated linked an emerging new view of the body politic with the materiality of the land. Sensitized and attuned to this connection between national identity and physical territory, young Chinese were then ready to take the next step, and those individuals who established native geological institutions in the years that followed soon learned that a new relationship with the land demanded a new way of relating to it as well.

TWO

Shaping the Field: Fieldwork and the Creation of the Modern Chinese Geologist

> But in the pursuit of knowledge, bandits could not scare him, barbarians could not stop him, and political turmoil could not sway him. . . . Today, all under heaven is in chaos again, but it is not as bad as in the late Ming. Yet the decay of scholarship is much worse, and the youth of today do not know hard work.
>
> —Ding Wenjiang, on Ming dynasty explorer Xu Xiake[1]

Throughout much of Chinese history, Confucian intellectuals took upon themselves the burden of responsibility for social harmony and the well-being of the state. For many Chinese youth, the failure of leaders both in and out of political office to respond to changing circumstances in the late Qing period was an indictment of their way of life and their ability to lead Chinese society. Through its criticisms of Chinese complacency, Zhou Shuren and Gu Lang's *Record of Chinese Mineral Resources* voiced an emerging desire for geology as not just state knowledge but a remedial form of collective "self"-knowledge incumbent upon all educated and patriotic Chinese. The men who would eventually establish modern geology in China, however, took this line of reasoning one step further and linked the geological investigation of Chinese territory to patriotic action at the level of the body and personal experience.

For Zhang Hongzhao, Ding Wenjiang, and their fellow geological pioneers Weng Wenhao and Li Siguang, positive fieldwork experiences abroad highlighted China's failure to understand its own territory. At the same time, fieldwork seemed to offer a way for these men to bring their newly acquired understanding of China and the outside world back home in a demonstrable and transformative way. If Chinese inattention to the physical nature of their own territory was a lapse of proprietary duty, then personal discipline could potentially redeem the group, and China's first generation of geologists looked to strenuous fieldwork not only to reinvent themselves but to foster collective renewal.

The cultural significance of geological fieldwork is hardly a new idea. But histories of geology have classically approached fieldwork as a manifestation of the epistemological and sociological presumptions that informed the development of geology in the nineteenth and early twentieth centuries. Most notably, in the context of Victorian England, fieldwork was an arena in which "gentlemen geologists" imprinted the earth sciences with the mores of public-school masculinity.[2] The great geological controversies detailed by Martin Rudwick, James Secord, and David Oldroyd all hinged on fieldwork as an empirical practice that united preexisting "elements of romanticism and tacitly pantheistic religion with those of robust, manly Christianity and the gentleman's love of the countryside and its sporting pursuits."[3]

In contrast, for Chinese eager to appropriate geology from abroad, fieldwork expressed not latent cultural values, but rather the possibility of renovating what seemed explicitly wrong about traditional Chinese ideals. Like modernizing intellectuals in other disciplines, the founding generation of Chinese geologists complained that "unawakened" peers had no real contact with actual things, whether in nature, the international arena, or even their own country.[4] However, many of those who clamored for national "self-strengthening" and modernization were themselves unwilling to engage in the manual aspects of investigating "things." Not so for Republican geologists, for whom "fieldwork" encompassed both *shidi diaocha/kaocha* (实地调查 or 实地考察, lit. "on-the-spot investigation") and *yewai gongzuo* (野外工作), or work in the outdoors, with an emphasis on exposure to the elements and physical labor.[5]

In fact, when China's early geologists began to fill in the newly perceived geological "blank" of the Chinese countryside, they found that fieldwork did as much to reshape their bodies and minds as it did to redefine their knowledge of the physical terrain. This dedication of the individual body to the body of the nation gave Chinese geologists a rare sense of possession over their own scientific labors that practitioners of other disciplines, such as physics or chemistry, often found it difficult to achieve. In their view, the geological knowledge they produced fit into international theories and norms, but its foreignness was amended by the unmediated contact of local scientist and native land.

Without a doubt, commitment to fieldwork was the cornerstone of the establishment of an independent native geological program in the years after the republican revolution of 1911. But it is less helpful to think of fieldwork as a "prerequisite" for the success of modern geology in China than to explore the complex ways that Chinese geologists' choices about fieldwork shaped what geology meant and how it developed.

For Zhang, Ding, Weng, and Li as individuals, fieldwork experiences abroad suggested a way to bridge the China of memory and the China of future promise, and as the dynastic era came to an end, these experiences inspired them to reimagine their roles as both scholars and political subjects. Their personal trajectories hinted at geology's potential to discipline the modern individual and introduced them to the epistemological possibilities of directly engaging the land. As these foundational figures returned home, they translated their overseas training into an ambitious plan for a native geological community. In the government-funded Geological School that they founded as a first step toward this goal, fieldwork was used both to attract students and teach them how to do science. This emphasis on empirical methods was aimed at producing effective geological practitioners rather than just faithfully transmitting existing geological knowledge. By the time geological pedagogy moved into the academic setting of Peking University (commonly shortened to Beida), it explicitly joined a broader intellectual movement to revitalize Chinese culture, and fieldwork became a program for creating not only better geologists, but better Chinese men, attuned to the needs of a changing China.

Fieldwork and the Cultivation of Geology Abroad

Like many other students who went abroad after the ailing Qing government abolished the largely classics-centered civil-service examination system in 1905, Zhang Hongzhao, Ding Wenjiang, Weng Wenhao, and Li Siguang found overseas education to be a personal as well as intellectual odyssey that allowed them to gaze at their homeland from a great remove while contemplating foreign ideas and peoples at close range. For this pioneering generation of modern Chinese geologists, who left during the Qing and returned to a new republic, the experience also channeled their general hope of strengthening the country into the specific goal of building a native geological research capacity for the new regime. Central to this was their exposure to geological fieldwork, which honed their technical skills and satisfied their desire to take concrete action.

As Zhang Hongzhao recalled in his autobiography, he was already a thirty-two-year-old *xiucai* (county-level scholar in the old examination system) in 1909, when he decided to start over and pursue the "new learning" of geology in the science college of Imperial Tokyo University:

> At this time I only knew that a great number of foreigners had investigated the geology of China, but I had never heard of any of my compatriots attending to such things. With so many [Chinese] nationals [國人*guoren*], not a one had explored the geology of a single portion of this divine land [神州 *shenzhou*] in any detail. Instead they let outsiders extend deep into the hinterland without their knowledge; it was shameful enough.[6]

Echoing Zhou Shuren's complaints from a few years earlier, Zhang's term for "hinterland," *fudi* (腹地, lit. "the belly of the country"), underscores the way that he saw unchecked foreign exploration as a violation of the national body and helps explain his willingness to "personally take responsibility" for studying the geology of China firsthand.[7]

In Tokyo, Zhang performed well in his studies, but the comprehensiveness of geological activities in Japan only exacerbated Zhang's sense of humiliation and urgency. Not only had his professors in Tokyo "left their footprints" all over Japanese territory, they also expected their students to conduct fieldwork whenever there was a school vacation.[8] These demands were so great that Zhang rarely had a chance to go home and visit his family, despite the relative proximity of Japan to the Chinese mainland. Finally, Zhang was encouraged by his advisor, Inouye Kinosuke (井上禧之助), to turn his attention to China for his thesis project, and he was able to return to his native Zhejiang for fieldwork during the summer before his senior year, in 1910.

Zhang's three months investigating the famous Hangzhou West Lake area were productive but bittersweet. In order to keep up with field training at school and live up to his family's high hopes, Zhang had missed his last opportunity to return home before his father passed away, and yet just half a year later he was in his native province conducting fieldwork. The trip was also a poignant reminder of how backward China was. Whereas geologists in Japan could find a clean and comfortable place to sleep in even the remotest regions in Japan, in China facilities were such that a researcher could get little rest after a hard day's work, and Zhang returned to Japan ill from the unclean conditions.[9] More importantly, Zhang had to work without the assistance of even rudimentary modern maps of the region. The best maps available were in the traditional style, without longitude and latitude, elevation or consistent scale.[10] Even as a student in Japan, Zhang could consult general maps of the whole country and check local reference materials before ever setting foot in the field, and his work was guided by accomplished geologists familiar with regional features.[11] In China, Zhang found no usable maps or books and had no other geologists to compare notes with. The study-abroad experience had deepened his identification with China, but it had also trained him to see his own country with different expectations, and he resolved to use the "joy within adversity" when "all that meets the eye is new knowledge and each beginning is an achievement" to rouse others to further action.[12]

Although Ding Wenjiang went abroad at age fifteen and followed a very different trajectory in his overseas education, he experienced a similar combination of attachment to and alienation from his native territory. While taking preparatory courses in Japan for two years (1902–4), Ding witnessed the ways in which Chinese students were constantly led to compare their new surroundings with their homeland and how discrimination encouraged a strong sense of "common interest" among students from all over China.[13] His most formative intellectual experiences, however, occurred in England, where he attended secondary school in the town of Spaulding in Lincolnshire. There he finished four forms in two years under the sponsorship of a local physician who had done missionary work in China.[14] Afterward he was accepted to Cambridge, but Ding found the lifestyle beyond his means, and he left after a single term.[15] Eventually he enrolled in Glasgow University (1908), where he completed a double major in zoology and geology, with a minor in astronomy. Of these subjects Ding's talents clearly shown in geology. The "broad-shouldered Chinese student of brilliant intellectual powers and of charming personality" was a favorite of Glasgow's geology chair, John Walter Gregory, and earned first-class certificates in geology during three terms. In 1911, Ding graduated with first-class honors and the Cowie Prize for fieldwork in geology.[16]

Rather than going directly home to Jiangsu Province after graduation, Ding took a circuitous route via Annam (modern-day Vietnam) and then Yunnan, Guizhou, and Hunan provinces. Over the course of three months in the summer of 1911, Ding traveled by foot and by boat, observing the geology of southwestern China for himself.[17] His published accounts of these travels were aimed at a nonscientific audience, and he gave few details of his geological investigations, but emphasized the ways that after nine years abroad fieldwork was a "revelation" that reintroduced him to the "real face" of the Chinese landscape.[18]

Both Zhang and Ding returned to China as soon as their studies were completed and looked for outlets to apply their newfound skills to local conditions. As we shall see in the next section, their efforts coincided with the fall of the Qing empire and cleared the ground for the Republic's future geological institutions. In contrast, Weng Wenhao and Li Siguang returned several years later, and though both men were similarly explicit about their eagerness to investigate Chinese territory, the political transition actually gave them reason to pursue higher degrees and broaden their fieldwork experiences abroad. Because they studied geology in Europe rather than Japan, this approach was partly determined by practical concerns such as distance and funding, but Weng and Li were also interested in diversifying their training to better locate Chinese geology within the international context.

After winning provincial support to go abroad in 1908, the only time that Weng Wenhao returned home from the University of Louvain in Belgium was between his undergraduate and graduate coursework, in the summer of 1911. At that time, funding from Zhejiang Province unexpectedly ended because of the political handover. Weng's father, a wealthy Ningbo businessman, was ambivalent about using family money to help him continue, but his mother secretly sold her trousseau to support his return to school, and he did not go home again until he had his PhD in hand.[19] For Li Siguang, whose studies at the University of Birmingham in England from 1913 to 1918 were funded by the Ministry of Education, visits home were specifically barred by national law.[20] Committed to staying abroad, Weng and Li both made the most of their novel opportunities.

Weng graduated from his undergraduate studies with highest honors in geology in 1911, after just two years, and he was immediately accepted into his department's doctoral program to work under Felix Kaisin.[21] He spent the summer of 1912 exploring the Belgian countryside, before concentrating his efforts on igneous strata at Lessines. After a year of focused fieldwork and microscopic analysis of collected specimens, Weng completed his doctoral thesis, *Contribution a l'étude de la porphyrite quartzifère de Lessines*, in 1913.[22] The thesis earned highest honors, and it was published in the University of Louvain's geological journal, where it garnered a great deal of attention for its careful treatment of Belgium's understudied igneous formations. Weng was not only the first Chinese geologist to earn a doctoral degree, but the first to make a significant contribution to the geological understanding of another country. More importantly, by mastering the skills of independent field research in a well-studied area under the exacting standards of a mature geological community, Weng prepared himself for the challenge of researching China's comparatively unexplored territory.

Li Siguang appeared to follow a more conventional route by choosing to survey existing research on the geology of China, but he also did so with a view to identifying lacunae and planning future fieldwork. After finishing his undergraduate work in geology at the University of Birmingham in 1917, Li devoted his summer to a comprehensive survey of foreign literature on Chinese geology, and produced a map of previously studied field routes.[23] Under the influence of his advisor, William Savage Boulton, Li expanded this project into a masters thesis that not only systematically analyzed available information about Chinese geology, but set out guidelines for future empirical study.[24]

Neither Weng nor Li limited their activities to schoolwork, and the two men threw themselves into life abroad, taking particular advantage of their proximity to some of Europe's classic geological field sites. Weng spent his first school vacation, in 1910, traveling in France and England, and he stayed on in Belgium for half a year after receiving his doctorate so he could take short field trips throughout Europe.[25] Li was unable to go to continental Europe because of World War I, but he spent several summers at English coal mines to supplement his official allowance. After completing his masters degree in 1918, he worked in Cornwall "to get a wide range of on-the-ground geological knowledge before returning to China."[26] When the war ended, Li did fieldwork in several mining regions of France and Germany before tackling structural and glacial geology in the French, Swiss, and Italian Alps.[27]

Though Weng and Li both turned down well-paid commercial opportunities in order to teach and do research in China, each delayed his return home in favor of additional fieldwork in Europe.[28] Fieldwork was not only a matter of scientific training, it was also a way of fitting China into a global frame of reference, and the chance for firsthand geological experience abroad was too precious to waste. Later, as Zhang Hongzhao and Ding Wenjiang began training Chinese geologists in Beijing, Weng and Li would contribute much of the foreign expertise needed to bring native geological education in line with Western standards in the 1920s.

Fieldwork as a Means of Establishing Geology at Home

For students like Zhou Shuren and Gu Lang, who returned from technical studies abroad in the last decade of the Qing dynasty, career options were generally limited. Though some returned students carved out creative paths to accomplishment (as Zhou Shuren would in later years) or parlayed their overseas experience into successes in officialdom, most students of mining, shipbuilding, and railway engineering who set out to change China's position in the world settled for minor government posts, teaching positions, or jobs at short-lived industrial and military ventures.[29] Fish out of water for more traditional careers, it was not always clear what these returned students could really offer, and it was frequently difficult to coordinate their efforts or integrate them into existing programs. Despite the intervening years and the myriad attempts to jump-start China's modernization at both the imperial and local levels, their range of options was not markedly different from those available to the first generation of Chinese students who studied technical subjects in America through the Chinese Educational Mission in the 1870s.[30]

When Zhang Hongzhao and Ding Wenjiang returned to Qing China in early 1911, their immediate prospects were also uncertain, but two key factors primed them for the political upheavals ahead. First, they were both committed to the idea that Chinese themselves had to take the lead in investigating the land, and second, they were both convinced that fieldwork was the most effective and practicable way to do this. While one or two other returned students had attempted to publish findings from firsthand observations of mining regions in the north, Zhang found it difficult to find a suitable Chinese audience for his thesis research.[31] He opted instead to try and gather support for a native geological survey that would work systematically to explore the territory, and he promoted his ideas in the Tianjin-based *Dixue zazhi* (地學雜誌, Earth studies journal) before he even graduated from Tokyo University.[32] In September 1911 he took the Qing court's returned students' examinations in Beijing and received a *jinshi* degree equivalent in *gewu* ("science"), but his brief experience teaching at Imperial Peking University's floundering Geology Division reinforced the sense that something more ambitious was needed to jump-start Chinese geology.[33]

As it happened, a week after results of the returned students' examination were announced, the October 10, 1911, Wuchang Uprising set in motion the unexpectedly rapid collapse of dynastic rule, and with the weight of former viceroy Yuan Shikai's military forces tipped against the imperial court, revolutionary leader Sun Zhongshan (孫中山, also Sun Yat-sen) officially declared the establishment of the Republic of China in Nanjing on January 1, 1912. Among many appointments on that day, Zhang Hongzhao was named the first director of a newly created Geological Section (地質科) under the Department of Mines in the Ministry of Industry, and it seemed as if his attempts to generate interest in geology through publication and networking had paid off.[34] Zhang sent out reports announcing the new section to provincial administrators and indicated his intent to train Chinese geologists, but could get little else accomplished. While previous generations of returned students often encountered resistance to new ideas, Zhang found the young Republican government open but indecisive, and too uncertain about its own survival to support geology with much tenacity. In April 1912, the national government relocated from Nanjing to Beijing, and the Geological Section survived the division of the Ministry of Industry into the Ministry of Agriculture and Forestry and the Ministry of Industry and Commerce, but support for its work was so shaky that Zhang left his post as section head in protest.[35]

Like Zhang Hongzhao, Ding Wenjiang earned a *gewu jinshi* at the returned students' examination in September 1911, but after the Wuchang Uprising, he kept to his original plans. He took up teaching duties at the Nanyang Middle School in Shanghai and continued to do fieldwork throughout the Yangzi River valley. Ding, who complained of the uneven quality of geological monographs published by Western amateurs in China, did not publicize his own work, preferring for the time being to use fieldwork as a private means to see his China with new, more competent eyes. Nevertheless, he had made contact with government officials at the examinations in Beijing and soon began to lobby for an official geological organization to sponsor a comprehensive study of Chinese territory, starting with the coal-rich regions of the northeast and the frontier areas of the southwest.

After Zhang's departure from the Geological Section in the winter of 1912, Ding was the natural choice to take over, but when he did so in February 1913, he quickly discovered that

> the [Geological] Section had a porter and two junior functionaries, none of whom knew any geology. The "section" was an administrative unit, but we didn't have anything to administer. I [Ding] repeatedly asked for travel [field-trip] expenses, but the Ministry said there were no travel funds.[36]

In fact, Ding recalled that when he complained to his friend Zhang Yiou (張軼歐), who had studied mining in Belgium before becoming head of the Department of Mines, his friend responded with a laugh, "That's exactly why I hired you. If everything was in place, what would we need you for?"[37]

With his work cut out for him, Ding set about laying out an action plan, which he distributed as a "Synopsis of Pilot Geological Investigations of the Ministry of Industry and Commerce" (工商部試辦地質調查說明書). In this sypnosis Ding described a fifteen-year program that would split Chinese territory into three tiers based on importance and accessibility. The first steps would require personnel to analyze available data (in multiple languages) and consult with local "elders" familiar with the special features and known resources of the landscape. With this preparation, "specialists with extensive knowledge and experience" could then "enter mountains and valleys with equipment in hand to undertake field investigations" using modern methods.[38] On the already ambitious assumption that a well-trained field geologist could measure and study about six thousand *li* on foot annually, Ding estimated that about fifteen hundred geologists could get the job done, with about three million yuan a year for salaries plus funding for travel costs, equipment, library resources, and publication.[39] Ding's estimates were in no way a serious request given the resources of the young Republic, but he hoped to convey the enormity of the task and the practical factors that had to be considered.

The plan incorporated Zhang Hongzhao's broad vision and helped Zhang Yiou convince the ministry to approve a Geological Survey in place of the inactive Geological Section. Even so, these moves did not represent a firm commitment, and the main challenges identified in the sypnosis remained: the survey lacked both money and talent.[40] The survey was charged with "planning and managing" all national geological investigations, yet the ministry earmarked fifty thousand yuan for start-up costs without committing to a regular annual budget.[41] When Ding Wenjiang assumed directorship in late 1913, he was still the only technical staff member, and as he was finishing fieldwork along the Zheng-Tai railroad line (from Hebei to Shanxi), the Ministry of Agriculture and Forestry and the Ministry of Industry and Commerce were amalgamated into a new Ministry of Agriculture and Commerce.[42]

The problem of personnel had been obvious to Ding since his frustrating days in the Geological Section, but with the encouragement of Zhang Yiou, he shoehorned a Geological School (地質研究所) along lines originally proposed by Zhang Hongzhao in 1912 into the Department of Mines in the run-up to the reorganization of the ministries.[43] Though he and Zhang Hongzhao were returned students themselves, they both knew that China's development of its natural resources could not rely exclusively on "pilgrims heading west to procure the scriptures." Not only was the cost of overseas education often prohibitive, but "it could not be denied that for those who have had many difficult years of study at the remove of many hundreds and thousands of *li*, China's geology was nothing more than hearsay or a vision glimpsed across a divide."[44] China required a natively trained corps of geologists personally familiar with problems of Chinese geology and willing to devote their service to national development. Fieldwork was essential.

Unfortunately, government austerity left little money for a school. Coincidentally, budget problems at Peking University had shut down its entire Science College, and its flagging Geology Department, which had only graduated two students in four years, was facing permanent closure.[45] The timing was right, and the university offered to lend its equipment and facilities to the Ministry of Agriculture and Commerce, cutting its own losses and minimizing costs for the new government Geological School.[46]

After the civil examination system was abolished in 1905, Peking University became the new domestic outlet for well-connected young Chinese with dreams of officialdom. In the early years of the Republic, the university was still seen as a privileged path to bureaucratic advancement rather than a serious academic institution, and students found it difficult to imagine how a degree in geology could further their political ambitions. The Geological School, in contrast, lacked the high-profile of Peking University, but it also dodged the air of decadence and intellectual indifference that the university had acquired. As a government agency tied to a new national survey, the school offered the tacit assurance of future employment and an explicit commitment to fieldwork.[47]

Zhang and Ding began recruiting students as early July 1913, and hundreds of youth who were disillusioned with the empty promises of Chinese educational reform turned out for the Geological School exams. Ding personally supervised entrance examinations in Beijing and Shanghai for middle school graduates with competence in mathematics, inorganic chemistry, English, and Chinese language and literature.[48] Given anticipated fieldwork demands, candidates also had to have physical strength, stamina, and the ability to endure strenuous labor.[49] Thirty-six students passed, but after personal interviews Ding whittled this number down to the twenty-seven he deemed truly qualified to handle the school's intensive three-year program.[50]

Those who enrolled were eager to put their modern-style schooling to use, so the school's founders fought hard to make sure that the program was both substantive and affordable. With no tuition costs and guaranteed housing, students at the Geological School had only to provide for board and personal expenses.[51] Similarly, the school provided all classroom materials and covered fieldwork costs so that its specialized curriculum would not be an undue financial burden for students.

Each school year was divided into three trimesters, and the first year was devoted to "general," or introductory topics, such as mineralogy, petrology, paleontology, current ideas in geology, geography, chemistry, zoology, surveying, and German.[52] Several of these courses, such as optical mineralogy and mechanics, had associated laboratory components, and Saturdays were reserved for outdoor surveying practice. Summer vacations were scheduled to include at least one long-distance field trip.

According to the original plan, the second school year added "fundamental," or advanced topics, including chemical analysis, structural geology, and historical geology, and Saturday excursions shifted to actual geological fieldwork.[53] By the third year, students would be expected to have a firm foundation in geological principles and they would be directed toward practical applications and field research. Besides advanced topics in petrology, ore deposits, physical geography, mineralogy, historical geology, and paleontology, coursework would include drafting, geological report writing, and photography, but the overall emphasis was to be on independent field research with both mineral resources and theoretical concerns in mind.[54] Ding and Zhang wanted geology to take root in China through firsthand experience and not textbooks or memorization.

The Ministry of Agriculture and Commerce, however, took a far more matter-of-fact view toward fieldwork and its own limited capacities. By October 1913, when classes began, Ding Wenjiang was occupied with research duties at the survey and preparing for an even longer trip through southwestern China in February 1914. With Ding in the field and the survey "active," ministry officials lost interest in geological training, and Zhang Hongzhao, who had taken over as director, had to fight to keep the Geological School open.

The first battle had to do with teaching staff, since Zhang had originally hired German geologist and mining engineer Friedrich Solger at great expense, and this fell through when Solger insisted on accompanying Ding on his Zheng-Tai railroad expedition. Once that was finished, Solger then left for Qingdao and ultimately returned to Germany as war clouds gathered in 1914.[55] Without Solger's assistance, Zhang was forced to take on the lion's share of first-year teaching at the Geological School, and even had to round up a few members of the ministry whose geological and mining expertise was going unused in their regular bureaucratic positions.[56]

Having improvised to keep the Geological School on course for the first year, Zhang still had to keep the ministry from shutting the whole project down completely. He argued that every time programs for students with modern training were aborted, vital talent was wasted.[57] Moreover, it was shortsighted to think that Ding's individual efforts could constitute a sustainable foundation for Chinese geology: "The Geological School was created to cultivate survey personnel. If there is a Survey but no surveyors, its strength is not real strength, and without real strength its actions are not real action . . . the so-called Geological Survey is merely a name."[58] The school's mission was not to teach geology as a subject, but to train skilled men for immediate use, and only the Geological School could provide the requisite field experience to make this a reality. Finally, after the Ministry of Education confirmed that it lacked qualified teachers for any comparable courses in geological study, the Ministry of Agriculture and Commerce relented, but the school was barred from accepting any further students, and it had to close as soon as the first class graduated.[59]

The second year of the Geological School brought with it a deep sense of mission. The school's existence was linked to the progress of one cohort, and students and teachers alike felt pressure to demonstrate the virtues of "Chinese students in a Chinese school under Chinese teachers studying Chinese geology."[60] During the school's administrative crisis, Zhang Hongzhao taught several first-year geological courses himself.[61] Beyond regular Saturday surveying, Zhang could only find time to lead one extended field trip in the summer of 1914, guiding his students through the Western Hills of Beijing as he had been guided through the Japanese landscape several years earlier.[62] By fall 1914, however, Weng Wenhao had returned with a doctoral degree in geology from the University of Louvain, and Ding Wenjiang was back from his Yunnan fieldwork. Both men took up teaching responsibilities and taught specialty courses to round out Zhang's introductory curriculum.

The arrival of Weng and Ding not only added breadth and depth to the school's teaching, it also reinforced the school's commitment to fieldwork, and Ding was especially eager to see greater integration of classroom and outdoor study. Despite differences in their overseas experiences, all three men felt that fieldwork was the critical element in geological training, upon which their efforts would stand or fall. To clarify concepts to Chinese students who were struggling to master an unfamiliar subject through English or German textbooks that used inaccessible examples from Europe and America, Zhang, Ding, and Weng "did not dare confine themselves to the common references on the shelves or the received theories of foreigners."[63] They believed that "if we want this discipline to move forward, the only choice is to shoulder [our] axes and go into the mountains, hacking through thorns and brambles and chopping down underbrush to further our knowledge of actual facts."[64] With the guidance of teachers who had studied abroad and the occasional visiting instructor, students learned to "climb Mt. Tai to study gneiss and ford the Yangzi River to observe alluvial deposits."[65] Fieldwork enabled them to fit the Chinese landscape into generalizable scientific categories without forgetting that "territories are not the same and many differences exist—only when one's shoes touch the ground can one see the reality beneath the surface."[66] Experiential learning not only bridged the gap between foreign concepts and natural phenomenon, it also heightened sensitivity to regional particularity and physical variation.

With its combination of geological theory and physical practice, fieldwork inspired "boundless interest" in the Geological School's student body.[67] In part, the reason was simply the beauty of mountains and lakes and the opportunity to explore distant, storied scenery that others could never glimpse firsthand.[68] But Zhang hoped his students would be able to use these "joys of the soul to conquer the pain of the body" intrinsic to research in the field, for "our country is now poor. If we desire to remedy this, we must begin with industry, but if we want industry to flourish, we must in turn start from geological investigation."[69]

Foreign powers, unfortunately, were all too aware of this. Midway through the second school year, on January 18, 1915, Japan issued its Twenty-One Demands, including possession of resource-rich Manchuria and Shandong Province, control over China's main coal deposits, and acceptance of Japanese "advisors" on China's military, commercial, and financial affairs. Upon threat of war, the Chinese government under Yuan Shikai (袁世凱) submitted to these terms in May, though Britain and the United States pressured Japan to eliminate the advisory clause. In the midst of this national emergency, Zhang Hongzhao quickly revised the Geological School's curriculum so that, by March 1915, students were studying mining, metallurgy, drawing, and mechanics, instead of just advanced theory.[70]

Group fieldwork was accelerated during term-time, and as a final project each student was assigned a district for independent research. China was in urgent need of both a detailed geological map of the entire country and a comprehensive catalog of the nation's mineral resources. Though the Geological School's students could not be expected to carry out such a monumental task, fieldwork guidelines were designed to maximize their contribution, for every student in the field was a pioneer. Students were graded on detailed field notes, to be taken on the spot and well-illustrated with pencil sketches, photographs, planar and sectional figures, longitude and latitude measurements, and verbal descriptions. They were also responsible for collecting and carefully labeling specimens for later analysis and preservation, with particular attention to ore samples. Finally, all field reports had to include several maps for each surveyed area, including a 1:50,000 topographical map and a 1:10,000 geological map (1:50,000 if detailed topographical maps were already available).[71] Instructors helped students process their raw materials in drafting and report-writing classes, simultaneously gaining intimate knowledge of student data and personally reinforcing good work habits.[72]

At the end of 1916, Weng Wenhao and Zhang Hongzhao published *Student-Teacher Studies of the Geological School* (地質研究所師弟修業記, *Dizhi yanjiusuo shidi xiuye ji*) based on group and individual field assignments over the course of the school's three-year program. Weng acknowledged that student work was "training" rather than "directed research," but reasoned that "there was no reason why it could not augment that which our predecessors have not yet attained and guide the path of inquiry for our successors."[73] To this end, Zhang and Weng did everything they could to synthesize student data and existing (foreign) geological research. They organized *Student-Teacher Studies* topically rather than geographically, covering systematics, igneous formations, structural geology, and mineral products across several parts of Zhili (Hebei), Shandong, Shanxi, Jiangsu, Zhejiang, Anhui, and Jiangxi provinces, with supporting maps and sections. In conclusion, they reflected on the experience of teaching in the field and offered speculative thoughts on broad questions, such as China's north-south divide, its geological history, and the relationship between geology and its natural resources.

Student-Teacher Studies was the first book-length, original contribution to Chinese geology written in the Chinese language.[74] Like the Geological School it commemorated and the survey activities that came after, this book displayed a timely mix of practical results and pedagogy, framing fieldwork as both the product and producer of China's newly minted geologists. When the Geological School ended in July 1916, eighteen students graduated with an advanced degree and three others received a basic diploma.[75] All of these advanced graduates entered the Geological Survey, which was reorganized in October 1916.[76] The revamped survey had three sections, a general affairs division headed by director Ding Wenjiang, a geology division led by Zhang Hongzhao, and a minerals division under Weng Wenhao. In its early years, however, it was essentially a continuation of the Geological School. The survey's own guidelines barred members from holding outside teaching positions, but with Geological School graduates as junior investigators and Ding, Weng, and Zhang working side by side, student-teacher relationships were built into the structure of the organization regardless of salaries or titles, and fieldwork retained the dual functions of "training" and "research."[77]

One of the strongest criticisms of modern education in Republican

China was its "revolving" nature, which simply reproduced itself without ever actually serving the nation. As late as 1924, a student at Nankai University complained in a now-famous quotation:

> The educators love to say: "The aim of education is to save the country." We like to ask, how? . . . Students of literature have become teachers; students of sciences have become teachers; students of commerce likewise have become teachers. You are teachers and we are teachers. . . . We ask what a middle school teacher does, and the answer is to teach the students English, mathematics, and other knowledge to prepare them for university education. Why university education? The answer is to get a diploma. Why a diploma? The answer is to become a middle school teacher. Thus men revolve around education, generation after generation. Can such a merry-go-round save the country?[78]

By emphasizing fieldwork as the key to understanding geology, the Geological School sidestepped this cycle and graduated students equipped to immediately take up research duties in the Geological Survey. The survey in turn served as an incubator, accepting fresh graduates from first the Geological School and later China's academic geology departments, and giving them practical opportunities to refine their skills. From the outset, China's foundational geologists asserted that "without men to investigate, there can be no talk of investigation," and throughout the Republican period geological education focused on producing neither teachers nor textbooks, but practitioners.[79]

Fieldwork as a Habit of Mind and Body

When the Geological School closed its doors in July 1916, most of its graduates were hired full-time by the Geological Survey, and Chinese geologists were set to begin organized study of their national territory. In order to guarantee future recruits and build on this momentum, Ding Wenjiang immediately initiated talks for a department of geology with newly appointed chancellor of Peking University, Cai Yuanpei (蔡元培, 1868–1940). Shifting geological training to an academic setting would put scientific education on a more permanent footing and embed it in the broader ideals of cultivating modern youth.

Cai, who had been minister of education in 1912 before resigning in disillusionment, was a Hanlin academy scholar but had also spent many years studying in Germany and France. He accepted the chancellorship of

Peking University in late 1916 in hopes of rehabilitating the school's corrupt reputation and establishing a true center of learning that would inspire changes in the entire educational system. At the heart of Cai's educational philosophy was a belief that education should prepare the individual for a constructive role in society through a tripartite emphasis on utilitarian education, ethical education, and aesthetics. With a strong commitment to academic freedom, Cai hired faculty from across the political, cultural, and social spectrum, inviting, for instance, both the radical champions of the vernacular movement and the staunchest defenders of the classical literary style.[80] A department of geology was, in Cai's view, a necessary component of utilitarian education, and he eagerly accepted Ding's proposal that the university resume geological instruction.

A few introductory classes were offered right away, and in the fall of 1917 Peking University reclaimed the facilities and equipment that it had previously lent to the Geological School and established a new Department of Geology,[81] with He Jie and Wang Lie as professors.[82] With the Geological Survey active and Cai Yuanpei at the head of the university, the department had little difficulty attracting students, and for decades after 1920, when its first eight students graduated, the Beida Department remained the survey's most dependable source of new geologists.[83]

Despite the advantage of continuity, however, the department did not immediately match the Geological School in quality of education, especially with regard to fieldwork. The school was a tightly knit, closed community of students progressing through the same classes at the same pace, and focusing on an urgent common goal. The department, by contrast, accepted new students annually, and, until 1920, it had to maintain a full complement of courses for all levels, with only two professors and insufficient facilities. By 1919, there were thirty-six students at three different stages of a geological program that was transitioning from a three- to a four-year curriculum and trying to introduce electives in course selection.[84] When students formed an independent study society in 1920 (see chap. 3), they complained that resources were so scarce that they rarely had any personal contact with teachers, and they had to fight over laboratory materials.[85] Faculty had no time to take students into the field and the university had no funds for geological trips. The department did not maintain formal fieldwork requirements and focused so much on in-class teaching that students found it difficult to keep up with reading and memorization.[86]

Though he was not officially connected to the Beida Department of Geology, as director of the Geological Survey, Ding Wenjiang had a vested interest in maintaining the highest standards possible for geological education.

He understood that He Jie and Wang Lie were overtaxed and underfunded, but felt that a reliance on textbooks and lectures was a giant step backward. As Ding explained at the graduation ceremony of the Geological School in 1916, "the greatest cause of our nation's weakness is that in general the upper classes do not have close contact with the [natural] world." Ding believed that once students learned to appreciate "the joys of nature," they could resist the temptations of wealth and status. Then, by climbing the great mountains of China's west and roaming where average men could not, young geologists would become "men of real abilities" who would make the nation flourish.[87] In his view, which was shared by Zhang Hongzhao and Weng Wenhao, fieldwork was necessary for both a solid understanding of geological principles and scientific method and an enlightened training of patriotism and character.

The active lifestyle of field geologists challenged deeply entrenched prejudices against the assertion of physical culture among intellectuals. So open was this disdain that Ferdinand von Richthofen opened his five-volume treatise on Chinese geology, with the view that geology was the least likely of modern sciences to ever take root in China, for

> the Chinese man of letters is sluggish and chronically loath to move rapidly; in most cases he simultaneously vexes one with his avarice and cannot free himself from native prejudices concerning decorum. In his view, to go on foot is demeaning, and the occupation of geologists a direct surrender of all dignity in the eyes of the world.[88]

Certainly this opinion reflected both Richthofen's frustration with Chinese officials and his own ethnocentrism,[89] but Chinese in many historical periods made similar complaints of their fellow scholars.

This attitude is often traced to passages from the traditional Chinese canon, such as Mencius's claim that "those who labor with their minds govern men, and those who labor with their muscles are governed by men."[90] It is debatable whether Confucius himself was as disdainful of physical cultivation as later adherents were, since he taught students charioteering and archery as part of their training as *shi*, or minor nobility.[91] Certainly, though, these physical skills were not matters of pride the way mental and moral attainments were, and Han dynasty scholar Wang Chong noted that though "Confucius could lift the portcullis of the northern gate, he never flaunted his vigor, knowing that the strength of sinew and bone was not as honorable as the strength of benevolence and righteousness."[92]

It is perhaps less important whether China's sages and philosophers truly disapproved of physical culture than that their teachings were historically assimilated in this way. As a modified form of Confucianism became the state-sanctioned path to officialdom,[93] intellectuals used physical values to separate their civilized (文, *wen*) pursuits of literature and art from the violent habits of martial (武, *wu*) men, and physical prowess was similarly employed to separate races by defining Chinese cultivation in opposition to barbarian brutality.[94] The class implications of the denigration of physical skill and exertion are clear in Mencius's quotation, and they were reinforced by the concentration of wealth, social status, and intellectual skill in the hands of scholar-officials. Even among intellectuals themselves, passages referring to the Sage-king Shun sitting "with hands folded in quietude and all under heaven in good order" could become pointed barbs suggesting that those who had to exert themselves to achieve a desired result were simply less competent.[95] It would have been shocking for traditional intellectuals to learn that Weng Wenhao considered it the highest praise to say that "even though there were several pack animals to transport luggage, . . . [Ding Wenjiang] walked for great distances, observing rocks and making sketches, hands and feet constantly at work."[96]

The privileging of the mental over the physical was also evident in art and literature. Though both *wen* ("civilized, literary") and *wu* ("martial, physical") heroes were popular in plays and fiction, *wu* figures were invariably subordinate to righteous *wen* characters and were often redeemed by a turn to scholarly pursuits. In China's famous romances, often referred to as *caizi-jiaren* (才子佳人, "scholar-beauty") stories, the protagonists were always fair, slender young men of literary skill who wooed women with poetry and brilliance rather than bravery or physical prowess.[97] Once women were allowed in Chinese theater, these romantic protagonists were frequently played by female actors who could better capture their softness and mildness of manner.[98]

In 1915, in an early volume of the magazine, *La jeunesse*, cultural critic Chen Duxiu mirrored this image of upper-class masculinity, but completely reversed its valuation. He wrote:

> Whenever I look at our educated youth, I see that they have not the strength to catch a chicken, nor mentally the courage of an ordinary man. With pale faces and slender waists, seductive as young ladies, timorous of cold and chary of heat, weak as invalids—if people of our country are as feeble as this in body and mind how will they be able to shoulder burdens and go far?[99]

The educated youth Chen described might still make desirable romantic matches for some, but their atrophied physical traits boded ill for the nation.

The Opium Wars (1839–42, 1856–60) paved the way for this shift in consciousness. China's losses seemed to illustrate how weak the nation was in comparison to Western powers,[100] sparking an interest among so-called Self-Strengtheners to adopt foreign technology (often referred to as *yangwu* 洋務) on the basis of Zhang Zhidong's *ti-yong* formula: Chinese learning for substance (體 *ti*), Western learning for practical use (用 *yong*).[101] This movement resulted in several attempts at military modernization, but it was only after China's unexpected defeat in the Sino-Japanese War (1894–95) that native thinkers began to attribute national weakness to the weakness of the Chinese people themselves.

One of the most influential of these thinkers was Yan Fu, who translated Huxley, Smith, Montesquieu, and Mill into classical Chinese and brought Spencerian social Darwinism to China.[102] In "On Strength," one of four essays published in the aftermath of the Sino-Japanese War in 1895, Yan argued that the "weak become the prey of the strong, and the clumsy submit to the crafty. . . . This is true not only of animals and plants but also of people."[103] For Yan, the unit of natural selection was the group (alternately the society, nation, or race), and to win the "struggle for existence," China had to increase the population's "fitness" by developing its "bodily vigor, intelligence, and moral virtues."[104] Yan did not limit his endorsement of sound physical culture to military drills and calisthenics; he also criticized footbinding and opium addiction, and urged physical education for women, because healthy mothers produced strong children.[105]

According to intellectual historian Benjamin Schwartz, this "affirmation of . . . physical virtues . . . [was] the most dramatic manifestation of the new transformation of values."[106] However, the humanists and educators who embraced this iconoclastic line were still quite instrumental in their motives.[107] Some, like the reformer and public thinker Liang Qichao, simply demanded: "With neither knowledge nor physical strength, how can the Chinese survive in a competitive world?"[108] In response, both military training and competitive sports filtered into modern educational reforms to build stronger bodies.[109] Others, like Cai Yuanpei focused on the body in the service of mental and moral advancement, claiming that "perfection of the spirit requires perfection of the body; if the body is weak, then how can thoughts and spirit flourish?"[110] In contrast, Ding and other pioneering Chinese geologists viewed cultivation of the body and perfection of the spirit as peripheral to the real business of directly experiencing nature.[111] They shrugged off the traditional aversion to physical exertion as counterproductive to their *intellectual* aims. For them, fieldwork was inherently both manual and mental labor, and its physical activities—whether accessing, measuring, sensing, or collecting—transformed landscapes into a natural laboratory for practice and discovery.[112]

Throughout his lifetime, Ding personally exemplified this ethos of muscular science in both his own fieldwork and his pedagogical approach. Two of his mottos—"when climbing a mountain, always go to the top; when on the move, always go by foot" and "take the long path, not the short path; take the mountain trail, not the flat road"—influenced several generations of Chinese geologists.[113] Weng Wenhao described Ding as constantly active in the field, with one hand at work with his hammer and the other taking measurements or making sketches, and Ding transmitted this energetic style to his students.[114] He carried his own equipment and slept, ate, worked, and rested with his students on the principle that fieldwork was always a collaborative endeavor.[115] As a teacher, Ding did not believe in leading students around and pointing out geological phenomena; instead he preferred to suggest problems and give students a chance to examine sites for themselves before influencing their observations.[116] Students could only learn by doing and by making their own mistakes.

For this reason, Ding Wenjiang could not stand aside and allow Peking University's Department of Geology to neglect fieldwork and fall back on textbook learning. Informally, Ding began his own search for qualified teachers to free more time for fieldwork in the department. While he toured Europe as part of Liang Qichao's unofficial Chinese delegation to the Versailles Peace Conference in 1919, Ding met with Li Siguang in England, and, stressing that "cultivation of geological talent is a matter of vital urgency," asked Li to teach at Peking University after finishing his masters degree at the University of Birmingham.[117] On the American leg of the same trip, Ding consulted with chief geologist of the US Geological Survey, David White, who suggested that Amadeus Grabau might be interested in working in China.[118] Grabau, whose pro-German sympathies led to his dismissal from the Geology Department of Columbia University in 1919, had been a beloved teacher to several Chinese overseas students, including mining engineer Wang Chongyou, and Ding was especially interested in gaining someone of Grabau's stature to teach paleontology and stratigraphy.

When the first graduating class from the Department of Geology at Peking University applied for positions in the Geological Survey in late spring of 1920, Ding seized on the chance to test their basic geological skills.[119] Appalled at the results, he sought out his good friend and leading figure in the New Culture Movement, Hu Shi, then a professor of history in the university.[120] Together they approached Cai Yuanpei with the examination results, and argued that both physical vigor and empirical habits of mind were basic to geological training. After several hours with Ding and Hu, Chancellor Cai approved their proposal to bolster the geological faculty and refocus on fieldwork. Amadeus Grabau and Li Siguang were hired immediately, and their arrival in the fall of 1920 ushered in a golden age that soon established Peking University's reputation as having the best geology department in Asia.[121]

The department did not yet have funding for fieldwork, but as soon as classes started, Li began to explore the Beijing surrounds. In October 1920, he took several first-year geology students with him to explore the Liuhegou Coalfield, and under his direction, they collected fossils and measured sections.[122] Early in 1921, professors' salaries in Beijing universities were several months in arrears and a general teaching strike was called. Li was active in these protests, but still took students into the field to survey area coal reserves.[123] Because these excursions were independent research trips rooted in Li's own concern for China's energy resources,[124] accompanying students learned how to approach general problems in the field rather than master pedagogical set pieces selected for specific geological demonstrations. However, this limited student participation to those few who were motivated enough to seek out extra opportunities, competent enough to be of actual assistance, and financially independent enough to pay their own way.

Li understood that a more comprehensive system was needed to give students fieldwork experience, and he repeatedly petitioned Cai Yuanpei for special funds. Finally, in November 1921, Cai invited Li to attend a University Advisory Board meeting, and the Department of Geology was granted a subsidy for geological fieldwork, with extra funding for improvement of laboratory facilities.[125] The effect of this subsidy was immediately felt and two large field trips were organized before the end of the year. University support also made it possible for Li to formally integrate fieldwork into his classes, making small group trips (which he personally led) a general requirement. Li also drew connections between field and classroom by teaching students to section specimens they collected and submit them to microscopic and chemical analysis.[126]

Amadeus Grabau's effect on fieldwork was more indirect, but no less significant. By the time he reached China, Grabau was so severely afflicted with arthritis that he required crutches to move, but his long history of publication in the United States showed a deep commitment to fieldwork, starting from his younger years leading trips for the Boston Museum of Natural

History and teaching at the Rensselaer Polytechnic Institute. This pattern continued through his professorship at Columbia University,[127] and over the course of his career in America he published several works which became standards in the field, including guidebooks of Eighteen Mile Creek and Niagara Falls,[128] *North American Index Fossils* (first edition 1906, with Hervey Shimer), and *Principles of Stratigraphy* (1913). As he was preparing for his move to China, Grabau completed a two-volume *Textbook of Geology* (1920–21), which brought the latest teaching methods of United States universities to Beijing and underscored the importance of strenuous fieldwork, stressing that "the man whose horizon is bounded by the walls of a city can never be a geologist. . . . The true geologist . . . goes directly to the earth and there begins his inquiries."[129]

When he first arrived in China, this enthusiasm for the vast new field of study available to him revived Grabau's health for a brief period, and he managed with some difficulty to conduct a nine-day expedition to the Kaiping basin, where he collected over forty boxes of fossils.[130] He also worked in the Tang Shan area of Hubei Province and several areas in the Beijing surrounds.[131] Grabau's return to health did not last long, but it was long enough to inspire his admiring Chinese students. Later when confined to a wheelchair, Beida students and Geological Survey members alike were untiring in their collection of materials for Grabau's research, proudly acting as his "legs" in the field.[132]

By the mid-1920s the Peking University Department of Geology was as well known for its field training as it was for the caliber of its lectures. By 1924, Ding, who had been so critical of the program a few years before, felt that Peking University surpassed most Western institutions outside of the United States in its attention to fieldwork.[133] The 1927 course schedule lists a full day of geological fieldwork each week for first- and second-year students, in addition to several hours of surveying practice and laboratory time. Upper-class students were required to participate in extended field trips during school vacations, and seniors researched their theses in small groups with the guidance of a faculty advisor.[134] Within a few years, the quality of native geological education convinced Ding that China need not automatically favor geologists trained overseas unless they had advanced degrees.[135]

Conclusion

In the face of imperialist interest in China's mineral and energy resources, many Chinese students studying abroad became aware of the urgency of mining and geology. What set Zhang Hongzhao, Ding Wenjiang, Weng Wenhao, and Li Siguang apart was less their recognition of China's need than their intuition—borne out by later experience—that fieldwork was the key to ownership and assimilation of geological knowledge. Lesser-known figures, such as Parkin Wong (黃波芹, MS Cornell University, 1914) and Wah Seyle Lee (李壽華, MS Stanford University, 1916) wrote meticulous compilations of existing research on Chinese geology, but they remained fixated on pinpointing China's "treasures" rather than understanding its territory.[136]

For Zhang et al., the materiality of the land reacted with their own physicality to form a personal bond that transformed both student and object of study. This direct experience brought abstract and unfamiliar concepts to life, and when China's geological pioneers set out to train future geologists at home rather than abroad, they used fieldwork to make both geology and territory sensible to their pupils. Other returned students, such as the mining engineer Wang Chongyou (王寵佑) (MA Columbia University School of Mines, 1908) and Wang Lie, were technically competent to teach basic geology in China, but only Zhang, Ding, Weng, and Li translated their overseas experiences into a program of education that linked the study of geology with self-cultivation.

As geological education in China moved from the narrow confines of the government Geological School to the academic setting of Peking University, the desired connection between training geologists and the training of a mentally and physically strong citizenry became even more explicit. In the popular press, Ding Wenjiang aggressively touted a "scientific philosophy of life" based on experience and logic, and Weng Wenhao complained that China's political ills could be traced to youth who were afraid of hard work.[137] Neither was able to change the shape of youth at large or effect the "government by good men" (好人政府) that they hoped for, but as teachers and mentors China's geological leaders insisted that men of science were properly men of action and integrity as well.[138]

By the early 1920s the Geological Survey of China was a tight-knit and highly productive organization that was as prepared to deal with the challenge of investigating China's vast territory with only thirty geologists as it was to tackle the technical problems of stratigraphy and structural geology. Over the course of the 1920s and 1930s the survey managed to organize and fund fieldwork throughout northeastern, central, and southern China by grasping fleeting opportunities amid the chaos of warlord and party politics.[139] However, the research efforts of the survey were shielded from the wider scientific community by their isolation from established scientific metropoles and had limited effect on curbing the imperialist tendencies of foreign geologists interested in China. Chapter 3 examines the ways that, through the establishment of the Geological Society of China in 1922, Chinese geologists learned to use newly acquired forms of scientific sociability to bring the products of their expanding fieldwork into the international scientific economy and thus claim a space for Chinese investigation of Chinese territory.

THREE

A Host of Nations: Cosmopolitanism and the Geological Society of China

> Nothing mattered more. Chinese history during the era of the first Republic was defined and shaped—and must ultimately be interpreted—according to the nature of its foreign relations. While few would dispute the contributions of what Paul Cohen has called a "more interior approach" to modern Chinese historical studies in the past two decades, there is no point searching for some uniquely "China-centred" historical narrative for this period. Everything important had an international dimension.
>
> —William Kirby, "The Internationalization of China"[1]

The first eight years of the Geological Survey of China coincided with World War I and its global aftermath. Not surprisingly, this took Chinese mineral resources out of European focus, and the national Survey made haste to expand its activities during the lull.[2] In 1914, geologist and former director of the Geological Survey of Sweden, Johan Gunnar Andersson, accepted Ding Wenjiang's invitation to join the Chinese survey as mining advisor, and with an infusion of Chinese personnel from the Geological School in 1916, the survey expanded to thirty researchers of varying ranks by 1921.[3]

Financing from the national government did not keep pace with the growth of the survey, but under the leadership of Ding Wenjiang, Zhang Hongzhao, and later Weng Wenhao, the survey made use of commissions from mining companies and government agencies at both the central and local levels to conduct broad-based geological investigations wherever possible. While the areas around Beijing received the most attention, the survey's activities often exceeded the effective reach of the national government and extended into warlord-controlled territories that gave little more than lip service to the central state. With its small but committed staff, the survey managed to complete fieldwork in several areas of northern and central China as well as the Yangzi valley region, and parts of Inner Mongolia. This track record and the promise it held for scientific integration of former Qing territories helped the survey weather the many regime changes that comprised the Beiyang government, including a poorly considered attempt to revive the imperial system (1915–16) under Yuan Shikai, and an ever-changing roster of regional warlords.

Though each shift of power in the Beiyang period brought a superficial reorganization of the survey's structure and priorities, the Geological Survey retained its core composition, and this allowed it to carve out a relatively independent path in spite of political pressures. By 1919, the survey had established both the *Bulletin of the Geological Survey of China* (地質彙報, *Dizhi huibao*) for its regular reports and two irregular series of more in-depth *Memoirs* (地質專報, *Dizhi zhuanbao*) for special reports such as Weng Wenhao's *The Mineral Resources of China* (1919). The next year, Amadeus Grabau came on board as chief paleontologist in the new paleontological division, and the Geological Survey was elevated to a position just below the ministerial level. In 1921, the survey even raised enough money for a new building to house its library and museum collections, and a number of China's leading cultural figures hailed the move an important step for Chinese science.[4]

For Chinese geologists themselves, however, several fundamental problems remained. Practical assignments helped the survey stay afloat and created valuable opportunities for fieldwork, but they also limited the time available for following up on basic geological questions. Survey researchers were careful to collect notes and materials with these underlying issues in mind, but data accumulated faster than it could be processed, and targeted surveys of coal and other resources kept pointing out new gaps in what was known of China's geological history and structure. Soon Chinese geologists' self-reproach shifted from shame that Chinese were not able to investigate their own territory, to the frustration that, despite their determined attempts, they were still not contributing a comprehensive new picture of Chinese geology to global science.

This concern that economic applications not overshadow scientific merit was important in the 1910s–1920s context of the New Culture movement, which idealized science and scientific thinking as a cure-all for China's ills, but it complicated Chinese geologists' relationship to the state. Struggling as they were to survive with limited financial resources and even shakier moral authority, many Beiyang politicians were dubious about the power of science to either forward their national agendas or help them stay in office, and support for geological activities was largely based on the expectation of short-term yields. Only a few of the survey's early champions considered the ways in which a less utilitarian approach might bolster the nation's legitimacy as a modern republic.

In point of fact, China's weak scientific standing continued to impinge upon geologists' ability to assert local research prerogatives, and the situation became more acute as foreigners began returning their attention to China and its outlying territories. Predictably, explorers such as Roy Chapman Andrews and Sven Hedin had their sights set on border regions, such as Yunnan, Mongolia, and Xinjiang, but organizations such as the International Geological Congress considered the whole of China a nagging question mark, and they remained quite open to the suggestion that an established colonial power should tackle the problem.[5] However much the founding of the Geological Survey reduced China's reliance on foreign geologists, the quietly growing body of Chinese geological work still fell outside of the international purview, and until China established a visible presence on the world stage, the survey could neither stem the tide of foreign geologists on Chinese soil nor force them to engage with local researchers.

When Chinese geologists finally began to bridge this gap through the activities of the Geological Society of China in 1922, the key was realizing how the need for scientific promotion and circulation could actually link their interests to the interests of others. Rather than demanding foreign respect for the sheer quantity and quality of its output, or initiating formal collaborations that would only reproduce existing power inequalities, the Geological Society played China's weaknesses for strengths. As Li Siguang observed at the twentieth anniversary of the Society, though China might suffer from "the absence of a geological legacy," it could provide locals and visitors alike with a forum free of the "formidable load of traditionalism often witnessed in the Western scientific world."[6] And if it could not keep foreign geologists with imperialist ambitions away, the Society could at least stage scientific events that would bring the international geological community close enough to be of service. In fact, as the Society became increasingly indispensable to foreigners interested in keeping abreast of China's geology, it not only spread awareness of Chinese research internationally, it also strengthened geologists' domestic position, attracting native enthusiasts in and out of the survey and forcing officials and industrialists to take notice of both practical and theoretical advances.

Through its flexibility, informality, and catholic vision, the Geological Society that emerged in the 1920s was less a professional society defining the discipline of geology than a scholarly-cum-diplomatic organ carving out a place for Chinese science at home and abroad. However, this cosmopolitan strategy only emerged after several false starts, and this chapter examines the evolution of Chinese ideas about the proper form and function of a geological society in light of shifting notions of how international science could serve the national interest.

The first half begins with Zhang Hongzhao's earliest published writings on the organization of Chinese geology. These discussions of a geological society arose in the shadow of foreign scientific models, and Zhang's desire for Chinese to both match the successes of geological societies abroad and escape the confines of their example reflected the uneasy political climate immediately before and after the republican revolution of late 1911. Implicit in his early writings was the assumption of a scientific struggle between nations that put responsibility for China's material weakness on the inability of Chinese to study their own territory. Later, the Peking University students who formed their own geological association in the aftermath of the 1919 May Fourth protests imagined themselves defending Chinese interests against rapacious outsiders by collectively pursuing geology and catapulting China to scientific parity with foreign powers.

The second half of the chapter then traces the actual founding of the national Geological Society and the development of its uniquely cosmopolitan style in the "golden years" of the 1920s. In contrast to the hotheaded students of Beida, Zhang and other leaders of the Chinese geological community understood that under the shaky new republic Chinese resources were too limited for an explicitly aggressive strategy. The Geological Society of China embraced international participation as a chance to use scientific sociability to discipline rather than dispute foreign activities. Under the direction of a tight-knit group of overseas-educated Chinese geologists, the Geological Society abandoned earlier models of competition and drew upon Beijing's thriving foreign community to create a meeting place for scientists from all over the world.

Zhang Hongzhao and the Blueprint for Chinese Geology

While studying in Tokyo, Zhang Hongzhao discovered not only modern fieldwork practices, but a thriving geological community that coordinated research with Japan's imperial policies. The example of both the Geological Survey of Japan (est. 1882) and the Geological Society of Japan (est. 1893) gave Zhang confidence that new scientific communities could be created by design and encouraged him to turn to community building to address China's geological needs.[7] From 1910 to 1912 he published three groundbreaking articles that expressed his evolving thoughts on both Chinese geology and China's shifting political circumstances.

All three of Zhang Hongzhao's articles appeared in *Dixue zazhi*, the journal of the Earth Studies Society (地學會, *dixue hui*).[8] The Earth Studies Society was founded in 1909 by the geographer Zhang Xiangwen (張相文) with the explicit goal of creating a cosmopolitan modern citizenry, and its relationship with the Qing court was complex. The Earth Studies Society was not a professional or research association, but a collective of literati who hoped to advance Chinese interests through the exchange of up-to-date information. On the surface, the Society was one of several programs that supported the constitutional assemblies promised by the Qing court in 1908.[9] However, several of its members, including Zhang Taiyan (章太炎) and Cai Yuanpei, were prominent revolutionaries, and its founders Zhang Xiangwen and Bai Yukun (白毓崑) both participated in local republican uprisings in 1912.[10]

The leaders of the Earth Studies Society may have been more politically radical than Gu Lang and Zhou Shuren a few years earlier, but they remained more conservative in other respects. Bai Yukun believed that, as part of the "original essence of our country, [the earth sciences were] especially able to arouse feelings of love and respect for the national territory."[11] While this reflected a new attention to the materiality of the land, it framed modern earth studies as an extension of existing traditions. Building a new citizenry meant arming educated Chinese with more accurate and detailed knowledge of world geography, modern methods, and local conditions. In the style of Wei Yuan and Xu Jiyu, the Society's founders felt at home in the national territory but wanted to reduce the "chasm" between China and the outside world.[12] They saw this as a problem of collection and dissemination and did not see themselves as building a research community committed to any particular science until the 1920s.

As a result, the early *Dixue zazhi* covered a wide array of modern "earth studies," including geography, geology, mining, arctic exploration, oceanographic and astronomical subjects, and land use. Each issue contained detailed travel accounts, personal observations of recent floods or earthquakes, reports of local agricultural, soil or climate data, and other firsthand contributions, but most of its technical articles were adapted from foreign periodicals and therefore focused on foreign areas. The only exception was a series of geological and mining maps of Zhili (now Hebei Province) produced by Kuang Rongguang from his own observations as chief prospector of the Zhili Provincial Bureau of Mines.[13] Unfortunately, though these maps (and an accompanying paleontological sketch) were the first original geological materials published by a Chinese investigator, they were very simple and sparked little more than patriotic pride, even among Earth Studies Society members.[14]

When Zhang Hongzhao made his first contribution to *Dixue zazhi* in 1910, he understood that it was not an academic journal, but he hoped to reach an audience of like-minded individuals with enough influence to create change. The journal was not only read by reform-minded literati and urban elite, but it was also monitored and excerpted by the official daily of the central government. Though the Earth Studies Society had revolutionary leanings, Zhang used *Dixue zazhi* to address his concerns about China's lack of geological research to the imperial bureaucracy.

In his first article, "The Workings of the Geological Surveys of the Countries of the World," Zhang used his knowledge of foreign geological institutions to offer a new spin on the familiar lament that the "competition between nations" for natural resources was holding China back.[15] Taking a page from the history of geology, he argued that national geological organizations had a proven track record of expanding national wealth, and the bulk of his multipart article consisted of detailed descriptions of national surveys, including personnel, budgets, primary activities, and major publications. He also offered his own opinions on the strengths and weaknesses of the various systems. Impressed as he was by the experience of the Japanese system, Zhang wanted to compare all available models and choose a strategy that best suited China's territorial extent and limited scientific resources. Unfortunately for the ailing Qing state, Zhang Hongzhao's article was published too late to be of much use to it. By August 1911, the political situation was so unstable that no further issues of *Dixue zazhi* appeared until January 1912, the first year of the new Republic of China.

Once the revolution was a fait accompli, Zhang continued to promote geological research, but his attitude toward the possibility of a native Chinese geological community shifted in line with the new possibilities and pitfalls of the young republic. In his first article, Zhang noted that the first geological surveys in Europe were preceded by active private associations such as the Geological Society of London (est. 1807).[16] However, he tried to deemphasize this and used the American example to demonstrate that, while geological societies made the work of national surveys easier, properly supported surveys could be successful even before a local geological society existed. Before the Xinhai Revolution, Zhang hoped this historical argument would preempt Qing concerns that the absence of a geological society might undermine a national survey. After the revolution, in his second article, "My Suggestions for China's Geological Investigations" (1912), a geological society could emerge as an actual desideratum, lamentably out of reach in China's present circumstances, but central to future development in the hands of local workers.[17]

This change reflected Zhang Hongzhao's new ambitions for Chinese geology as a whole. While the Qing were in power, the best way for Zhang to justify geological research was to point out the immediate benefits for the empire. Despite its weaknesses, the standing of the Qing as rulers of China was not in question and natural resources would easily be in their purview. Zhang maintained an air of objectivity, describing different foreign geological surveys and hoping that Qing bureaucrats would make an informed choice, or indeed *any* choice rather than inaction. Though he did not deal with this issue in writing, Zhang was all too aware that there was a dearth of qualified Chinese personnel, and that the Qing would most likely hire foreign geologists to carry out any necessary work. However, his goal was to harness geology for advantage of the country as a whole, and discussion of a native geological community or other long-term plans would only discourage the already lethargic dynasty.

Once the new republic began, Zhang faced an untested and resource-poor national government that required the support of local officials and ordinary citizens to generate change, but also espoused lofty ideals of modernization and self-determination. As a result Zhang's writing emphasized persuasion over raw facts, balancing both danger and opportunity to make a case for a truly Chinese geology. Zhang would later recall that he pitched his second article in *Dixue zazhi* at provincial leaders and the educated public, as well as the central government.[18]

Geology was central to both the survival of the Republic and its status as a Chinese nation. It was humiliating for Zhang that foreigners monopolized scientific understanding of Chinese territory, and Zhang advocated geological investigation as part of owning the land, or, as he put it, fulfilling the "proper duties of a landlord."[19] Zhang likened China to the heart of Asia and appealed to national pride to encourage Chinese to be responsible to Asia's "limbs," Russia, Japan, India, the Philippines, and Indochina.[20] Moreover, all civilized nations understood the value of contributing to geological knowledge; if China continued to shirk this duty, it would forfeit its claim to "civilization."[21] With these new arguments at his disposal, Zhang asserted that China was in urgent need of both scientific principles and practical knowledge. He turned to the history of geology again, this time highlighting the ways that, in Europe, geological societies had always preceded national surveys and laid the foundations for practical advancements with theoretical research.[22] Though he still insisted that a national survey was China's most urgent geological priority, Zhang was no longer simply advocating getting the task of surveying done; he was proposing building blocks for an independent local geological community.

This new emphasis was even more apparent in Zhang Hongzhao's third article in *Dixue zazhi*, "Discussions on the Proper Duties of a Geosciences Society."[23] It is unclear whether Zhang saw this article as just a proposal for reform of the existing Earth Studies Society or as a blueprint for a new organization, but the use of "geosciences" instead of "geology" underscored his belief that China was not ready for a narrowly professional geological organization concerned with licensing or boundary keeping. "Discussions" was written in Nanjing, after Zhang's appointment as director of the Geological Section under the Ministry of Industry.[24] The geosciences society he laid out was a learned association that would mediate between official projects and private (scholarly or community-based) initiatives in the earth sciences. In contrast to the existing Earth Studies Society, which provided a neutral forum for interested parties to exchange as much information as possible on geography, geology, and related disciplines, Zhang envisioned an organization that would advise the government and educate the general population. "Within the scope of the geosciences," he proclaimed, "the work of benefiting society and guiding the government should all be part of the [Geosciences Society's] agenda."[25]

These objectives would be met through a combination of four types of activities, to be carried out as fully as opportunity and resources allowed: (1) research, (2) lectures, (3) field trips, and (4) compilation. The research component of the Society's activities included both pure scientific work and studies in support of government and industry such as urban planning, river management, and mining. Lectures were to be held during school holidays in spring and fall and were designed to make geological research accessible to a broad range of the public. Zhang did not expect the Geosciences Society to organize field trips itself, but saw for it a consultative and supportive role. He suggested that members of the Geosciences Society gather before field parties set out in order to discuss sites, research objectives, and methods. Then members could pool resources to help supply books and equipment, with only the expectation that field findings would be shared with the group at large.[26] In Zhang's view, publication activities should also be aimed at bridging the gap between research needs and nonspecialist concerns, and he gave textbooks and standardization of terminology priority.[27] If possible, the Society should possess a geosciences library and museum to aid in this work.

The Geosciences Society envisioned by Zhang as a bridge between bureaucratic demands and technical specialists never materialized. Adapted from traditional study and mutual-aid societies to address China's modern scientific needs, it was unable to define its place in the shifting world of the new republic and contradictions abounded.[28] The Society was to give advice to both field-workers and policy makers, but it had no way of certifying the professional or technical qualifications of its own membership, who Zhang referred to simply as "scholars."[29] The Society's mandate was intended to encompass the entirety of China and yet members were to meet face-to-face to discuss the plans of individual field projects, an impractical scheme at best.

Fundamentally, Zhang's idea of a geosciences society as the Chinese answer to other nation's geological societies was unworkable because the government itself did not have a clear scientific mission.[30] According to Zhang,

> that which the administration cannot accomplish, the Society can step forward and carry out. For those things which the administration must do, the Society should offer advice and discussion to expedite them, though it should not be required to serve as the administration's agent.[31]

However, the newly instituted Geological Section bounced between ministries with little genuine support,[32] and in the absence of a coherent government program of research this careful demarcation of administration and Society had little practical meaning.

China was neither ready for a professional geological society in the Western mold nor an amateur-activist geosciences society in the Chinese gentry tradition. The existing Earth Studies Society actually petitioned the government in the summer of 1912 to help it establish local branches that would collect field data for compilation by a central branch in the national capital, but the government was unwilling to support an outside agency on such a scale.[33] By fall of that year, Zhang landed on the wrong side of political infighting and was moved from the Geological Section he had founded to an assignment monitoring regional flooding in Hunan.[34] Through the efforts of his replacement, Ding Wenjiang, the dream of a geological school and survey did not die, but a geological society remained elusive.

The Geological Society of Beida

In the first issue of the *Annual of the Geological Society of the University, Peking* (國立北京大學地質研究會年刊), Jiang Menglin, a prominent educator and Chief of General Affairs at the University, wrote that Chinese geological workers "emphasized empirical work in the field in order to gain immediate utility in the short term but few had time to study basic principles." Luckily the "far-sighted" students of Beida's Department of Geology hoped "to study that which fieldwork could not cover" and founded an association to make these findings known to the public at large.[35] Professor Jiang noted the paucity of trained geological "talent" in China, but by 1920, when he helped open the first general meeting of the Geological Society of Beida, the number of Chinese geologists had already increased markedly from 1912.

In the years between Zhang Hongzhao's first discussions of a geological society and the founding of this student organization, the Geological School of the Ministry of Industry and Commerce had graduated 21 students, most of whom qualified to serve in the recently founded Geological Survey. The survey, which replaced the Geological Section in 1913, was as understaffed and underfunded as its predecessor until this new tide of young geological workers began active duty in 1916. The Ministry of Agriculture and Commerce (reorganized in 1914 from the Ministry of Industry and Commerce and The Ministry of Agriculture and Forestry) was disinclined to continue the Geological School, and pressed the Ministry of Education to take over training responsibilities after the first graduating class completed its studies.[36]

This move suited the new Chancellor of Beida, Cai Yuanpei, and his vision for modernizing Chinese higher education quite well. As a personal friend of both Zhang Hongzhao and Ding Wenjiang and an original member of the advisory board of the Earth Studies Society, Cai firmly believed in the importance of geology and geography for both utilitarian and moral education. He reestablished the Peking University Department of Geology (previously closed for lack of interest) in 1916 with two professors and equipment inherited from the Geological School.[37] Though courses and materials (especially textbooks) were few, the formation of the department brought a steady trickle of students into geology and afforded a continuity that the Geological School had been unable to provide. By the time the first class of eight students graduated in 1920, Amadeus W. Grabau of Columbia University and Li Siguang (known abroad as J. S. Lee) of the University of Birmingham had joined the faculty and the freshman class included over twenty students.[38]

The arrival of Grabau and Li greatly strengthened and invigorated the faculty, and the student body responded in kind by taking initiative in extracurricular affairs. In the fall of 1920, a group of seven friends led by Yang Zhongjian and Zhao Guobin rallied their classmates to establish a Geological Society at the university. As rising sophomores these seven leaders were familiar with the geological community at Beida but junior enough to maximize the impact of a new society on their training. They entered Beida in 1919, just months after the May Fourth Incident rocked campuses across China's major cities and student activism was at its apex.[39] The May Fourth Incident, ostensibly a response to the Versailles Treaty, which granted control of Germany's leasehold in Shandong Province to Japan instead of returning it to China, was the crystallization of several years of "new thought tide." It is often understood as the focal point of the broader May Fourth, or New Culture, Movement, which spanned the period from 1915 to the early 1920s.[40] The May Fourth period on campuses was marked by a proliferation of student organizations, which sprouted "like spring bamboo after the rain."[41] Most of these were devoted to social or political issues, but a few centered on the sciences. The Geological Society of Beida was preceded by a Mathematical Society and soon followed by a Chemical Society.

The May Fourth context of the Geological Society explains some of the student group's lofty ambitions. Though explicitly social or political views were aired in other fora, such as student papers, debates and protests, the Geological Society nevertheless had its eye on the national stage.[42] According to Zhao Guobin, "routinely when we discuss various aspects of Chinese geology, we aren't even as familiar [with them] as certain foreigners; shame of this measure could not be greater! When classes are over and my colleagues get around to this subject, there are none that do not sigh. Consequently, we thought to found a group to study geology outside the classroom."[43] The object of the Society was to support independent student research for collective betterment, but the ultimate goal was to raise public awareness of geology and to wipe away the humiliation of foreign control over China's geological exploration.[44] The first general meeting was held on October 10, 1920, in honor of National Day, and the Society saw its mission as a patriotic responsibility.[45]

The students of the Beida Geological Society were keenly aware that theirs was the first geological society in China:

> At the time [of the Society's founding] everyone felt the urgent need for putting China's geology in order. In the public sphere there still were no research organs—excepting the government's Geological Survey—and since our school is the highest seat of learning in the entire nation, even though [our] knowledge is shallow, it seemed that we should quickly found a society for mutual aid in research.[46]

Like Zhang Hongzhao, students associated a geological society with more basic or theoretical study than a national survey had leisure to support. This view was reinforced by Yang Quan's 1915 article "Learned Societies and Science" in the journal of the Science Society of China, which declared that "universities were the sole mother of science" and described scientific societies as the only sustenance for research after graduation.[47] Whether one's formal duties were in the government survey or in the classroom, a separate space was needed for abstract discussions and the exchange of ideas.

Before 1920, when Amadeus Grabau began teaching paleontology and Li Siguang lectured on historical and structural geology, Beida students were largely limited to courses on economic geology, metallurgy, and mining. These were supplemented by general geology and mineralogy but a large proportion of students' energies were devoted to foreign languages and other preparatory subjects.[48] The scientific curriculum was taught entirely from Western language textbooks and publications, and students described their schoolwork as "enough to make a person's head spin."[49]

Despite the added work, students were extremely eager to push beyond economic geology to a more fundamental mastery of geological principles, and new courses offered by Grabau and Li only whet their appetites. Five of the seven students who led the effort to create the Beida Geological Society were in the Prospecting Division of the Engineering College (rather than the Geology Department of the Science College), and for them the Society represented an even more precious opportunity for "research," or cooperative study of geological theory with students from other disciplinary backgrounds.[50] "Amongst other purely scientific disciplines," in Zhao Guobin's words,

> geology has the greatest ambition . . . geology is the crystallization of the myriad sciences. One who researches geology must simultaneously understand chemistry, physics, zoology and botany, paleontology, mineralogy, petrology, geography, sedimentology, etc. If you lack just one of these you will never become a really good geologist. As for applications: if you are someone that studies anthropology or ethnology, you must first study geology; as for the future of the mining industry, it also has a great influence, and geologists must count as the vanguard of the mining industry.[51]

This sense of importance was reinforced by the involvement of professors and administrators, who took unprecedented steps to attend Society activities and provide informal settings for discussing geological issues. In particular, faculty involved with the establishment of the Society viewed its gatherings as an opportunity to stress "principles" over "applications." Amadeus Grabau used the example of chemistry to explain that advances in a basic science would propel its sister applied science. Geology Department Chair Wang Lie urged students not to rush into applied aspects of geology but to use scientific method to organize existing geological knowledge.[52] Professor Wang Fuwu even suggested that principles were important because the Society had a responsibility to educate the educated public about geological concepts as tools of patriotic citizenship.[53]

While professors were moderate in their actual expectations and encouraged students to simply start with whatever they could manage right away or choose one discrete topic in geology to examine closely,[54] Society members had grander ambitions. The Society aimed to foster research among its members, help popularize geology in society, and wash away the shame of yielding China's geology to foreigners. In its first public announcement, the Society proposed to achieve these objectives through: (1) purchase of maps and books pertaining to geology for shared study; (2) fieldwork; (3) organization of lectures by specialists in geology; (4) publication of a regular journal; and (5) compilation and translation of geological material.[55]

Over the course of its first year, the Beida Geological Society organized six lectures by professors and guests from the national Geological Survey on topics ranging from basic elaborations of coursework to new techniques from abroad and ongoing research in China. Under university auspices, the Society also petitioned provincial Bureaus of Industry, mining companies, and other groups for donations of local specimens and reference material.[56] The Society gradually built up a small nonlending collection and library, which included journals from the American Geographical Society as well as student contributions and donations from railway companies, steel companies, and natural history collectors.[57] Though spotty, this material had a range and specificity unavailable in the university's own library, including detailed maps of certain mines, accounts of private explorations, and handbooks on methods.[58]

These activities were reported in the Society's *Annual*, but a regular journal remained out of reach. The *Annual* contained transcripts of lectures, Society business, and miscellaneous reports, but planning for a journal of original student research continued despite a lack of resources.[59] Fieldwork was similarly constrained by problems of scheduling and funding, but the Society's ongoing efforts led to greater integration of fieldwork into the regular curriculum, and students were encouraged to publish their individual field notes in later issues of the *Annual*.[60] Compilation and translation, however, remained elusive. Students frequently shared translations of short texts and discussed technical terms, but the stated goal of standardizing these translations and producing both popular and specialized textbooks was totally beyond the wherewithal of a student organization.[61] Access to foreign geological works was limited, and students lacked the time, experience, or authority to settle thorny linguistic and geological issues on their own.

The Geological Society of Beida was a curious mix of modesty and ambition. Its membership, which never topped forty and was dominated by underclassmen, freely admitted its "lack of knowledge and experience." Students organized the Society precisely because of the inadequacies of their formal training. Yet, far from being daunted, students were eager to take on tasks that were obviously beyond their means. However quixotically, they felt that the burden of popular education and Chinese pride lay on their shoulders, and their elders took them seriously. This spirit reflected broader trends of patriotism and activism among China's May Fourth era youth, but it also hearkened back to older Confucian traditions of scholarly responsibility for the welfare of state and society. Beida geology students were very aware of the small number of geological specialists in China, and understood that with continued effort they would quickly become local authorities.

The formation and tentative first steps of this student society flagged both a sense of urgency about China's need for underlying geological knowledge and an awkward adolescent stage in the development of the native geological community. Students and teachers constantly referred to Chinese geology as "sprouting," and indeed by 1920 the demand for informal, social contexts for geological exchange that Zhang Hongzhao anticipated in 1912 was beginning to emerge. Like Beida's students, geologists and interested mining specialists in the survey, mining companies, provincial industrial bureaus, and academe thirsted for access to new information and desired opportunities to interact beyond the confines of their own professional organizations.

Founding of the Geological Society of China

In December of 1921, Yuan Fuli (袁復禮) and Xie Jiarong (謝家榮) of the Geological Survey expressed interest in an organization to "strengthen contacts amongst domestic geological workers" and "initiate scholarly communication and exchange with foreign geological societies."[62] Yuan and Xie had both recently returned to China from study in the United States at institutions with active student geological associations and established paths of professionalization.[63] Earlier that same year, donations from two of China's largest mining companies funded a library and museum in the newly designed buildings of the Geological Survey at 9 Bingma Si (兵馬司).[64] Amid this flurry of activity and interest, Zhang Hongzhao—whose vision for Chinese geology provided the blueprint for both the Geological School and the survey itself—considered a geological society "the natural step in [China's scientific] evolution."[65] Director of the Geological Survey Ding Wenjiang and associate director Weng Wenhao encouraged Yuan Fuli and Xie Jiarong to draw up an English-language constitution for the proposed society.[66] Amadeus Grabau, in his capacity as both Peking University professor and chief paleontologist of the survey, edited their draft and Weng provided a Chinese translation.

These drafts were presented to twenty-six Chinese and foreign geologists gathered in the rooms of the new library on January 27, 1922, Chinese New Year's Eve. The geologists Zhang Hongzhao invited were all friends who worked together in the survey or taught in Beijing universities, and they were eager to form a Society that would reach beyond their small circle.[67] Ding Wenjiang presided as details of the constitution were hammered out and a committee was selected to nominate candidates for Society office after the five-day holiday. A week later, a second meeting was held, and attendees approved the proffered Constitution on February 3, 1922. Officers were elected, and Zhang Hongzhao was named the first president of the Geological Society of China in recognition of his longstanding sponsorship of Chinese geology.[68]

As Davidson Black of the Rockefeller Foundation-sponsored Peking Union Medical College noted, the Geological Society of China was "the first non-medical scientific association initiated and organized wholly by Chinese investigators for the advancement of one of the pure sciences."[69] Unlike the Earth Studies Society that preceded it, the Geological Society was aimed at trained professionals and designed to facilitate discussion of original research. Though the Science Society of China antedated the Geological Society as well, its major function was to popularize scientific thinking and support general scientific activity rather than foster contributions in any specialized field.[70] A few research-oriented organizations such as the Anatomical and Anthropological Association existed in Beijing before the Geological Society, but these were organized by foreigners working temporarily in China.[71] Several native scientific associations formed during the Nanjing Decade (1927–1937) under the Guomindang, but until that time, the only organizations comparable to the Geological Society were the Chinese Astronomical Society (est. 1922) and the Chinese Meteorological Society (est. 1924), earning the Geological Society the title of "older brother" among China's scientific organizations.[72]

The constitution of the Geological Society defined two types of membership. Geologists and other scientists interested in geology, such as meteorologist Zhu Kezhen, could become fellows of the Society with two nominations from existing fellows and the approval of the Society's council. Through a similar process, college students in geology or allied sciences could become associates of the Society, but they would not be eligible to hold office. Nationality was not mentioned as a consideration until 1924 when corresponding and honorary memberships were created for foreign nonresidents.[73] During its first full year, the Geological Society had sixty-nine fellows and nine associates, including twenty-three foreigners.[74]

Society offices were also open to all nationalities, though Amadeus Grabau and J. G. Andersson were the only foreigners ever elected. Besides a president, two vice presidents, a secretary, and a treasurer as councillors ex officio, two to six other councilors could be elected at the recommendation of the passing council.[75] Officers and one-third of the council were elected yearly, though council terms were three years each. The council had to approve all business before it reached the Society at large, and amendments to the Society's Constitution required a two-thirds general vote.

Elections were held at each year's general meeting along with the presentation of annual reports, discussion of scientific papers, and transaction of occasional business. According to the constitution, the Council of the Geological Society could convene special meetings and excursions between annual meetings.[76] These irregular gatherings at the Society's headquarters in the Survey library quickly became one of the most influential factors in the Society's rapid expansion, and important papers read at meetings were published in *The Bulletin of the Geological Society of China* (中國地質學會誌, *Zhongguo dizhi xuehui zhi*, hereafter *BGSC*). In later years the council also sponsored group field trips in conjunction with annual meetings and special events.

The limiting factor of the Geological Society's activities was funding, but the Society managed to become a hub of Chinese geological life without sacrificing its independence to political or business interests. Unlike the Geological Survey which was an official agency funded by the Chinese government (with crucial supplements from the China Foundation and the China Medical Board of the Rockefeller Foundation), the Geological Society did not rely on government monies.[77] The Society's regular income depended on membership dues and sales of its *Bulletin*. Fellows paid five dollars annually and associates paid two dollars. Lifetime membership cost fifty dollars and several members made special endowments for medals and specific scholarship funds. Corporate memberships were created in 1932 for institutions that donated at least 400 dollars, bringing the Geological Survey, Academia Sinica, provincial surveys, and various academic departments under the banner of a single voluntary organization.[78]

The legitimacy of the Geological Society came from its broad membership and not from any explicit connection to the government. As the "Aims of the Geological Society of China" suggest, the Society functioned best as an alternative to formal research organs:

> This society affords an opportunity for full and free discussion of the principles and problems of our science, such as is not possible in our official organization [the Geological Survey of China] where of necessity attention must largely be centred on detail.[79]

The survey was not only constrained by attention to "detail," but it was also limited by its own resources and tethered to the practical agendas of the national government. In contrast, Roy Chapman Andrews of the American Museum of Natural History's Central Asiatic Expedition spoke for many foreigners in describing China as "a field for investigation which is unrivalled in importance and interest."[80] As an unstructured space for discussion and exchange, the Geological Society of China was flexible enough to encompass shifting research opportunities and concerns. Its meetings and publications thus tapped the "importance and interest" of Chinese territory to simultaneously expand the reach of Chinese geologists and increase their theoretical sophistication.

The Bulletin of the Geological Society of China

The founding of the Geological Society harnessed some of this international interest in China's geology by giving local geologists, resident foreign geologists, and temporary scientific visitors a meeting place under native auspices. On their own, however, Society meetings were still too ephemeral to have a systematic influence, and more was needed to fully address the marginalization of Chinese researchers. At the beginning of the 1920s they still struggled to reach an audience that was qualified to evaluate and utilize their work, and they needed an outlet that was thorough and credible enough to have a tangible impact.

Domestic Chinese newspapers, which had previously carried the bulk of writing on topics such as earthquakes, topographical features, and mineral resources, were largely uninterested in more technical reports.[81] Though it carried the first geological maps ever published in Chinese, *Dixue zazhi* was committed to preparing its readers for the demands of modern citizenry, and it preferred broad digests of eye-opening geographical information to detailed reports of original research. By contrast, the scientific journals that various universities began publishing in the early years of the Republic were narrow in focus and often pedagogical in nature, while regional administrative journals and company mining reports limited themselves to economic resources and practical applications. Until 1919, even the Geological Survey of China could only circulate its work-in-progress informally and then publish results of practical significance in the ministry gazettes (*gongbao*).[82] This left little scope for elaborating on the more theoretical aspects of survey research, and it did not do much to attract those scattered Chinese who were beginning to get involved in geology, physical geography, or mining outside of officialdom.

After 1918, Chinese geologists who wanted to engage a more scientifically or technically aware readership could publish in *Science* (科學, *Kexue*), the journal of the Science Society of China.[83] But *Science* catered to generalist tastes, and in its first decade it concentrated on introducing various new disciplines to China's reading public.[84] Though *Science* did aspire to be a showcase for Chinese scientific advances, many of the geological articles it published dealt with pedagogy, international institutions, metallurgy, or historical overviews of the earth sciences.[85] Several other journals, such as *The Journal of the Association of Chinese and American Engineers* (est. 1920) and the *Magazine of Natural History* (博物雜誌) (est. 1919) covered various aspects of mining and related fields in the early 1920s, but none of these targeted a specifically geological audience.[86]

Chinese who had studied geology or mining overseas were sometimes able to address knowledgeable readers outside of China, but this was not always the case. The full scientific results of Zhang Hongzhao's undergraduate thesis, which marked the first extensive geological fieldwork conducted in China by a Chinese researcher, were never made public, despite Zhang's influential position.[87] Others were more fortunate, but publication abroad did not guarantee an audience. Wang Chongyou, who studied with Amadeus Grabau at Columbia University, put out a *Bibliography of the Mineral Wealth and Geology of China* in England and America in 1912, but this was merely a compilation of existing literature, and its impact was so limited that two other Chinese students earned mining and geology degrees in the United States based on comparable work, and both were unaware of Wang's earlier efforts.[88] Neither Parkin Wong's "The Mineral Resources of China" nor Wah Seyle Lee's "Bibliography of the Geology of China" ever saw the light of day, however, and later geologists in China had to compile their own bibliographies to meet the demands of research.[89]

With the exception of Ding Wenjiang, who published preliminary resource reports in his capacity as director of the Geological Survey, those Chinese who managed to publish in foreign periodicals were often limited to prosaic reporting on the state of Chinese mining or superficial overviews.[90] Wang Chongyou wrote a few short reports for the *Bulletin of the American Institute of Mining Engineers*,[91] and Weng Wenhao contributed "The Geology of China" to Samuel Couling's *Encyclopædia Sinica* in 1917, but none of these articles pushed any boundaries.[92] When, in 1921, Li Siguang turned his 1918 master's thesis into "An Outline of Chinese Geology" in Britain's *Geological Magazine*, he made a challenging and original synthesis of existing knowledge on Chinese geology available to a Western audience, but forced his compatriots once again to turn to a foreign journal for insight into the geology of their own territory.[93]

Interestingly, while Chinese researchers struggled to find appropriate ways of circulating work on the geology of their homeland, outsiders had a wide variety of options. Some studies of Chinese geology, like those of American geologist Bailey Willis and Swedish explorer Sven Hedin, were backed by foreign institutions and therefore automatically reported by their sponsors. But there was no shortage of specialized journals for less prominent geologists to publish in. Like popular magazines that saw China as distant and exotic, geological and mining journals were eager for news and experiences of Chinese territory, which was seen as terra incognita from the standpoint of modern geology.

More to the point, Ferdinand von Richthofen and Raphael Pumpelly's late nineteenth-century explorations had intimated that China possessed vast stores of mineral wealth, especially coal and iron, and readers were eager for detailed information.[94] Tantalizing collections of "dragon bones" from Chinese apothecaries hinted at important paleontological mysteries in China's fossil record,[95] and the tectonics of Asia were ever more important for a structural understanding of Europe.[96] Problems of mountain building, isostasy, glaciation, and stratigraphic correlation created a ready-made place for China in existing geological debates, and Western audiences welcomed everything from tidbits of mining news to speculative papers at international geological congresses.

Foreign geological workers with only short-term experience in China could speak as experts in diplomatic and consular briefs, company mining reports, and geological society meetings all over America and Europe.

Western residents in China also put out expatriate publications, like the *Journal of the Royal Asiatic Society North China Branch*, to exchange information with each other and other scientific centers.[97] Japanese geologists in China even established their own research institute in Manchuria and had several publications devoted to the geology of China.[98]

Information on Chinese geology peppered the proceedings and journals of scientific societies outside of China and were widely scattered through specialized journals in countless languages.[99] Because (with very rare exceptions) Chinese-language material was unintelligible to all but Japanese geologists, this network of publication and exchange functioned with scant reference to the Chinese geological community or the Chinese government.[100] It seemed as if Chinese workers could only participate if they were willing and able to operate exclusively through foreign journals, and even so, readers reacted more favorably to reports by their own countrymen or by scientific notables than by Chinese themselves.

An alternative solution was to publish Chinese geological work in China but in Western languages, rather than Chinese. In fact, in a book review of a Japanese study of Chinese geology, Eliot Blackwelder—who had accompanied Bailey Willis on the Carnegie Expedition and displayed a lifelong interest in the geology of Asia—wrote:

> Since it is printed in the Japanese language and characters, Mr. Ishii's paper on the Yang-tze Valley will be of little use to nearly all geologists outside of Japan and China. . . . Although there may be some compelling reasons unknown to the reviewer, . . . it would be hard to defend on general grounds, the printing of technical scientific papers in any language which is not in more or less general use in the scientific world. . . . Nearly all educated Japanese and Chinese read English, . . . so that even a paper intended largely for local use in Japan would be quite as intelligible to its readers if presented in one of the more important European languages and it would at the same time be available for foreign students in general. A popular summary in Japanese might be appended for the edification of the few who read only the mother tongue.[101]

The suggestion went unheeded in Japan, which continued to publish its major geological journals exclusively in Japanese until the late 1940s, but against this backdrop, it is not surprising that the Geological Survey of China had hoped to make a splash with its bilingual *Geological Bulletin* in 1919.

As a national institution with both accomplished natives and distinguished foreigners in its employ, the survey *Bulletin* quite naturally included reports written in both Western languages and Chinese, and its founders

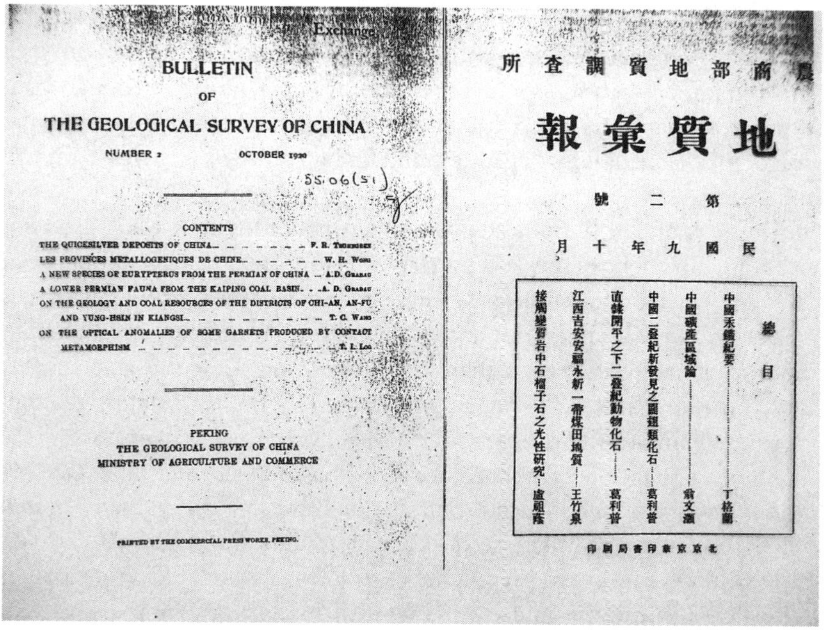

BULLETIN
OF
THE GEOLOGICAL SURVEY OF CHINA

NUMBER 2 OCTOBER 1920

CONTENTS

THE QUICKSILVER DEPOSITS OF CHINA ... F. R. Tegengren
LES PROVINCES METALLOGENIQUES DE CHINE ... W. H. Wong
A NEW SPECIES OF EURYPTERUS FROM THE PERMIAN OF CHINA ... A. D. Grabau
A LOWER PERMIAN FAUNA FROM THE KAIPING COAL BASIN ... A. D. Grabau
ON THE GEOLOGY AND COAL RESOURCES OF THE DISTRICTS OF CHI-AN, AN-FU AND YUNG-HSIN IN KIANGSI ... T. C. Wang
ON THE OPTICAL ANOMALIES OF SOME GARNETS PRODUCED BY CONTACT METAMORPHISM ... T. L. Loo

PEKING
THE GEOLOGICAL SURVEY OF CHINA
MINISTRY OF AGRICULTURE AND COMMERCE

PRINTED BY THE COMMERCIAL PRESS WORKS, PEKING.

農商部地質調查所

地質彙報

第二號

民國九年十月

總目

中國汞礦紀要……丁格蘭
中國礦產區域論……翁文灝
中國二疊紀新發見之圖翅類化石……葛利普
直隸開平之下二疊紀動物化石……葛利普
江西吉安安福永新一帶煤田地質……王竹泉
接觸變質岩中石榴子石之光性研究……盧祖蔭

北京京華印書局印刷

2. English and Chinese covers of the *Bulletin of the Geological Survey of China*, showing the different orientations of text.

hoped that by mixing scripts, it would embrace the widest audience possible. In fact, since Western languages were printed horizontally and bound on the left, and Chinese was traditionally printed vertically and bound on the right, the layout of the *Geological Bulletin* seemed ideally balanced and egalitarian. The journal had two complementary "front" covers, and papers were grouped by language category, with every article appeared in both a full original version and a Chinese or English abridgement. Pages were numbered in Chinese numerals in the one case and Arabic numerals in the other, so neither section took precedence, and the two halves of each issue extended quite organically from their respective covers inward, conjoined at the center and equally available to all bilingual readers.

For most foreign readers, however, this linguistic segregation hid the research of Chinese geologists in plain view. The layout made it simple for Western geologists accustomed to colonial surveys to start from the English cover and assume that the only scientifically productive researchers in China were the foreigners whose publications were printed "up front." Unlike abstracts, whose brevity often hinted at the richness of the full text, the abbreviated English versions of Chinese papers in the Western half of the journal seemed to stand on their own, and by not pointing to the comprehensive original reports at the unintelligible "back" of the *Bulletin*, they failed to impress outsiders with the caliber of native research. Though visually neat, the Janus-faced *Geological Bulletin* was in fact a physical representation of the deeper failure of Chinese and foreign geological circles to mingle and intersect.

This was of course quite opposite the spirit of the new Geological Society of China, whose primary purpose was to foster engagement between local and foreign geologists, and, where possible, among researchers across academia, officialdom, and industry. As a nongovernmental organization, the Society had more leeway than the survey in pursuing this goal, and its leaders quickly saw the advantages of a journal for both circulating geological research and fortifying the initial enthusiasm of its membership.

All of these factors came together in the pages of the *Bulletin of the Geological Society of China*, which began its run at the end of 1922 and immediately extended the influence of the Society into new topics and spaces. Early volumes set several precedents for the Society's relationship to other geological institutions, and each issue displayed the growing diversity of both Chinese interest in geology and foreign interest in China. But the real success of the *BGSC* hinged on the early decision to capture the outward-facing atmosphere of Society meetings in print and publish primarily in Western languages.[102]

Geological Society meetings were polyglot affairs, and though all official business was conducted in English, foreigners were encouraged to speak in their native tongues, with English translation as needed. Because Chinese geologists were all trained in English or another foreign language, they were discouraged from using Chinese whenever possible, which Li Siguang noted was "not without a sacrifice to our own convenience." As he explained it years later, the Geological Society "adopted the leading Western languages, principally English, as the official means of communication" because, as hosts, Chinese geologists did not want foreigners to feel "shy" or "less at home" among Chinese speakers.[103] This bit of altruism turned out to be a savvy move, attracting a large number of foreigners to Society meetings and making its *Bulletin* a viable forum for non-Chinese geologists.

Unlike Western-sponsored journals in China, the *Bulletin of the Geological Society of China* was not bound to any one foreign language because it was not aligned with either a foreign metropole or a particular expatriate community.[104] The majority of its papers were in English but contributors could submit in any leading scientific language, such as French or German. If in the past foreign geologists had lacked motive or opportunity to publish in China, the *BGSC* provided both. The combination of China's vast field for exploration and Beijing's cultural attractions lured the most famous geologists working in Asia into the *BGSC*'s orbit, and the journal's linguistic flexibility meant that, for the first time, geologists of almost any nationality could publish on the spot instead of scattering their results across foreign journals. This accommodative policy allowed the widest range of geological workers to reach the greatest concentration of other experts on Chinese geology without loss of fidelity or impact.

No geological journal in the Americas or Europe could boast original research in so many different languages. Within a few years the *BGSC* became such an important source for Chinese geology and its bearing on geology at large that non-Chinese workers from relatively marginal scientific communities used the *BGSC* to advertise their own scientific activities. Swedish, Russian, and Japanese geologists were among the most active groups studying China, but linguistic challenges often made their national journals inaccessible, and Anglo-European geologists frequently overlooked their China studies. By publishing in the *BGSC*, Swedish, Russian, and Japanese geologists not only publicized their own work, they also raised awareness of these local scientific communities and made bibliographies of their compatriots' research available in English, French, or German.[105] Even English-speaking geologists from colonial areas, such as Hong Kong, saw the *BGSC* as a means of broadening their audience.[106]

Naturally, the benefits of exposure were still greatest for Chinese geologists for whom the *BGSC* was a primary vehicle. Over half of the papers printed in the *BGSC* in its first decade were by Chinese researchers, and these Western language studies reached almost two hundred libraries in over twenty countries and territories by 1929.[107] Though the Geological Survey and the Geological Society both exchanged their publications with foreign geological organizations to enlarge the periodical collection of China's geological library and expand the circulation of Chinese journals overseas, the Society's strategy of publishing Chinese reports in Western languages without consideration of national origin enabled Chinese papers to "travel" wherever the *BGSC* was found. Essentially, by joining the crowd rather than standing apart, Chinese papers were finally able to share the limelight.

Because the Geological Society was organized as an alternative to the utilitarian strictures of survey work, it also challenged the stereotype that Chinese geologists were only interested in practical applications by encouraging theoretical speculation, and the *BGSC* went so far as to redistribute the more applied papers given at Society meetings to other publications like *The Oriental Engineer*. This gave survey geologists publishing through the Society a dedicated space to explore problems beyond the purview of government research and helped them cooperate with geologists at a growing number of academic and research institutions. The *BGSC* also published the thesis work of Chinese studying abroad, which brought new ideas and techniques back to China and demonstrated that Chinese interest in geology did not stop at China's borders.

At the Geological Society's twenty-year Jubilee, Li Siguang noted:

> Our society, though [it] carries the usual banner of a national organization, is in fact international in character. You need only turn over the pages on which the names of our members are printed. They come from all lands, and actively participate here in a fraternal atmosphere that, as far as I can gather, is seldom so widely and deeply felt in many of the similar organizations in *other* countries.[108]

In its range of topics, diversity of contributors, and linguistic flexibility, the *BGSC* mirrored the international geological community in China and projected the cosmopolitan image of Chinese geology all over the world, yet it was still but a static representation of the Geological Society itself.

Host of Nations

The "Aims of the Geological Society of China" called for "intercourse with the scientific men in all parts of this country by bringing them together at intervals in our large meetings." As predicted, the exchange of ideas was both a "benefit to all participants" and "a factor in the advancement of the scientific life of the nation," but not necessarily in the ways originally imagined by Zhang Hongzhao or the students of Beida. In 1922, the native geological community was still largely sequestered in Beijing, and from its inception, the Society focused more on intercourse between Chinese and foreign geologists than among the already close-knit Chinese geologists themselves. During its first decade, the Geological Society of China established itself as the primary meeting place for all geological workers on Chinese soil. Instead of trying to wash away the "shame" of foreign achievements in Chinese geology through competition or exclusion, the founders of the Geological Society cultivated an inclusive vision of Chinese geology and enrolled foreign scientists in its service through participation in Society functions.

From its first general meeting on March 2, 1922, the Geological Society drew on the support of foreigners to strengthen its position as a center for Chinese geology. The Society, which had no official connection to the national government, derived its legitimacy from the reputations of its charter members and the acknowledgement of the principal foreign scientific men in Beijing. These guests dominated the program of the inaugural meeting and spoke not just as individual researchers but as representatives of their overseas colleagues and home institutions.

In welcoming new members, Society president, Zhang Hongzhao, took the first step in joining Chinese and foreign geological efforts and set the tone for the entire ceremony:

> We are all students of nature. We are friends and teachers of one another. So long as we are willing to learn and are united in our efforts, neither the lack of money nor that of political peace will hinder us from our progress. The presence of so many distinguished guests here tonight shows that our efforts have not been unappreciated.[109]

Some speakers, such as Davidson Black of the Anatomical and Anthropological Association and the Rockefeller-sponsored Peking Union Medical College, echoed this sentiment directly. "China, from the standpoint of a worker in the natural science[s], offers a field of the richest promise to investigations," said Black, and "I can assure the Geological Society," he continued, "that it may count on the heartiest support and cooperation, whenever possible, among the members of its sister organization in whose name I tender our warmest congratulations and good wishes."[110]

Others, such as Roy Chapman Andrews, leader of the American Museum of Natural History's Third Asiatic Expedition, and Eduard E. Ahnert (Эдуард Эдуардович Анерт), director of the Geological Committee of the Russian Far East, seized the occasion to affirm their own interests. Andrews, in praising Chinese geologists for "the finest type of scientific spirit—that of cooperation," urged them to continue this policy, "for more and larger expeditions are being sent to this country from foreign lands every year."[111] Ahnert, who traveled all the way from Vladivostok to attend the meeting, commended Chinese geologists to the loftiest of ideals, reminding them that Russian geologists had not only "consecrated the best years of their lives to the study of those regions which form . . . a natural if not political entity with the territory of China," they had, "like all true scholars, freed themselves from the bias of political separatism, and . . . always accorded the freest facilities to the scholars of other lands who desired to make a study of the vast territory of Russia."[112] These men, according to Ahnert "have sincerely at heart the success of geological study in China, by all who are qualified to undertake it" and he hoped to find like-minded men of science in China.[113]

With their tactful attempts to protect a scientific "Open Door" in China, Andrews and Ahnert lent consequence to the founding of the Geological Society of China that even the highest praise could not. The Society, by bringing together foreign researchers with competing interests, secured a place for Chinese geologists while driving home the mutual advantages of openness and cooperation in the study of Chinese geology. Chinese geologists served notice through their new Society that they were now at the heart of an international project, rather than on the periphery of an imported science.

The Geological Society reinforced the image of Chinese geology as a common endeavor and demonstrated the productivity of Chinese researchers at its first annual meeting in 1923. Thirty-two papers were presented at the three-day conference, fourteen by foreign geologists and eighteen by Chinese.[114] The Chinese papers showcased some of the best work of the past several years, presented publicly for the first time. These included Li Siguang's foundational studies of *Fusulinidae*, Weng Wenhao's early work on the tectonics of northern China, Ding Wenjiang's classic survey of Yunnan stratigraphy, Yuan Fuli's geodetic research, Wu Kuang's application of radioactivity in geology, and Wang Chongyou's physical analysis of coal composition. Foreign submissions ran the gamut from system-building studies on the Devonian and Carboniferous of China to preliminary reports of ongoing research in the Gobi, with provocative papers on the Hipparion fauna of Shanxi Province and prosaic summaries of recent advances in the Far East by Russian paleontologists in between.

"While this society is essentially a Chinese scientific institution," wrote Weng Wenhao in the *BGSC*, "we have always welcomed scientific cooperation from all workers on the geology of China."[115] The minutes of the First Annual Meeting record that of seventy-seven active members of the Society, twenty-seven were foreign residents in Asia, including "ten Americans, five Swedes, three natives each of Russia, France, and Great Britain, two Japanese and one native each of Belgium, Czecho-Slovakia and Austria."[116] With the creation of corresponding memberships in 1924, the Geological Society selected "geologists and paleontologists of note" residing in each country that had contributed to Chinese geology.[117] In the first cohort of corresponding members America was represented by C. D. Walcott and B. Willis; Austria by L. V. Loczy; Great Britain by J. W. Gregory and A. Smith-Woodward; France by M. Boule and E. de Margerie; Germany by M. Schlosser, E. Tiessen, and F. Solger; Japan by B. Koto and M. Yokoyama; Russia by W. A. Obrutchev; Sweden by T. G. Halle and C. Wiman; and Indochina by H. Mansuy.[118] Other corresponding members were intermittently elected to maintain this representative diversity and welcome famous guests to the Society.

While this roster of corresponding members might not have been impressive compared with that of the Geological Society of London, the display of foreign interest encouraged new researchers coming to China to pay heed to the Society and the local geological community. It also helped to cement the position of Chinese geologists with their own national government, which, as late as the summer of 1922, still seriously considered dissolving the Geological Survey for budgetary reasons. This threatened closure was met with significant opposition from the leaders of several Chinese universities and later the directors of China's major mines. Both groups argued for the importance of geology to China's development, but their most persuasive point was that geologists were the only scientists in China whose institutions were on par with international standards, a fact whose "general acceptance by Chinese and foreign specialists" was demonstrated by the international participation of geologists in both the Geological Survey and Geological Society.[119]

Under these circumstances, the Geological Society walked a delicate line, between erasing difference, as it did by publishing all reports in its *Bulletin* in Western languages, and celebrating diversity, as it did in its face-to-face meetings. In the one case, the intent was to allow Chinese work to speak for itself wherever the journal was read, and in the other it was to invite researchers from all over to take part in the local community and speak on its behalf. Annual meetings were good examples of how this international cachet was put on display.

At the Geological Society's annual dinner, on January 13, 1923, toastmaster and incoming president of the Society, Ding Wenjiang selected "speakers . . . from the various nationalities represented in the society, and their theme was the contribution of these nations to the upbuilding of our knowledge of Chinese geology."[120] Zhang Hongzhao spoke for Chinese geologists and Wang Chongyou for Chinese mining men. J. G. Andersson, from Sweden, wittily boasted about the efforts of the "Germanic races," including Germans, Austrians, and Scandinavians. Weng Wenhao who received his doctorate at Louvain, spoke in the name of the "Latin races" of France, Belgium and Italy and recounted the "grave dangers and . . . formidable obstacles" their geologists overcame in China. Frederick Morris, geologist to the Third Asiatic Expedition, fancifully likened American milestones in Chinese geology to "the isolated stone obos, heaped by the pious pilgrims on every eminence in Mongolia" as guides to navigation, and, because the Russian contingent was still en route, Ding Wenjiang took up their great labors in the Northeast on their behalf. Mr. Tung (董常), a Chinese mining geologists who studied at Imperial Tokyo University, represented

Japan, noting that "to mention all the Japanese who at one time or another investigated the geology of China, would be equivalent to giv[ing] you a list of Japanese geologists, for every one of them has added to . . . Chinese geology." Davidson Black of Canada described British geological work in China, and Amadeus Grabau concluded by saying that "China is the land of promise to the man of science, but the foreigner who enters it must come with the desire for service which actuated the pioneers, if his work is to have lasting value."[121]

This exercise, performed with jokes, anecdotes, and teasing rivalry, exemplified how the Geological Society constructed "Chinese geology" as an international collaboration, and it was far more important to have every major national group represented than to nitpick about who represented what. Instead of the dry bibliographies that early overseas students pieced together from chaotically scattered foreign sources, the Society brought the idea of cumulative advancement to life and portrayed "Chinese geology" as a coherent field, though no such unity existed historically. Geological Society activities encouraged foreign geologists to frame Chinese geology as a scientifically valuable part of world heritage, and to credit Chinese geologists with providing an unprecedented and indispensable opportunity for them to meet each other on friendly terms.

By hosting international guests, the Geological Society projected an image of modern China as both cosmopolitan and magnanimous, and this image was solidified through social activities as well as scientific exchanges. Annual meetings were arranged in split sessions, with informal lunches ("tiffins") given by various geological organizations in China, such as those at Beida or Qinghua College (清華學校). These lunches introduced conference participants to local personnel and facilities, but were also meant as entertainments, with presentations by students, short speeches, and sometimes special treats, like desserts made in the accurate likenesses of fossils which guests were asked to identify and label.[122] The evening sessions of the annual meeting always ended in a smoker hosted by the Geological Society and participants often lingered until midnight over coffee and cigars discussing the finer points of the day's presentations or exchanging field stories.

The most anticipated social event of each year's conference, however, was the annual banquet thrown by the Geological Society. These dinners brought together paper presenters, members of the Society, their families, and celebrated guests from all walks of Beijing life. The American Ambassador to China was a frequent guest throughout the 1920s, as were the president of Yenjing University, the chancellor of Peking University, the

Chinese Minister of Education, and several well-known artists, missionaries, businessmen, military men, and society women. Despite the scientific occasion, a local newspaper marveled that an eavesdropper would "listen in vain for echoes of the controversy over the Sinian System, or the discussion on pleochroism . . . which swayed these men a few hours earlier."[123] Until after dinner drinks were served and the gathering was called to order for special toasts and speeches, geology was not discussed, and in this context speakers were more likely to "twit" one another or tease "Professor Grab-all" than rehash geological debates.

Language consciousness played a key role in the Society's social exchanges. While guests were relaxing over dessert at the third annual dinner of the Geological Society of China in 1925, acting chancellor of Peking University, Jiang Menglin offered an admiring toast to the Society in Chinese, explaining to his very international audience that "as a representative of a National institution he felt . . . he should use the national language."[124] No offense was taken by Ding Wenjiang, whose witty English repartee made him the evening's natural toastmaster, and none by president of the Society Weng Wenhao, whose eloquence in French was legendary.[125] Ding and Weng were also representing China that night, but they did so as Chinese men of science rather than Chinese officials, and they freely peppered their speech with English, French, German, and Japanese in the spirit of camaraderie and mutual understanding. For scientists like Ding, Weng, Li Siguang, and later Yang Zhongjian, facility with foreign languages and manners was a mark of open-mindedness, not servility,[126] and the cosmopolitan reputation of the Geological Society rested as much on the personal appeal of its leaders as on its publications and geological reports.

These core figures in the Geological Society exerted their influence through the society council, which had the authority to convene special meetings for geological business, as deemed necessary. Though a third of the council positions were open at every annual meeting, reelections were the rule and the Society's leadership rotated amongst a handful of the most influential Chinese geologists, with Amadeus Grabau and J. G. Andersson occasionally holding tenure as well. The expansive powers of the council gave the Geological Society great flexibility, and under its direction the Society was able to seize fleeting opportunities for hosting foreign scientists and spotlighting developments in Chinese geology. Over time, this pattern of occasional meetings turned the Society's cosmopolitan approach to the advantage of Chinese geologists by bringing famous foreign researchers into the Society's compass and disseminating timely information that would otherwise be unavailable.

The first of the Society's occasional meetings was convened only a month after its inaugural ceremony, on April 15, 1922, to "welcome the Japanese paleontologist, Dr. Hayasaka Ichirō (早坂一郎), who recently came to study the geology of northern China with his students of the Tohoku Imperial University."[127] Dr. Hayasaka delivered a lecture on the geology of Japan to the gathered audience, with translation by Li Siguang. This was the first presentation by a foreign researcher on the geology of his home country to a predominantly Chinese audience outside of a pedagogical setting. It was an especially important milestone because Japan's older geological community often took upon itself the responsibility for describing the geology of China to non-Asian audiences, and had not previously acknowledged the rapid development of China's own geological institutions.[128]

The next meeting, held on May 26, 1922, was timed for Davidson Black's return from a brief stint with the Third Asiatic Expedition. Black gave an informal talk on his observations in the field illustrated with lantern slides, and allowed Chinese geologists a glimpse into the day to day workings of the American Museum of Natural History's highly publicized motor vehicle exploration of Mongolia. J. G. Andersson then discussed the theoretical aspects of his survey research on the Cenozoic of China, and Li Siguang rounded out the meeting by sharing the work on Pleistocene glaciation that he had recently published in England.

On September 29, 1922, a "meeting was especially held to welcome the American geologists who had just returned from their successful research in Mongolia" and this turned out to be a great success for both the weary adventurers and their Chinese hosts.[129] Ding Wenjiang praised Roy Chapman Andrews for continuing the American tradition of disinterested research and told a story about Bailey Willis's famous Carnegie Institute expedition in China. Ding jokingly mentioned that "a prominent member of our foreign community assured me of his unalterable conviction that Mr. Carnegie must have kept back all the information of an economic value for his own personal use," but if any unspoken concerns lingered about the Third Asiatic expedition, they were soon dispelled.[130] Expedition geologists Charles Berkey and Frederick Morris projected pages directly from their field notebooks onto a large screen and allowed audience members to sample other sections during the question and answer period.[131] This was followed by a description of paleontological finds and lively audience discussion of possible correlations between the strata of Mongolia and China proper that lived up to Beijing's growing reputation as "the center of scientific life in Asia."[132]

The vigor of scientific exchange in the city was demonstrated once more at the last meeting of the year, on November 6, 1922. Eduard Ahnert outlined the work of Russian geologists in the Russian Far East bordering China, and during discussion Frederick Morris brought up points of comparison with his experiences in Mongolia.[133] Next Amadeus Grabau presented his study of Devonian brachiopods, indicating that at that time China was "the center of radiation" from which brachiopods migrated into western America and central Europe. Weng Wenhao closed the program with an account of the proceedings of the thirteenth International Geological Congress in Brussels and reported reactions to the Chinese contributions placed on the program.[134] Chairman of the meeting, Li Siguang, pointed out that this was the first time that the government had sent a Chinese geologist to the Congress.[135]

During its first year, the Council of the Geological Society exercised its powers five times, and called four general meetings in addition to its inaugural ceremony. Though the council had no more coherent strategy than to capitalize on the heavy traffic of foreign geologists in Beijing, these early meetings established the Society as a point of convergence for all researchers interested in the geology of China and demonstrated the power of providing an ecumenical forum for scientific exchange. Before 1922, Chinese and foreigners alike had to go abroad to learn of other investigations within China or its neighboring territories. The reversal of this trend by Dr. Hayasaka and E. Ahnert suggested that it was no longer politic for geological visitors seeking data in China not to share information in return.

The vitality of the local geological community made this practice mutually beneficial and the Third Asiatic Expedition of the American Museum of Natural History (one of five Central Asiatic Expeditions) used the Geological Society as a platform to generate interest in their fieldwork. After the success of the Society's 1922 meetings, Roy Chapman Andrews agreed to open and close his annual field seasons in Mongolia at the Geological Society. Whether or not these meetings were open to the public, he always drew capacity crowds to the Library of the Geological Survey and attracted front-page media coverage in the foreign press. When the expedition suddenly announced their return to China for the 1925 season and the council called a meeting later that same day, "more than one dinner party was cancelled to be replaced by the greater attraction of the intellectual feast spread for all in the West City by the Geological Society of China."[136]

In the past, it took several years before major expeditions in China, such as those of the Carnegie Institute, Ferdinand von Richthofen, and Sven Hedin, organized, analyzed, and circulated their results. News took even longer to reach China because these findings were published abroad.[137] With the Central Asiatic Expedition, the Society not only hosted team members at the beginning and end of every field season; when possible it held special meetings for updates from the desert. The Chinese geological public was brought up to date on behind the scenes work done each winter at the distant Museum of Natural History in New York and got to see the newest field specimens right away. Andrews once said, "I think that all of the Expedition would feel very badly and would feel the expedition had been a failure in many respects if we were not asked to give our first report to the Geological Society. It has become one of the things to which we all look forward to during the Expedition [*sic*]."[138] Chinese geologists were given the chance to discuss observations and interpretations while expedition scientists were still processing their data, and as the relationship between the Geological Society and the expedition grew, Society members were even made privy to the researchers' upcoming plans.

The open mood of Geological Society meetings gave Roy Chapman Andrews a chance to win over his Chinese colleagues and protect his access to the Gobi. He was able to publicly dismiss rumors that the expedition was expelled from Mongolia because it was making military maps and inciting natives against the government.[139] Citing the same supranational scientific motives that shaped the Society's welcoming attitude, Andrews claimed that the American Museum of Natural History had "no interest whatever" in the politics or economics of the country it explored.[140]

The history of the Central Asiatic Expedition in China suggests that it was much more effective for Chinese geologists to use engagement rather than pressure to penetrate foreign activities. By serving as host rather than rival or antagonist, the Geological Society gained privileged access to expedition research and raised its own standing in the intellectual life of Asia. The American Museum of Natural History voluntarily courted the Society's favor and announced its donation of specimens—including one of the expedition's famous dinosaur eggs—to the Museum of the Geological Survey at a special meeting of the Society.[141] Contrastingly, when the Chinese rights' recovery movement reached a fever pitch and the Chinese government and the Committee for the Preservation of Ancient Objects demanded that duplicates of all paleontological finds be deposited in China *before* any material could leave the country, the agitators only earned China the expedition's rancor and derision.[142] Chief paleontologist Walter Granger complained that even he could not know what would count as a duplicate until all collected material was processed and studied in New York. The Chinese government, he contended, clearly did not understand science, and when its demands did not diminish, Andrews packed up and went home.[143] Though the Geological Society could not intercede directly, the successful reopening of negotiations for the following year was made possible in part by the expedition's record of disclosure at Society meetings. From both the American and Chinese perspectives, Zhang Hongzhao's idea of a geosciences society that could accomplish what the government could not was at least partially realized.

Few other foreigners had the large expeditions or longstanding relations with the Geological Society that Roy Chapman Andrews had, but the precedent he set during the Society's special meetings in 1922 prompted foreign geological investigations of all sizes to bring their work in progress before the Society. There, discussions of process, raw data, and interim conclusions set up an informal but voluntary scientific partnership between Chinese and non-Chinese that shortened the distance between fieldwork and the site of knowledge production.[144]

For native geologists, the Geological Society's occasional meetings also provided an opportunity for them to share their most sophisticated work with domestic audiences, rather than simply "exporting" it to more established geological outlets abroad. Li Siguang's notes on Pleistocene glaciation from *Geological Magazine*, which he presented at the Society's third general meeting, prefigured his later theories of quaternary glaciation and laid the groundwork for the most polarized debate in Chinese geology.[145] Weng Wenhao, who spoke after Dr. Hayasaka at the second occasional meeting of the Geological Society, received feedback on his paper for the upcoming International Geological Congress and provided his Chinese colleagues with a more accessible English version of his French original.[146]

The flexible timing of special meetings also enabled Chinese delegates at international conferences to report back on the proceedings. They reviewed responses to Chinese papers, summarized important advances in the field, and communicated disciplinary news. Similarly, Chinese students returning from advanced studies abroad presented the results of their research and alerted local colleagues to new approaches and techniques. In the resource-poor environment of Republican China, society meetings amplified the findings and experiences of individual geologists to benefit the community as a whole and maximize collective impact.

It did not take long for Geological Society members to apprehend their growing scientific reputation, and soon they were drawing geologists to Beijing who had no scientific business in China but supported the Society's efforts. The most famous of these were Alfred Lacroix (France), Hendrik Brouwer (Netherlands), and Bailey Willis, whom the Society hosted on their way home from the 1926 Pan Pacific Science Congress in Java. The Geological Society of China and the Peking Natural History Society (whose

President at the time was Weng Wenhao) invited Willis and Lacroix to give public geological lectures, and several participants of the Pan Pacific Congress attended a Geological Society meeting in their honor. Li Siguang thus had the opportunity to deliver a sweeping view of continental drift before Willis, one of the world's leading opponents of mobilist theory. Li's paper, "The Fundamental Cause of Evolution of the Earth's Surface Features," analyzed available literature on widespread continental movement and offered a dynamical mechanism for global tectonic features based on stresses created by changes in the earth's rotational speed. The presentation generated heated discussion from Willis and Brouwer and brought one of geology's most critical debates to a Chinese audience. Bailey Willis, though unconvinced by Li's arguments, was impressed by his ambition and marveled at his ability to work on such a topic "under considerable difficulty in a place practically isolated."[147] Later visitors to the Geological Society were also surprised at how productive and informed its membership was, not fully appreciating that their own visits were a key factor in maintaining the vibrancy of China's geological community.

Conclusion

When the Geological Society was established in 1922, the Geological Survey had only been active for six years and just a handful of students had graduated from the Geology Department of Peking University. Textbooks and reference materials were all in foreign languages and researchers interested in the geology of China could only consult the works of foreign geologists and explorers. Western geologists were fascinated with China's territory because, despite being inhabited by one of the oldest continuous civilizations known to man, it still seemed totally obscure from a geological standpoint. What was known suggested vast mineral riches, and this was supported by traditional Chinese complacency about their empire as "vast in territory and abundant in all things." Increasingly, too, fossil evidence suggested that China might have been the center of mammalian radiation, from which modern mammals—including man—evolved and migrated to the rest of the world.

Western naturalists, geologists, and adventurers flocked to China from the mid-nineteenth century onward. They traveled to China, collected, sketched, measured, annotated, and then left, saving their insights and conclusions for audiences at home eager to find scientific and romantic possibility in a faraway land. For Chinese geologists just learning to be professionally competent and struggling with scant financial and material resources, China was not simply a distant field or a blank on a map. For them, China *was* home, and, as the hot-blooded May Fourth students of Beida lamented, it was a matter of guilt and shame that they had to depend on the questionable motives of outsiders for knowledge of its physical structure and history.

Instead of railing against this reality and trying to claim Chinese geology for themselves, China's most established geologists, who had gone abroad to study from the West, opted to invite foreigners into their midst. This was not a capitulation, but a canny move that turned China's vulnerability to foreign exploitation into a windfall. Foreign geologists who hoped to catalog China's mineral wealth or uncover its scientific value had to work in China, and by embracing rather than excluding these researchers, Chinese profited from their proximity. By welcoming foreign scientists into the Geological Society of China, Chinese geologists were able to enhance the reputation of their own geological community and broadcast native research through borrowed scientific networks. Though political difficulties and inadequate resources prevented Chinese geologists from large-scale explorations at the frontiers of China's territory, such ventures were the forte of foreign expeditions, and the Geological Society allowed Chinese to access data they could not have obtained themselves.[148]

The Geological Society did not co-opt foreign geologists so much as enlist them, for Chinese geologists did not engage outsiders at the expense of their own scientific development. Rather, they used both publications and public occasions to elaborate the idea of "Chinese geology" as a historically multinational project, in which all parties were invested and from which all benefited. The Geological Society represented Chinese *geologists*, but since Chinese *geology* belonged to world science, it opened its doors to workers from all nations in order to promote knowledge. Because foreign contributions to Chinese geology were so widely scattered, with many national rivalries and little coordination, the Geological Society's cosmopolitan stance made it valuable to geologists from everywhere. Over time, the Society became an "obligatory passage point" to which foreigners, who had previously only extracted data from China, voluntarily imparted new ideas and methods.[149] Given the size of Chinese territory, the burden of geological work could have crushed China's small corps of native geologists. By hosting rather than competing with foreigners, Chinese geologists bought themselves time to meet international expectations and, even more importantly, gained confidence to weather political storms at home.

FOUR

The Nanjing Decade: Geological Expansion and the State

The Master said, "The gentleman is not a utensil." 子曰:君子不器。

—*Analects* 2:12

From the moment the charismatic revolutionary Sun Zhongshan was pressured to yield power to the military leader Yuan Shikai in 1912, the idea of a democratic Chinese republic was already more fantasy than fact. Yuan manipulated the provisional parliamentary system and promulgated a new constitution that made him president for life before declaring himself the Hongxian emperor in January 1916. Predictably, Yuan's reign was brief, and the secession of several southern provinces inaugurated over a decade of warlordism, in which a shifting array of regional forces fought over ravaged pieces of Chinese territory. After Yuan's death in June 1916, a new parliamentary government was established in Beijing, but disagreements over which constitution to uphold and heated debates about China's participation in World War I created so much chaos that the national assembly was dissolved within a year. Frustrated, General Zhang Xun marched on the capital and tried to reinstall the young Manchu heir, Puyi, as emperor. Though this restoration collapsed within days and a new government formed in Beijing under the nominal banner of republicanism, actual power rested in the hands of competing military-political cliques, with presidents and cabinets of the Beiyang government in constant flux. None of these regimes could effectively control the domestic situation, but for warlords with national ambitions, domination of the political apparatus in Beijing not only made foreign loans and customs revenue available, but it also conferred widespread legitimacy for the purposes of treaty obligations and international diplomacy.

As Chinese geologists struggled to accommodate the uncertain political scene of the early 1920s, they also made use of international participation to create an image of Chinese geology that was more coherent than immediately warranted. However, this image successfully tapped into the real concerns and interests of both local and foreign researchers, whose coordinated efforts quickly gave substance to the Geological Society's aspirations for Chinese geology. Notably, this transformation took place within a context that embraced differences, instead of pitting them against one another, and the Geological Society made it possible for scientists trained in different places and committed to different goals to share an intellectual space without compromising their own individual or institutional sensibilities.

Such a solution was not an option for those vying to determine China's political future, and even the idealistic Sun could not avoid employing divisive tactics in his push for a unified China. With the backing of different militarists, Sun Zhongshan established three separate "national" governments in Guangzhou to challenge what he considered an unlawful Beijing regime. None of these efforts was able to extend its power beyond the far southern coast, but from 1917 until his death in 1925 Sun repeatedly pressed for a military campaign to reincorporate north and south China under the newly reorganized Guomindang (Nationalist Party, GMD). When Jiang Jieshi (蔣介石, Chiang Kai-shek) finally launched the party's Northern Expedition in July 1926, Chinese around the country saw Jiang as fulfilling Sun's patriotic vision, and they greeted the National Revolutionary Army with high hopes for the end of warlord depredation.

The Northern Expedition's victories in south China were spectacular, and within nine months its armies had taken the entire Yangzi valley. However, these successes exposed deep rifts within the Guomindang itself. Left-leaning GMD and United Front Communists led an inland offensive up through Hunan and then into Jiangsi and Hubei, where they declared their own government headquarters at Wuhan in early 1927. Meanwhile, Jiang Jieshi took his Huangpu Military Academy graduates and their troops up the southern coast to Shanghai, where he captured China's modern financial center and liquidated urban Communist operatives in a bloody April purge. The GMD left tried to dismiss Jiang, but he joined with rightist ("Western Hills") leaders to establish a government seat in Nanjing on April 18, 1927.

Meanwhile, Stalin and Trotsky's power struggle in the Soviet Union inspired the Chinese Communists to attempt a takeover of the Wuhan government, and the first United Front collapsed as Guomindang from across the political spectrum joined against their former Communist allies. Wang Jingwei's leftists, Hu Hanmin's Western Hills faction, and Jiang Jieshi's military centrists all recognized the Nanjing regime in February 1928, and with the help of entrenched regional militarists in the northwest, Jiang pushed the warlord Zhang Zuolin out of Beijing. The Northern Expedition officially came to a close after Zhang was assassinated by the Japanese, and his son Zhang Xueliang pledged allegiance to the GMD at the end of 1928.

By January 1929, the Guomindang flag flew across both China proper and Manchuria, and an uncontested national government was up and running in Nanjing. This inaugurated the period of greatest centralization during mainland China's Republican period and was heralded as the consummation of the 1911 revolution, but the legacy of turmoil persisted and plans for a strong modern China faced daunting structural challenges. With the end of the military phase of Sun Zhongshan's revolutionary roadmap, the government was charged with guiding a period of political tutelage according to Sun's Three Principles of the People (democracy, nationalism and livelihood). Once the Chinese people learned to exercise their rights as modern citizens, party domination would theoretically yield to democratic process. However, there was no clear time frame for this period of tutelage, and it was often used to justify dictatorial measures by the state.

The Nanjing government was also hampered by economic weakness that crippled its ambitious plans for "reconstruction" (建設 *jianshe*). Following Sun's modernizing vision and eagerness to restore national pride, the GMD hoped to strengthen China through roads and rail, electricity and running water, public education, industrial development, and scientific planning. Unfortunately, the central government had few reliable sources of income and heavy debts inherited from its predecessors. Though it managed to recover tariff autonomy from foreign powers and establish the salt gabelle as a revenue-making institution, at no point during the Republic of China's Nanjing Decade (1927–37) could the Finance Ministry ever collect central tax revenues from more than half of China's provinces. Land reform was stunted by the government's reliance on landlords for political and economic support, and most agricultural and land taxes were "delegated" to local and provincial governments for lack of adequate personnel. Trade and manufacturing taxes hobbled the nascent industrial sector, and as the effects of worldwide depression belatedly reached China's shores in the 1930s, the Republic floated on a rising tide of bond issues and domestic debt.

The Guomindang regime was often corrupt and inefficient, but the root problems behind these fiscal issues remained political and military. On the domestic front, between 1927 and 1937, the Nanjing government faced repeated insurrections from warlords co-opted during the Northern Expedition, and GMD factions challenging the center's legitimacy. Even mainstream elements loyal to Jiang were sharply divided into "cliques" that jealously vied for power. Portentously, the Chinese Communist Party (CCP)—freed from the constraints of the United Front—openly instigated urban uprisings, and a splinter group under Mao Zedong and Zhu De began organizing rural soviets in Jiangxi and Hunan. These diverse forms of internal opposition hampered tax collection and routinely raised military expenditures to over 50 percent of central government revenue throughout most of the Nanjing Decade. The tax burden on farmers in the provinces was multiplied by the layering of local and central authorities, and de facto independence of regional powers rendered many of Nanjing's programs and policies ineffective.

Internationally, China was largely sheltered from the direct effects of worldwide depression by its small industrial presence, but widespread financial instability overseas stepped up Japanese expansionism and weakened Western response, exacerbating existing threats to Chinese territorial sovereignty. Japanese militarists not only engineered the Manchurian Incident of 1931 that led to the creation of the puppet state, Manzhouguo (Manchukuo), but they also adopted a policy of progressively "separating" north China from Nanjing control. As these moves accelerated toward war in 1937, the question of China's stance toward Japan divided domestic politics even further.

Faced with enemies within and without, Jiang Jieshi's agenda emphasized centralization and consolidation wherever possible, and modernization programs often doubled as attempts to enhance Nanjing's influence. In this situation, although scientific initiatives were assumed to be apolitical, they were expected to fall in line with the center's vision of national reconstruction. Despite changes in rhetoric that sounded a more modern note, Jiang was essentially a leader in the old Self-Strengthening mold who saw science as a means of fortifying the state rather than interrogating nature.

This instrumentalist attitude made the Nanjing regime a difficult partner for China's developing scientific community. For geologists at the survey, which had outlasted several Beiyang regimes and maintained a strong institutional identity despite the uncertainties of the warlord period and the Northern Expedition, Jiang's hands-on, ends-driven approach was awkward but manageable. Survey geologists were no strangers to the nation-building cause, but they were accustomed to a great deal of autonomy, and they welcomed the government's new attention to science and technology with a strategy of wary flexibility. Though officially a government agency, the survey never relied exclusively on the state for funding and repeatedly won support from private sources, including both Chinese entrepreneurs and international grant-making agencies. As a technical organization, the survey selected and trained its personnel outside the normal bureaucratic channels, and its cohesion and efficiency depended upon student-teacher bonds rather than state policy. With Weng Wenhao at the helm, the survey not only steered its own course for improving science and nation, but it also eventually nudged broader Guomindang policies in more effective directions.

At less established geological institutions, however, the Nanjing regime's narrowly practical agenda felt more intrusive. It forced geologists to sharpen both their institutional cultures and their approaches to serving the national interest. For independent-minded leaders like Li Siguang, who had no qualms about exploring unconventional geological interpretations at his newly founded Research Institute of Geology (地質研究所)," scientific patriotism had little to do with toeing the GMD's party line, and pursuing practical geological applications did not entail forsaking theoretical sophistication. In fact, challenging accepted orthodoxy seemed central to serving China's scientific interests, and as Li took steps to distance himself from Jiang's peremptory brand of nation building, it became apparent how differences that enriched the Geological Society could seem defiant to the state.

Expansion of the Geological Survey

Throughout the 1920s, the Chinese geological community was busy showing the world that even if the political situation was a shambles and the country divided, Chinese scientists were united in their desire to take responsibility for understanding their own territory. In addition to ramping up publications, building local and foreign networks, and expanding research facilities, the 1926 discovery of Peking Man at Zhoukoudian boosted China's image as a modern scientific nation even further, and put the survey at the helm of an international investigation that aspired to rewrite the evolutionary history of all mankind.[1]

These activities and the practical utility of the survey's coal and iron studies raised the domestic profile of geology, and as the GMD's National Revolutionary Army moved northward, provincial surveys were set up to encourage exploration of local resources. Once the Guomindang established its national government in Nanjing, it reorganized the Ministry of Agriculture and Commerce into the Ministry of Agriculture and Mines, and the Geological Survey negotiated a revision of its constitution. Under the constitution of 1928, the survey abandoned its original sections of geology, ore resources, and publishing, and separated its research and support functions. Focused research departments were established for physical geology (comprising mineralogy, petrology, and stratigraphy, etc.), economic geology, and paleontology, with publication and outreach handled by distinct administrative, library, and museum divisions.

After years of struggling to stay afloat amid the political turmoil, economic scarcity and military disturbances of the warlord era, the Geological Survey saw the Nanjing regime as a new beginning. A central government committed to modernization and development promised stability and growth, and the survey laid out a bold plan for the diversification of geological research in hopes that officials would recognize the scientific potential of unification. The survey's existing objectives had been practically oriented: (1) 1:1,000,000 scale geological mapping of the entire country; (2) resource surveys focused on iron, coal, and other industrial ores; and (3) paleontological and mineralogical studies in support of stratigraphy, structural geology, and economic geology. These projects remained priorities, but under the new administration, the survey called for increased emphasis on regional geology of the southeast, exploration of frontier territories, and geochemistry and geophysics.[2] The survey would collaborate and coordinate with both provincial organizations and the Academia Sinica's Research Institute of Geology to reach these goals and extend its fieldwork into understudied areas in the south and west of China.[3]

The Ministry of Agriculture and Mines was impressed by this plan to launch a coordinated geological program and approved the accompanying budget. But the government's desire for geological results outstripped its financial resources, and it could not honor its target appropriations. Survey director Weng Wenhao made repeated inquiries and finally secured a loan of 1,000 *yuan* a month from the Research Institute of Geology of the newly created Academia Sinica, "the highest academic research institution of the Republic of China under direct command of the Nationalist Government."[4] However, the Academia Sinica's organization and mandate were unstable until the 1930s, and, faced with its own financial troubles, its loans were short-lived.[5] Finally at the end of 1928, after the Ministry of Agriculture and Mines reneged on an emergency remittance of just 1,000 yuan, Weng sent a letter to the ministry explaining that the survey was on its last legs. Even without the salaries of the survey's leading investigators (who had all agreed to forego their wages) basic operating costs topped 4,000 yuan a month.[6] Foreign experts such as Grabau and Johan Gunnar Andersson were not only working without pay, but they were also advancing personal savings to help fund publications.[7] In the interest of salvaging previous work, the survey's thirty-five members had been "attending to their duties on an empty stomach"—even going into the field whenever possible—but they could not continue for much longer, and morale was disastrously low.[8]

While Weng acknowledged the government's periodic enthusiasm for geological research, he argued that instead of proposing new research enterprises, it should first disburse the funds it had promised to institutions like the Geological Survey. The survey not only had the qualified personnel and the track record to provide real results, but it was also already well-established in the international scientific community and could advertise the government's commitment to scientific programs.[9] What disturbed Weng most was the ministry's complete unwillingness to face up to financial realities. Research proposals could be designed to fit a very restricted budget if only the ministry would commit to a considered amount and make good on it.[10] During its years under the Beiyang government, the survey was at least able to anticipate a deficit, and throughout the warlord era it worked for mining companies and local construction bureaus to support research wherever conditions allowed. In contrast, the Nanjing government's centralizing and bureaucratic bluster had lulled the survey into trusting its overly optimistic allocations, and subsequent funding problems took the survey by surprise.

The sum total of irregular ministry funds received by the survey in 1928 amounted to less than 10,000 yuan for the entirety of 1928, and this fiasco taught the survey that it could not depend on the central government for maintenance let alone expansion. That year, the Geological Survey eked by on funds from a preexisting grant of 35,000 yuan from the China Foundation for the Promotion of Education and Culture, which managed American Boxer Indemnity remissions. Meanwhile, Weng Wenhao and Ding Wenjiang scrambled to find nongovernmental support and used their personal influence to secure a larger, renewable grant of 50,000 yuan from the China Foundation for the next several years.[11]

On the domestic front, six prominent mining companies donated money for a major museum and library expansion to show their support for the survey's contributions to Chinese society.[12] A few Guomindang intellectuals who wanted to preserve the preeminence of Beiping ("northern peace," formerly Beijing, or "northern capital") as a research center also formed the National Academy of Beiping under the Ministry of Education, and in March 1929 the survey negotiated an annual grant of 24,000 yuan for an institute of geology at this new academy.[13] Weng Wenhao was named director of the institute, and for all intents and purposes the Beiping Institute and the Geological Survey, which completely shared their facilities and personnel, were a single entity.[14] By the end of the 1920s, both academic and industrial interests saw the broad research agenda of the Geological Survey as an important part of building national prestige, and they made every effort to intervene when the Nanjing regime fell short.

These measures safeguarded basic survey operations against the government's financial instability, but survey members were eager to launch more ambitious ventures and expand the range of their geological research along the lines they had previously outlined. As soon as the discovery of a hominid tooth at Zhoukoudian was made public in October 1926, Davidson Black of the Peking Union Medical College approached the Rockefeller Foundation with a proposal to fund a full-time vertebrate paleontologist to study the remains of any anthropoid material found at the site. Weng Wenhao matched this suggestion with a pledge to provide survey personnel for topographic and geological study and contribute as much as possible to the working expenses of the project.[15] Though the survey's initial two-year cooperation with the Rockefeller Foundation was very limited and only sponsored work directly connected to hominid evolution, the Zhoukoudian excavation developed into a major focus of survey activity in 1929 and provided a successful model for the survey's independent financing of several later projects.[16]

By the end of 1928, two hominid mandibles were found and Davidson Black felt confident ascribing Peking Man to a new genus *Sinanthropus*. Given the bounty of associated mammalian fossils and the extent of the Zhoukoudian deposits, Chinese paleontologists wanted to continue excavating the caves but the survey was financially hamstrung and could not assume responsibility for such a large-scale investigation on its own.[17] Davidson Black mediated negotiations between Weng Wenhao and Rockefeller representative Roger Greene, and in February 1929, an agreement was reached to fund a Cenozoic Research Laboratory as a permanent division of the Geological Survey of China. With over US $80,000 for the first three years alone, the Cenozoic Laboratory extended the survey's work far beyond the narrow scope of the Peking Man investigation to broadly support the study of geology and paleontology in the Tertiary and Quaternary periods.[18]

This landmark collaboration was a vote of confidence in the administration and staff of the Geological Survey, which persevered throughout the political and fiscal difficulties of 1926–28 and produced talented vertebrate paleontologists like Yang Zhongjian, Pei Wenzhong (裴文中), and Jia Lanpo (賈蘭坡) in remarkably short order.[19] Unlike the American Museum of Natural History's Central Asiatic Expedition to Mongolia and Sven Hedin's Sino-Swedish Expedition to Xinjiang, which ran into bitter public opposition from the Committee for the Preservation of Ancient Objects and the Federation of Chinese Academic Societies of the late 1920s, Chinese intellectuals welcomed the Rockefeller-sponsored Cenozoic Laboratory with open arms.[20] The constitution of the laboratory stipulated that "all material collected shall entirely belong to the Geological Survey of China" and nothing would be exported out of the country. The Cenozoic project disavowed any interest in cultural artifacts, which would be handed over to Chinese authorities, and all papers produced in the laboratory were to be published in the journals of the survey or the Geological Society of China.[21]

These conditions were very much in keeping with the Rockefeller Foundation's express purpose of fostering science in China,[22] but they were the specific product of personal ties established through the Geological Society and other academic associations. Survey members were an integral and irreplaceable part of the Peking Man investigation, and when the first skullcap of what Davidson Black dubbed *Sinanthropus pekinensis* (now recognized as Homo erectus) was discovered, noted American geologist George Barbour wrote:

> The whole circumstances of the discovery reflect the highest credit on the methodical persistence of the staff of the Survey, which has worked loyally through many months of extreme difficulty and political disorganization. . . . The credit for the recent discoveries goes to Mr. W. C. Pei, a young geologist on the staff who has supervised all the recent excavations. One has to visit the 'cave' at Chou Kou Tien [Zhoukoudian] to appreciate his achievement[23]

Julia Strauss has pointed out that the Republican era in China was characterized by "strong institutions in weak polities," and that the success of individual government organizations "hinged on two closely related variables: insulation of the organization from external pressures, and general perceptions of the organization as successful in meeting its goals."[24] Strauss focuses on institutions like the Ministry of Finance and the Sino-Foreign Salt Inspectorate, which deal with what she calls "core" state-building activities, but the Geological Survey adopted a similar strategy. By proving itself effective and reliable in the face of political vicissitudes, the survey became one of "the only government institution[s] . . . able to get financial support from private sources."[25] In turn, these private arrangements buffered the survey from excessive government intrusion, while expanding the survey's capabilities and making it increasingly indispensable to the Nanjing regime.

The Rockefeller collaboration was an ideal blueprint for how fiscal independence could allow the survey to support state interests without sacrificing organizational autonomy. Instead of funding specific research projects, like mining companies and local government offices often did, the Rockefeller established a new disciplinary center that both encompassed the Peking Man investigation and fulfilled an existing institutional need. It also guaranteed a regular multiyear budget that allowed for continuity and planning, and provided independent oversight to ensure that money could not be diverted away from research. Finally, it took account of both China's nation-building and national-pride-building goals by reducing China's dependence on foreign facilities or specialists and helping keep data and personnel in the country. Bolstered by its success with the Cenozoic Laboratory, the survey returned to its earlier approach of setting its own research agenda by actively seeking private sponsors for a seismological station, a fuel laboratory, and a pedological department by the end of 1930.

Earthquakes had been a topic of interest for Chinese geologists for over a decade, but the Geological Survey had neither the instruments nor the expertise to conduct modern seismological studies of its own. Even though the first paper presented by a Chinese geologist at an International Geological Congress was Weng Wenhao's "L'influence seismogenique de certaines structures géologiques en Chine" (1922), it was based entirely on historical records and field observations.[26] Weng published several other studies of seismotectonics and earthquake precursors, but at the Fourth Pan-Pacific Science Congress in May 1929, it became clear that China was the weak link in the emerging web of seismic stations across East Asia.[27] This was all the more embarrassing to Chinese geologists because earthquakes had been recorded in China as early as 1831 BC and the first seismoscope was invented by the royal astronomer of the Eastern Han dynasty in AD 132[28].

In the 1920s the only two active seismic stations in China were operated by French and German authorities in Shanghai and Qingdao, and Weng felt that even if China was unable to make any theoretical contributions to seismology in the immediate future, it had to record and publicize accurate measurements to maintain its international reputation.[29] Moreover, Chinese seismic activity was widely distributed geographically and accounted for one-third of the world's continental earthquakes. With a modern facility in north China, triangulation with Shanghai and Qingdao could pinpoint the epicenters of domestic earthquakes without relying on foreign seismic stations.

Seismology was not a national priority for the Ministry of Agriculture and Mines, but Weng Wenhao published a small volume on earthquakes in late 1929 and convinced Lin Xinggui, a prominent lawyer, to donate a tract of land on Jiufeng Mountain, in the western suburbs of Beiping, for a seismic station.[30] According to the negotiated agreement, Lin would also provide 2,000 yuan for construction costs and provide housing for staff in a temple adjacent to the station.[31] However the land and any structures on it would revert to Lin if the seismic station were ever inactive for longer than a year, and the survey used this clause to press the national government for ongoing support. The station was outfitted and staffed using discretionary funds from the China Foundation, and the survey bought a large Wiechert seismograph, a radio receiver, and astronomic clocks.[32]

Li Shanbang (李善邦), a graduate of the Peking University Physics Department, was sent to Shanghai for training, but he received little assistance from Father E. Gherzi of the Siccawei (徐家匯) Observatory. He returned to Beiping in June 1930 and figured out how to use his new equipment from books and manuals. The first seismological readings were recorded and verified in September 23, 1930, and within a year tremors as far as Mexico and New Zealand were detected. Li and his assistants spent several months checking their data against other stations and learning to interpret different signals, and Li studied seismology in Japan before the Manchurian Incident of 1931.[33] In August 1932, two state-of-the-art Galitzin-Willip seismographs arrived, and recordings were published in the bimonthly *Seismological Bulletin* for international exchange.[34] The station also put out an irregular journal containing comprehensive analyses of notable earthquakes, technical advice, and discussions of theoretical issues such as crustal composition. The Jiufeng Seismic station was continuously active until the Lugouqiao Incident in July 1937, which precipitated full-scale war between Japan and China.

Like the seismic station, a fuel laboratory was contemplated long before 1930. Chinese territory was generally thought to be coal-rich and petroleum-poor, but estimates varied widely, and the Geological Survey considered it vital to analyze existing fuel sources in order to rationalize both prospecting and utilization of fossil fuels.[35] By the 1930s, domestic production satisfied less than 1 percent of China's demand for gasoline and even less of its kerosene consumption, making fuel research vital not only for China's industrialization but for its basic economic solvency.[35] But though the central government was extremely interested in boosting fuel supplies, it still deemed the startup cost of a laboratory facility prohibitive until Jin Shuchu (金叔初, Sohtsu G. King) donated 15,000 yuan for the project. A wealthy businessman who was also a serious student of natural history and an active member of the Geological Society of China, Jin Shuchu became involved with the fuel laboratory in late 1929 through his friendship with the leaders of the Geological Survey and his nephew's interest in chemical engineering.[37] In the end, Weng Wenhao decided to name the fuel laboratory after Jin's father, Jin Qinyuan, and Jin's nephew, Jin Kaiying, volunteered to go to the United States to study petrochemistry for the survey.[38]

When it opened on National Day (October 10) 1930, the total cost of the new three-story Qinyuan (Sin Yuan) Fuel Laboratory was almost double Jin Shuchu's donation, but the survey used his generosity to justify its use of China Foundation funds to make up the difference.[39] This opportunity allowed the survey to revamp all of its laboratory facilities, not just those for coal and petroleum testing, and helped it balance its strengths in the field with comprehensive benchwork and analysis. In effect, the fuel laboratory became the geochemical and geophysical wing of the survey, staffed by technical specialists but available to the entire organization. Besides space and equipment for petrochemical research, the fuel laboratory also specialized in micropaleontology and housed a microscopic study, a mineralogical and petrographical laboratory, a spectroscopic laboratory, a photography room, and special areas for instruments and balances.[40]

In addition to the analysis and classification of native fossil fuels, the fuel laboratory conducted extensive experiments on the feasibility of gasoline production from coal and shale using various methods of low-temperature distillation and hydrogenation. The processes were not cost-effective, however, and when fuel research moved to Nanjing with the survey in 1935, work focused on cracking vegetable oil as an alternative fuel source. After the outbreak of war in 1937, a small branch of the laboratory stayed with the survey to conduct chemical analysis but most staff members moved to industrial research facilities and synthetic gasoline factories for the duration of the conflict.

The last major addition to the Geological Survey in 1930 was the soil laboratory, or pedological research department. Because Chinese viewed agriculture as the basis of their civilization, the analysis of regional soils was a part of Zhang Hongzhao's original proposal for the survey in 1912, and soil science was listed under the ore resources section in the survey's first constitution in 1916. Nevertheless, due to a lack of funds and technical personnel, no progress was made beyond the collection of a few scattered soil samples, and even prominent debates about the origin of northern China's spectacular loess formations did nothing to change this.[41] In 1928, Dr. Neustruev of the Institute of Soil Science of the Soviet Union contacted Weng Wenhao to make arrangements for the completion of a soil map of Asia by 1930, when the Second International Congress of Soil Science would be held in Leningrad. Neustreuv not only highlighted China's lack of soil data "on the principle of the so-called climatic soil types" which was becoming the international standard, but he also suggested that Soviet scientists conduct large-scale fieldwork to investigate Chinese soils.[42] Soil quality and land utilization were also topics of concern at the 1929 Pan-Pacific Science Congress in Java, and in 1930 the China Foundation finally called upon the Geological Survey to form a department exclusively devoted to soil science.

The soil laboratory was officially established in September 1930, with Weng Wenhao himself as director. The China Foundation commissioned the Geological Survey to conduct a detailed soil study of the entire nation and committed 50,000 yuan annually to the project from 1931 to 1940.[43] The survey used the foundation's startup grant of 20,000 yuan in 1930 to start construction on new facilities for the soil laboratory and hire three recent college graduates with backgrounds in geology, agricultural science, and agrochemistry.[44] Robert L. Pendleton, of Johns Hopkins University, arrived in Beiping in February 1931 for a one-year stint as chief soil technologist of the Geological Survey to train Chinese staff and organize the soil laboratory's research program. The soil laboratory began publication of its *Soil Bulletin* in 1931 with "Soils of China" by Charles F. Shaw, a soil scientist on sabbatical from the University of California at Berkeley who spent his sabbatical year making preliminary studies in association with noted agricultural economist John Lossing Buck at the University of Nanjing.[45] In 1933 James Thorp, of the US Bureau of Soils, was invited to lead a three-year national soil survey, and with the addition of new Chinese graduates trained overseas in soil science, the work of the soil laboratory began in earnest.

The soil laboratory focused on soil mapping, soil classification, soil morphology, and fertilization and usage studies. A national soil map was constructed in 1935, and by 1936, when the first edition of the *Geography of the Soils of China* was completed, the soil laboratory had sixteen scientists and technicians on staff.[46] In addition to the *Soil Bulletin*, the laboratory also published a monograph series and the *Soils Quarterly* (1941–48). Though the soil survey was strongly influenced by the Soviet and American emphasis on field research, it also cooperated with both the fuel laboratory and several agricultural bureaus on physicochemical analysis of Chinese soils. The work of the soil survey was greatly complicated by the absence of uncultivated soil in most of China proper, a shortage of trained surveyors, and the physical difficulty of accessing many areas in the frontier regions. However, the economic value of soil research kept the soil laboratory active through the War of Resistance, and from 1936 onward the laboratory was completely staffed by native scientists.[47]

The survey's remarkable expansion in 1930 not only reflected its reputation for industry and achievement, but it also demonstrated the willingness of wealthy urban Chinese to support national scientific programs. In return, the Geological Survey was eager to benefit the general public more directly, and in 1930 it took on the task of compiling the first national atlas of China.[48] Ding Wenjiang convinced Shi Liangcai (史量才), general manager of Shanghai's premier daily newspaper, *Shen Bao*, to commission an up-to-date national atlas to commemorate the paper's sixtieth anniversary in 1934. Shi agreed to pay all production costs and provide 2,000 yuan a month to cover research expenses as well. Ding brought Weng Wenhao onto the project, and Weng put the survey's library and research materials at the disposal of Zeng Shiying (曾世英), who handled the actual compilation. As a result, Zeng joined the survey in 1930 and established a separate department for drafting and analysis of mapping data.[49]

When the *New Atlas of the Republic of China* (*Zhonghua minguo xin ditu* 中華民國新地圖, or *Shenbao Atlas*) was published in April 1934, it featured seven general maps of China and forty-four detailed political and topographical maps. The general maps of population, climate, linguistic distribution, agricultural production, mining output, and transportation were compiled by disciplinary specialists, such as historian and philologist Fu Sinian (傅斯年), geologist and meteorologist Zhu Kezhen, linguist and musicologist Zhao Yuanren (趙元任), agriculturalist Zhang Xinyi (張心一), and geologists Xie Jiarong and Hou Defeng (侯德封).[50] These topical maps were designed to give readers new images of the nation as a coherent, connected whole, and to drive the point home, regional maps were divided by quadrants rather than provinces, replacing local political ties with a unified mathematical scale across China's vast territory.[51] This somewhat cumbersome layout was justified by rigorous updating of elevation data, geographical coordinates, and place names using almost eight thousand foreign and domestic maps, many of which were previously confidential. Information was augmented as needed with field measurements, and the atlas addressed both the representational inaccuracies of existing Chinese maps and the out of date political geography of foreign sources (whose names and boundaries were often holdovers from the Qing or Ming dynasty).[52] The atlas also included insets of major Chinese cities and was the first publication to include a comprehensive index of place names and geographical coordinates. The *Shenbao Atlas* and subsequent revisions remained the standard reference map of China until the 1950s, with several guides and translations published in foreign languages for international use.

Though the Geological Survey saw the unification of China after the Northern Expedition as an opportunity to place geology at the disposal of the state, the central government was too politically divided and economically weak to take the lead in scientific development, and the survey channeled its nation-building energies elsewhere. Forced to seek its own financing, the survey achieved most of the new research priorities that it had outlined for the Nanjing administration in 1928 through independent funding and private initiative. The process of finding nongovernmental support not only kept the survey free from the administration's narrowly utilitarian focus, but it also allowed the survey to respond creatively to unexpected opportunities like the discovery of Peking Man. In so doing, it subscribed both international organizations and domestic industry to the national geological research program, and allowed the survey to represent Chinese geology as a functioning whole. In turn, these successes shielded the survey from bureaucratic control, and allowed it to act on behalf of, rather than at the behest of, the state for much of the Nanjing Decade.

The Geological Survey and Nation Defense Planning

In the early years of the Nanjing Decade, the Guomindang made significant strides toward recouping Chinese sovereignty in Manchuria. The Japanese Guandong Army's attempt to thwart Guomindang extension into Manchuria by assassinating warlord Zhang Zuolin backfired, and Zhang's son, Zhang Xueliang, became a staunch supporter of the Chinese national cause. The Nanjing government was determined to break Japan's stranglehold over the Manchurian economy and began building its own rail and communications network across the northeastern provinces of Liaoning, Jilin, and Heilongjiang. Emboldened by international acceptance of the Guomindang regime, Chinese officials refused to renew several Japanese forestry and mining concessions and encouraged native agriculture and industry. The Geological Survey focused much of its fieldwork on the region as well. These rights recovery initiatives and the worldwide economic downturn of 1929 weakened Japanese holdings in the region, and Chinese seemed set to halt foreign encroachment on their strategic northern border.

Circumstances changed drastically on September 18, 1931, when officers of Japan's Guandong Army took matters into their own hands and staged an attack on the Japanese owned South Manchurian Railway as a pretext for taking over Manchuria. They immediately blamed China for destroying Japanese property and "retaliated" against Chinese forces in the area. Within a few months of the Manchurian Incident, known simply as 9–18 to Chinese, the Guandong Army had occupied most of Manchuria, and launched an invasion at Shanghai.[53] In order to defend Japanese interests against "Chinese provocation," radical militarists deposed the moderates in the Japanese cabinet in December 1931 and supported Guandong Army advances. With extremists in power, Japan established Manzhouguo as a puppet state in February 1932, flouting not only the Washington Conference of 1922 and the Kellogg-Briand Pact of 1928, but also diplomatic intervention from the United States and the League of Nations. In 1933, the league formally censured Japan's continued expansion into Rehe (Jehol), and Japan withdrew from the league, shattering the dream of peaceful internationalism and opening the door for even bolder incursions into north China after 1935.[54]

In real terms, Japan's invasion of Manchuria was not only a shocking blow to Chinese pride and self-determination, it was an economic disaster. The loss of Manchuria meant the loss of "one fifth of China's territory, which included half the nation's railways, four-fifths of its iron production, and . . . a tenth of its customs revenues."[55] Chinese had also looked to Manchuria's pristine soil for much needed farmland, and in the years before 1931 its soybean production had provided one of China's few profitable agricultural exports. Japan's occupation of Manchuria cut off long-standing trade routes connecting China and the Soviet Union, and left Beijing and Tianjin vulnerable to future attack.

For Jiang Jieshi personally, however, the Manchurian Incident was a blessing in disguise. Though he successfully established a central government in Nanjing in 1927 and incorporated Beijing and Manchuria by the end of 1928, Jiang's hold on Chinese politics was continually challenged by both ideological opponents and disgruntled warlords. In May 1931 dissatisfaction with Jiang was so great that Wang Jingwei's leftists and Hu Hanmin's Western Hills faction (themselves bitter enemies) united with Guangdong and Guangxi militarists to form a separatist government in Guangzhou and demand Jiang's resignation. In the midst of this power struggle, Japan's sudden aggression gave Jiang an opportunity to prove his indispensability, and he retired from both civilian and military duties in December. A new government formed in Nanjing, but without support from Jiang, it collapsed within a month, leaving the country incapable of facing the Japanese threat. Grudgingly, Wang Jingwei returned to Nanjing as premier, but all power rested with Jiang, who assumed control of a newly created Military Affairs Commission and dominated the Guomindang from January 1932 onward.

The Manchurian Incident drove home the inevitability of war with Japan, but Jiang was eager to avoid conflict until China had the military and industrial capacity to defend itself. Chinese Communists capitalized on anti-Japanese sentiment to win over the public and challenge Guomindang rule, but, though thankless, Jiang's policy of nonresistance bought China time to mobilize. Behind the scenes, preparation for war became Nanjing's top priority, and this new clarity of purpose pushed geology to the forefront of Chinese economic policy. The Geological Survey not only led the way in the search for vital iron and coal reserves, it also provided a critical model for the organization and management of industrial development.

The first half of the Nanjing Decade, from 1927 to 1931, had been marked by grandiose planning and embarrassingly meager results. Iron and steel works were located in areas with no coking fuel, massive automobile factories were planned before highways existed, and power plants were announced without financing or distribution networks. The National Reconstruction Commission charged with coordinating industrial development had little money of its own and could not attract foreign investment. In the exasperated finance minister's words, the commission's endless stream of "pet projects . . . involving huge expenditures" were all "unrealizable because of the known shortage of funds," but when put in charge of a new National Economic Council in 1931, he could do no better himself.[56]

A change was clearly necessary to face the Japanese challenge and after private meetings with scores of the nation's most respected experts in economics, sociology, history, science, and engineering, Jiang Jieshi tapped Weng Wenhao to lead a secret National Defense Planning Commission (國防設計委員會, NDPC) in 1932. The NDPC was to be under Jiang's direct control, and he was prepared to back it with his own financial resources, but he wanted to find a political outsider with both an economic vision for Chinese development and a track record of success to run the organization. Weng had resisted earlier attempts to lure him into government, but after the Manchurian Incident he became more vocal about national affairs and joined his friends Ding Wenjiang and Hu Shi in founding the weekly journal *Duli pinglun* (獨立評論 *Independent Review*).[57] He gave Jiang several private lectures on China's geology and strategic resources, and Jiang became increasingly convinced that Weng had the qualities he was looking for. By the end of the summer, Weng agreed to behind-the-scenes leadership of the NDPC, with the understanding that British-trained economist Qian Changzhao (錢昌照) would handle everyday duties so Weng could remain in Beiping as director of the Geological Survey.[58]

William Kirby has pointed out how effectively Guomindang industrial policy after 1932 was dominated by the NDPC, which, unlike its predecessors the National Reconstruction Commission and the National Economic Council, made significant advances in "producer industries and national defense."[59] Kirby stops short of calling the post-Manchurian Incident regime a "technocracy" but attributes many of its developmental successes to the creation of "an enduring civilian government in which engineers—those who wanted to build things, get things done—would prevail over economists, who knew how little the government could afford."[60] He also links the National Resources Commission (資源委員會, NRC), which emerged from the NDPC in 1935 and guided China through its eight-year war with Japan, to the Society of Chinese Engineers and credits the NRC's achievements to the Society's ethos of technical professionalism. Engineers constituted a large portion of the NRC's personnel and were vital to the execution of its industrial plans, but the NDPC/NRC absorbed the culture of the Geological Survey rather than that of the Society of Engineers. Unlike the Society of Engineers, which never had to function within the confines of the government, the survey was adept at protecting its technical dominion from bureaucratic interference. More importantly, the survey's awareness rather than defiance of "how little the government could afford" was precisely what allowed it to "get things done" under difficult circumstances and made Weng Wenhao an ideal leader for the NDPC.[61] What the NDPC envisioned, Jiang Jieshi nurtured, geologists fueled, and engineers helped bring to life, but it was Weng's experience guiding the Geological Survey, his firsthand knowledge of China's territorial resources, and the professional reputation of China's geological elite that made the NDPC the "strong institution in a weak polity" par excellence.

In his first *Duli pinglun* article, published while he was discussing national development with Jiang, Weng explained why Guomindang reconstruction had become its own worst enemy. Reconstruction was such political gold that there was greater incentive for proposing new (and often competing) industrial and public works projects than for implementing existing ones. "Projects are started and then midway the responsible bureau is reorganized or the leadership changes, overturning everything so you have to start over from scratch and repeat what has already been done."[62] While earlier schemes failed for lack of planning, suddenly planning was a cottage industry of its own, and the public no longer trusted the government's ability to execute its increasingly improbable ideas. "Every day I hear of several plans," wrote Weng, "but the substance of these plans invariably has no connection with reality."[63]

Because planning had taken on the character of political performance, high-ranking officials raced to submit new proposals and were more concerned with "writing introductory essays and inscribing dedications" than analyzing actual conditions or fiscal realities.[64] For Weng, this was not planning at all. Real planning relied upon technical expertise, on-site research, and the freedom to conduct serious study of practical problems. Nothing could be done without an accurate assessment of available resources and local conditions. At the very least, planning required enough continuity and coordination to produce reliable, time-tested results. Citing the old Chinese saying, "a seven-year ailment can only be cured by three-year old herbs," Weng suggested that China's age-old problems demanded several years of uninterrupted expert planning to be effective.[65]

The Geological Survey had weathered countless regime changes and government reorganizations to create a reputation for just this kind of cumulative research and planning. However, Weng lamented that bureaucrats enamored with planning as just another political tool ignored geological surveys because they did not "want to bother reading them and would rather be satisfied with their silly fantasies." As a result a sulfuric acid factory was opened in a province with no sulfur ore, and politicians still had "the haughty audacity to say that ores are available everywhere."[66] Though the Chinese had long believed their country was "vast in territory and abundant in all things," careful geological surveys had made it clear that relative to its size and population, China was actually mineral-poor in many key areas and its fabled coal and iron reserves were greatly overestimated. By clinging to this outdated faith in China's mineral wealth, politicians squandered precious development funding on extravagant blunders and missed opportunities to exploit China's actual resources.

The government's disregard for available data in planning new programs was not limited to technical information. Weng's criticisms plainly echoed his concerns from 1928, when the Geological Survey first came under Nanjing's jurisdiction and discovered the danger of relying on its financial projections. In his memos to the Ministry of Agriculture and Mines, Weng was very clear that the ministry's inability to "weigh its income before making expenditures" (*liangru weichu*) and failure to prioritize among desired projects made it impossible for the survey to carry on its research even though detailed plans had been submitted and approved. Though the survey was in dire straits by the end of the year, Weng was less concerned about obtaining the promised sums than about assuring accurate figures and timely remission in the future. For Weng, it was a refusal to deal with "real circumstances" that interfered with the survey's own tradition of effective management.[67]

Fortunately, though government interest in geology changed with the political weather, the survey had developed a clear sense of China's most pressing geological questions through more than a decade of field research and international exchange, and it was determined to follow its own program of growth. Within two years, the survey successfully funded, staffed, and equipped all of the projects outlined in its initial plan to the Nanjing regime and took on new initiatives like the soil laboratory as well.[68] At the end of 1930 the survey reported its activities to the newly established Ministry of Industry (which replaced the Ministry of Agriculture and Mines) as fait accompli rather than items for approval and explained the importance of each new survey branch to China's economic development as part as an overall plan to rationalize use of territorial resources.[69]

The survey was successful when it separated itself from bureaucratic inefficiency, or "scrambling," and Weng brought this same attitude to the NDPC.[70] He also imported the Geological Survey's stress on independent recruiting and expertise. All candidates for the survey had to pass stringent entrance exams and work through several stages of probationary status and apprenticeship before becoming full-time researchers.[71] There were no hires from the regular civil service pool and new recruits were immediately indoctrinated in the survey's culture of fieldwork and personal loyalty. Though not geologists per se, the NDPC's technical personnel were imbued with the values of the new Chinese intellectual, as embodied by Weng Wenhao and Ding Wenjiang, and expressed in Ding's 1923 "Debate of Science and Metaphysics." For Ding and other leading intellectuals, the "scientific philosophy of life" represented logical thinking, healthy skepticism, strict empiricism, and results orientation rather than empty talk and wishful thinking. By merging scientific method and physical vigor, geologists typified the ways that technical proficiency supposedly cultivated superior moral values, and in many ways, the NDPC under Weng constituted the "government by good men" proposed by Ding in his writings in *Nuli zhoubao* (努力周報, *Endeavor Weekly*). In a *Duli pinglun* article entitled "My Views Are Just These," Weng pulled all these ideas together, equating scientific professionalism with both national achievement and good citizenship.[72] In the aftermath of the Manchurian Incident, intellectuals who had previously eschewed government sought ways to make tangible contributions to society, and Weng's personal reputation as both a geologist and an organizer of men attracted China's best and brightest to the NDPC. Under the guidance of Minister of Education Zhu Jiahua (朱家驊), who was also trained as a geologist, technical education was vastly strengthened at the secondary and university levels, and in the 1930s the NDPC/NRC became the leading employer of scientific and engineering graduates in China.[73]

Weng Wenhao not only stamped the NDPC with the Geological Survey's reputation for efficiency and integrity, but he also built upon the survey's existing proposals and brought geology to the core of China's economic plan. In addition to fuel and oil surveys, the NDPC took important steps toward coordinating the efforts of the national survey with provincial organizations by rationalizing the division of labor and organizing collaborations to maximize available staff.[74] This was suggested by the survey in 1929 and helped move Chinese geological fieldwork both southward and to the interior in preparation for the anticipated Japanese invasion.[75] After the loss of mineral-rich Manchuria, the survey's turn to the southern provinces of Yunnan and Jiangxi and the central province of Hubei for ore and coal, helped bring much-needed cash to the Nanjing government. In fact, sale of China's exceptionally rich tungsten (Jiangxi) and antimony (Hunan) ores to Germany funded almost all of the NDPC's industrial and infrastructural projects, including the Central Steel Works, Central Machine Works, and Central Electrical Manufactory.[76] The successful exploitation of China's unique mineral resources rather than its more obvious but weaker coal and iron reserves vindicated Weng's faith in surveying and strategic development, and tied the Chinese state even more tightly to geological expertise.

Ironically, though the National Defense Planning Commission's semi-autonomous style was heavily influenced by the Geological Survey, the survey itself became less and less independent as the NDPC grew. Once the NDPC began implementing its first Three Year Plan in April 1935, it changed its name to the National Resources Commission. Like the NDPC, the NRC remained a secret government organization until 1938, but as industrial development moved into its active phase, the NRC hired large numbers of engineering personnel and required more and more attention from Weng.[77] Despite his heavy commitments at the NRC, Weng still personally directed the survey's activities, and in the winter of 1935, the Geological Survey, which had been one of the only national organizations to maintain its headquarters in Beiping, finally moved its main offices to Nanjing.

As usual, this relocation was largely funded by the survey itself through private donations, but a significant portion was paid for by the NDPC/NRC. The Geological Survey built a new fuel laboratory, soil laboratory, seismological observatory, library, and museum in Nanjing and moved most of its specimens, reference materials, and analytical equipment south.[78] The old Beiping complex became the survey's Beiping Branch and maintained its joint status as part of the National Beiping Academy.[79] All of its original buildings and divisions were kept, though research was scaled down, and the Cenozoic Laboratory remained completely intact because of its reliance on the Zhoukoudian site and the Rockefeller sponsored Peking Union Medical College.

The Chinese geological community had been concentrated in Beiping since its founding, and in the early 1930s there were several attempts to widen the geographical base of geological recruits by establishing new academic departments and regional research organs, but Beiping's dominance continued unabated until the end of 1935. Even then, several of the survey's leading researchers, such as economic geologist Xie Jiarong and paleontologist Yang Zhongjian, chose to stay at the old survey and remain affiliated with Peking University. However, once the survey's headquarters moved to Nanjing, the capital became the center of Chinese geological research, and the entire organization became linked not just with China in general, but the Guomindang government in particular.

In terms of the survey's activities, little was compromised. Arguably, the hard-won distance from government bureaucracy that Ding Wenjiang and Weng Wenhao established for the survey in its first two decades of operation was always just a mechanism for accomplishing more and better serving the national interest. On the surface, the survey's new relationship with Nanjing allowed it to do just that by following through on more of what it proposed than ever. After 1935, the survey had major outposts in both northern and southern China (including specialized laboratories at both facilities) from which it could systematically survey resources and stratigraphy in previously inaccessible interior areas. It was also empowered to coordinate China's economic geology research across institutions, and it successfully led a comprehensive standardization project for both terminology and mapping conventions. Nevertheless, Weng's successive political advances, to secretary general of the Executive Yuan in November 1935 (and minister of economic affairs in January 1938) increasingly blurred the boundary between survey and state. Ding and Weng's ideal of technical expertise in government infused Nanjing with new life but cost the survey itself the flexibility and independence that had made it unique.

The Research Institute of Geology and Space for Theory

Of the new crop of geological organizations that arose as the Guomindang consolidated power in the Northern Expedition, the only one that could compare to the Geological Survey of China in prestige and personnel was the Academia Sinica's Research Institute of Geology, under Li Siguang. Both organizations were national in scope and headed by internationally recognized geologists committed to strengthening China at home and abroad, but the focus of their research varied widely, and they charted very different courses during the Nanjing Decade. While the survey was an independent-minded institution that worked more and more closely with the central government as the Japanese crisis intensified, the Institute was a Nanjing-sponsored creation that increasingly diverged from the regime as it pursued its own intellectual and political vision of the national interest.

The idea for the Academia Sinica can be traced to Sun Zhongshan's 1924 proposal for "a central academy as the highest research organization of the country with a view to effecting a national reconstruction through scientific studies."[80] The Research Institute of Geology was one of four original institutes established at the Academia Sinica's inception in January1928.[81] Its mandate made no mention of natural resources, focusing instead on "the advancement of geological science," and Li Siguang was chosen as director on the basis of his strong record of theoretical research.[82]

Li's career in geology began with a desire to help China develop its mineral reserves, and in 1913 he chose to attend Birmingham University because it was located in one of England's famous coal and iron regions. However, Li's interests quickly shifted from mining and surveying to underlying geological problems, and before returning to China in 1920 he conducted a systematic review of existing geological work on China to help identify promising new areas of investigation. In China, Li joined the newly revitalized Peking University Department of Geology and taught petrology, physical geology, and structural geology. Li quickly developed an enthusiastic following among Beida students, to whom he emphasized the importance of physics and chemistry, the necessity for rigorous scientific thinking, and the primacy of fieldwork. He was also a strong supporter of the student-run Geological Society of Beida and frequently lobbied university administration for new equipment and programs.

While at Beida, Li was briefly affiliated with the Geological Survey but, like his colleague and friend Amadeus Grabau, he opted to focus on teaching and research, becoming especially interested in fusulinoids in the early 1920s. Fusulinida, a kind of benthic foraminifera, were bottom-dwelling single-celled organisms that formed rice-shaped calcareous shells (known as tests) and evolved to great complexity in worldwide Carboniferous and Permian seas.[83] Li found fossil fusulinoids widely distributed in limestones associated with north China's coal measures, and in 1923 he devised an innovative graphical method for identifying and correlating individual species.[84] He subsequently published several studies of test morphology and development that established fusulinida as index fossils for China's Paleozoic strata.[85] Because different fusulinida species each thrived under restricted environmental conditions, these fossils offer clues to the ancient marine setting, often indicating water depth, salinity, temperature, etc., and thereby facilitating hydrocarbon exploration (both coal and petroleum). Li was part of a pioneering generation of micropaleontologists in the 1920s and 1930s that laid the groundwork for the use of foraminifera as tools in fuel prospecting, but though he was very concerned with the implications of his work for finding coal in China,[86] he focused his own energies on using fusulinida to interpret stratigraphy, paleogeography, and paleoclimatology. He was particularly struck by the radical difference in fusulinoid forms in north and south China, which suggested an ancient land barrier between the two regions, and he speculated on the tectonic shifts that might have accounted for these. In 1931 he earned a doctorate in science from Birmingham University for his classic 1927 study, "Fusulinidae of North China," and the Research Institute extended his comparative work to south China throughout the 1930s.

Li Siguang's work on fusulinida exemplified his belief that research on fundamental scientific problems could have both theoretical and practical benefits, and that China's need for fuel and minerals could be as productively approached through the study of scientific concepts as through simple surveying. Besides his micropaleontological work, Li famously presented his early tectonic ideas before American geologist Bailey Willis at the Geological Society of China in 1926, and pursued a wide range of topics in structural, regional, and—most controversially—glacial geology in the pages of both Chinese and foreign geological journals, like the *Bulletin of the Geological Society of China* and the London's *Geological Magazine*.[87] These ideas, which would develop into a unique school of "geomechanics" (地質力學) in the late Republican period and early Communist era, all reflected Li's growing interest in geophysics as an explanatory framework for the rest of geology.

Just as president of the Academia Sinica, Cai Yuanpei, chose Li to be director because of his commitment to theoretical problems, Li cherry-picked his institute researchers for their intellectual curiosity. High-powered figures like Weng Wenhao, Ding Wenjiang, Zhu Jiahua, and foreign geologist Teilhard de Chardin signed on as special research fellows,[88] but most of the new institute's active staff were recent graduates of Peking University who had impressed Li with their potential for advanced work in both field and laboratory. Li took care to assign each research fellow to projects that would maximize his particular talents, and several young investigators were sent abroad for specialized study,[89] but the Research Institute's overall research program clearly bore the stamp of Li's own interest in micropaleontology and geophysical problems.

Li frequently had to defend the special character of his research institute against the strictly utilitarian demands of Nanjing officials, who periodically tried to take control of China's geological work (before lapsing into disinterest again and leaving the survey to function as unofficial coordinating agency). As Li reiterated to the Ministry of Industry in 1931, after several rounds of frustrating meetings,

> The Research Institute [of Geology] of the National Academia Sinica is not subject to your regulations. . . . The enterprise of geological research takes fieldwork as its basis [and] because China's territory is vast, we worry that we cannot cover all of it even if we all work together. . . . But, in their mutual relations [organizations] should not place restrictions upon each other. . . . The Ministry of Industry's regulations are for the amalgamation of its subordinate bureaus, but collaboration and division of labor in research is very different.[90]

In Li's view the Academia Sinica pointedly created a "research institute" rather than a "survey," and though the Research Institute was happy to coordinate with other bodies and share mapping and surveying duties, these were not its primary responsibility.[91] According to Li, "[the Research Institute's] work should particularly emphasize discussion of geological theories. . . . its purpose is in resolving special geological problems and not in being satisfied with collection or verification of data."[92] Though central government officials chided Li for neglecting economic priorities, he insisted on the relevance of geological principles for solving China's fuel crisis,[93] and the survey encouraged the Research Institute's pursuit of theoretical research as a vital complement to its own work. In an open letter to China's geological establishment in 1929, the survey stated that it had long hoped to pursue physical geology and geophysics, which were "the most modern and most advanced areas of geology," but since the establishment of the Academia Sinica, it was convinced these topics would be more appropriately studied by the Research Institute of Geology.[94]

Unfortunately, the institute did not have the resources to make much headway at first, and it spent its early years tackling financial problems and inadequate facilities. Though geology was endowed with the largest natural science institute of the Academia Sinica (with a staff of twenty-nine), when it opened in January 1928, it had just 20,000 yuan with which to equip itself and run the year's research operations. The central government originally allotted the Academia Sinica 2 million yuan for start-up and research expenses, but predictably, remitted only a quarter of that amount (in devalued bonds), and, unlike the Geological Survey, the institute was not at liberty to seek outside funding.[95]

With such a small budget, the Research Institute could not find offices in the new capital, where real estate prices were soaring, and it joined the engineering, physics, and chemistry institutes in Shanghai, settling in the northern district of Zhabei.[96] The institute moved to the French Concession in the west of the city within half a year, and hoped to build a new building to house a display room, library, and special laboratory for optical research, high-temperature, and high-pressure studies. However, money for construction was not forthcoming, and the institute had to scrap immediate plans for fuel research and geophysical work. According to its 1929 report, books were tucked in every room, "specimens were piled in a stable across the street," and the roof was "hopelessly" leaky.[97] The institute finally found offices near Soochow Creek at the end of 1929 and managed to get special funding for new buildings, but Guomindang officials opposed the permanent establishment of Academia Sinica institutes outside of the capital, and construction was limited to a few renovations.[98] In early 1932, institute buildings were damaged when the Japanese attacked Shanghai, and in September, the institute temporarily set up in the Academia Sinica's main offices before occupying its permanent quarters near Rooster Crow Temple in Nanjing the following fall.[99]

Despite these difficulties, the Research Institute pushed forward with paleontological benchwork and fieldwork on problems related to structural geology and stratigraphy. Institute members even managed to conduct preliminary gravimetric experiments with a new torsion balance and perform analyses of rock elasticity.[100] Work was divided between four research groups, each led by a research fellow specializing in stratigraphy and paleontology, mineralogy and petrology, applied geology, or dynamic geology (including structural geology and geophysics). Research associates and research assistants were assigned to groups on a project basis so they could gain experience across a range of topics and methods, and researchers in the field frequently collected data for multiple groups to maximize coverage. In 1930, Li Siguang also instituted biweekly colloquia that allowed researchers to share findings, problems, and techniques.[101]

The Academia Sinica charged the Research Institute with overall geological investigation of south China, and the institute took on topics such as crustal movement of the lower Yangzi valley, intrusive and extrusive igneous rocks on the southern coast, and "practical" problems concerning porcelain clay (kaolin), silting at the Yangzi delta and hydrogeology of the capital region. The institute also maintained a policy of helping local government bureaus with geological issues and organized several collaborations with regional surveys. Li Siguang gave group leaders a great deal of latitude in selecting specific field sites and analytical methods, but he personally oversaw the work of institute members and demanded that they submit daily written progress reports.[102] As the institute established permanent facilities and expanded its equipment and library,[103] researchers were encouraged to undertake increasingly ambitious projects, but the institute remained best known for its elaborations of Li's own theories, including global tectonics, micropaleontology, and quaternary glaciation.[104]

Li Siguang's "Glacial" Reception

In his study of the Academia Sinica, Chen Shiwei analyzes the way that the central academy tried to defend the intellectual autonomy of "pure research" against the utilitarian demands of the Guomindang regime, and how the ultimate acceptance of "applied" science as a patriotic duty during the Japanese crisis shaped the status of the sciences in Republican China. Proponents of "science for science's sake" saw applied science as a concession that compromised scientific professionalism and association with the government as a surrender of academic objectivity. Others, like Weng Wenhao and Ding Wenjiang, considered technical cooperation with the state to be an expression of professional ethics and insisted that disciplined scientists could remain impartial, even in government service.[105]

Li Siguang, however, was neither a purist nor a technocrat, and he struck an independent path both in his own research and as leader of the Academia Sinica's Research Institute of Geology. His attraction to theoretically challenging problems, far from being highbrow, was a product of his faith in the basic utility of unraveling geological nature. It was no accident that his work on fusulinida, structural geology, and paleoclimatology all had implications for resource prospecting, and he vigorously drew attention to China's need for modern fuels. Li was convinced that scientists could serve the national interest in multiple ways and that "applied" science was simply the most obvious.

Just as critically, Li saw geology as a way for Chinese to question the authority of international claims about their territory while pressing the boundaries of scientific knowledge. In his 1939 book, *The Geology of China*, Li wrote that

> comparative study of common tectonic types recognized in China naturally brings the whole world into our view; and in order to pursue the argument to a logical end an exploration into the borderland between Geology and Geophysics becomes inevitable.[106]

But his work was intended to make other borderlands visible as well, and during the 1934–35 lecture tour of England (from which *The Geology of China* was compiled), Li defied convention by opening his treatment of Chinese geology with a description of Tibet. "China," explained Li,

> embraces an expanse of land-mass in eastern Asia which half-encircles the stupendous massif of Tibet, or "the roof of the world," forming a staircase, as it were, that generally steps down towards the shelf-seas bordering the Pacific.[107]

After the death of the thirteenth Dalai Lama in 1933, the Chinese were concerned that Britain had imperialist designs on Tibet,[108] and decades later Li Siguang recalled that

> at the time most people in western Europe invariably thought that Tibet was not fully part of China, and to correct this mistaken concept (whether intentional or unintentional), I purposely gave the Tibetan plateau first place among China's natural regions.[109]

Tibet, Li informed his British hosts, was not only part of China, but the ancient kernel from which the familiar coastal regions extended.

Li also made it a priority for his research institute to raise the profile of Chinese geology, explaining in the first general report of the Academia Sinica that "to investigate this kind of [theoretical] problem, amass many sources, and develop an unbiased view will not only contribute to global science but win international respect."[110] During the Nanjing Decade, Li became increasingly convinced that eastern China (both north and south) had experienced several rounds of glaciation during the early Quaternary period (i.e., the Pleistocene epoch, ~1.64 million to 10,000 years ago), and his insistence on this possibility in the face of foreign skepticism raised delicate questions of impartiality and expertise for the Chinese geological community.

While the Geological Society kept relations between native and foreign investigators cordial and brought Chinese research to the ears of international scientists, Chinese geologists still faced an unspoken double bind common to all incipient scientific communities.[111] If native scientists disagreed with established foreign experts on significant matters of interpretation, their mastery of the science was called into question, but if native scientists did *not* disagree with established foreign experts on real, scientifically relevant issues, they were disparaged as sterile and derivative. This was particularly problematic in the field sciences, where assertions, by their very nature, were rarely reproducible or testable in the way that questions in the laboratory sciences could be. By pushing a contentious theory and tenaciously revisiting the evidence, Li was in effect challenging both foreign geologists and his own compatriots to put scientific "due process" to the test.

Li Siguang first proposed the possibility of Pleistocene glaciation in north China in 1922 at the third meeting of the Geological Society of China, where J. G. Andersson and others expressed serious doubts.[112] It was generally assumed that unlike Europe and North America, China proper had no Pleistocene Ice Age because the depth of loess formations proved its colder northern latitudes were too arid to support glaciers. In Andersson's view, the evidence Li presented (including U-shaped valleys, erratics, and striated boulders) was therefore intriguing but inconclusive, and he challenged Li to explain undisturbed deposits nearby, which contraindicated glaciation in the region.[113]

Li's own confidence in Pleistocene glaciation in China was shaken by these objections and he did not pursue the problem again until he noticed landscape features suggestive of glacial action during a Peking University field trip to Lushan (Mt. Lu, Jiangxi) in 1931.[114] Li returned to Lushan for further study the following summer, and in 1933 he delivered "Quaternary Glaciation in the Yangtze Valley" as the presidential address of the Geological Society's tenth annual meeting.[115] In his paper, which was supported by several other institute papers on Lushan, Li presented topographic and depositional evidence for what he identified as three distinct waves of alpine glaciation in the Lushan area, focusing on hanging valleys, flat-bottomed valleys, striated rocks, and boulder clays (large, angular, unstratified rocks in a fine matrix.) Reaction among foreign geologists was even more negative, and Davidson Black, normally known for his mild manner and friendliness to Chinese colleagues, likened acceptance of Li's views to a "crisis of turning day into night."[116]

In part Black's reaction might have been related to his deep involvement in the Peking Man project and his insistence that Asia was the original center of mammalian dispersal. Both China's great faunal and floral diversity and its plausibility as the cradle of human evolution were generally attributed to the absence of extensive Pleistocene glaciation, and Li's extension of ice coverage as far south as the Yangzi valley and as recently as 14,000 years ago was a serious challenge.[117] On the other hand, the strong objections Li faced were in a sense basic to the problem of glaciation itself. Arguments for glacial action often rested on phenomena that was difficult to explain otherwise, rather than on incontrovertible positive evidence. In Europe and North America Louis Agassiz's original proposal of a worldwide Ice Age had been extremely controversial for just this reason.[118] Even American geologist George Barbour, who, along with Teilhard de Chardin, staunchly opposed the idea of Pleistocene glaciation in China, admitted that whatever the correct interpretation of the phenomena, no simple explanation would suffice to explain the available data. However, the camps were clearly divided. As Weng Wenhao observed, "foreign scientists still can't help but have a degree of habitual disdain for Chinese scientific research,"[119] and Li considered it a matter of pride to defend his "rather harassed and intimidated idea."[120]

The question of Pleistocene glaciation was about more than just ice ages and boulder clay. Foreigners were more familiar with "type localities" of alpine glaciation, especially in the Alps and Rockies, and it was difficult for Chinese researchers who might have supported Li's theory to claim first-hand experience of glacial phenomena. Yang Zhongjian, though a rising star in the Chinese geological community, did not feel that he was in any position to disagree with "experts" like Barbour, de Chardin, and Erik Norin (of Sweden).[121] Li himself visited the Alps after graduating from Birmingham University, and he went back with glacial geology specifically in mind during his1935 lecture tour, but few others in China had similar opportunities.[122] Aware of this problem, Li explained, "I have, therefore, gone into numerous details which are almost superfluous to those who are familiar with countries known to have been glaciated," and he tried to add credibility to his findings by discussing them "in the order in which they happened to come under [his] notice."[123]

Questions of glaciation were not an important component of the Geological Survey's research agenda, and the survey tried to remain neutral in this debate, but it was in an awkward position. Its members, who worked in close collaboration with foreign specialists (especially at Zhoukoudian) and developed their expertise in the loess-covered north, leaned toward the view that Pleistocene China had remained unglaciated at low altitudes, but they also took great pride in China's right to an independent scientific opinion.[124] In the spring of 1934 the survey organized a joint Yangzi expedition led by Li in the hope that a consensus could be worked out in the field.[125] The trip, which included George Barbour, Teilhard de Chardin, Erik Norin, Yang Zhongjian (Geological Survey) and Yu Deyuan (Research Institute), passed pleasantly, but neither side relented.[126] No agreement was reached beyond a recognition that the problem remained difficult and unresolved, though the burden of proof landed on Li's shoulders.

In Li's view, foreign geologists did not approach the possibility of Pleistocene glaciation with an open mind. After the 1934 joint expedition to the Yangzi valley, Erik Norin told Li in private that had he seen the same phenomena in his native Sweden, he would have immediately ascribed them to glacial action,[127] leading Li to complain that "so radically this explanation [of the phenomena] differs from the 'orthodox' view that 'scientific skepticism' has tended to do all that it can to keep such a disturbing interpretation in the background."[128] Teilhard de Chardin, on the other hand, thought it was Li who was too caught up with pet theories.

Li kept his research institute team working on quaternary glaciation and gradually expanded his claims to cover Huangshan (Anhui), Tianmushan (Zhejiang), and Huaiyangshan (Hubei).[129] In 1936 Hermann Wissmann of Germany accompanied Li to several field sites and became convinced of Pleistocene glaciation in China, but though he published in support of Li's ideas (using paleoclimatological data), the outbreak of war with Japan silenced further debate until after 1949.[130]

Li was the first major Chinese geologist to take issue with an established foreign position, and by promoting Pleistocene glaciation amongst foreign geologists in China and in England, he made a point of showing that Chinese were both qualified and willing to interpret their own territory without the "permission" of foreigners.[131] Li displayed similar independence dealing with the Nanjing regime, and unlike the "pure science" group, Li was unafraid of being politically vocal. In fact, Li's involvement with politics actually antedated his interest in geology, and his aloofness from the Guomindang was many ways more political than Weng's entry into officialdom.

By 1905 Li Siguang was already a founding member of Sun Zhongshan's Revolutionary Alliance, which he joined while studying in Tokyo at the age of sixteen.[132] He returned to his native Hubei with a technical degree in nautical engineering in 1910, and after the Wuchang uprising of 1911, became financial advisor to the provisional local military government. Sun appointed him Hankou city construction planner in January 1912 and the following month Li was named both Secretary of the Hubei Revolutionary

Alliance and Minister of the Hubei Ministry of Industry.[133] Upon the official founding of the new republic, Li's ministry became a provincial department and he continued to work on industrial, agricultural and commercial development as department head until Yuan Shikai assumed presidency of the Republic. Feeling that the revolution was likely to fail in spirit, despite of the ouster of the old Manchu regime, Li resigned from his posts in August 1912. Yuan's political maneuvering in Beijing and the collapse of Sun's second revolution in 1913 led Li to take a government scholarship to study mining in England, and he did not accept another political appointment until after 1949.[134]

In contrast, Weng Wenhao had no record of political activity during his student days in Belgium and no party affiliation prior to his recruitment by Jiang Jieshi in 1932. Even then, Weng was reluctant to enter government and only did so to foster more effective ties between research and national defense in the wake of the Manchurian Incident.[135] His ability to attract a wide range of China's best scholars and experts—who had hitherto eschewed government—into the National Defense Planning Commission was in fact directly related to his reputation for being patriotic but apolitical.[136] The May Fourth generation of Chinese intellectuals idealized scientists as truth seekers impervious to the temptations of money or fame, and Weng was seen as an exemplar of these scientistic virtues whose "shining integrity lifted him above ordinary politics."[137] Though personal ties to Jiang eventually pulled Weng into nontechnical areas of the bureaucracy, and he considered it a scientific duty to contribute to nation-building applications, Weng shared the "pure science" contingent's aversion to partisanship, and he tried to maintain as much separation between government service and party politics as possible.[138]

Given their different backgrounds it is not surprising that Weng and Li reacted very differently to the Nanjing regime and to the Manchurian crisis of 1931. As an opponent of the Beiyang government, Li had welcomed the Northern Expedition and been sufficiently optimistic about the completion of Sun Zhongshan's unfinished revolution to follow the government south and help found the Academia Sinica. But his experiences in the opening years of the Republic had made him wary of believing political promises, and he soon became disillusioned with the violence and corruption of Guomindang factional struggles. As early as 1928, Li warned against the new leadership relocating to Nanjing and disposing of internal enemies behind closed doors while ignoring the clear threat from Japan and Russia to the north.[139] He was therefore appalled to find that Jiang's response to the

Guandong Army's aggression in Manchuria was as focused on the eradication of Communist dissenters as it was on industrial growth, and he openly supported student protests against the government's "conciliatory" policy toward Japan.[140]

Like Weng Wenhao and Ding Wenjiang, Li Siguang was part of the small group of friends whose meetings to discuss current events launched the *Duli pinglun*, but Li opted not to participate in the journal itself.[141] Instead Li joined Cai Yuanpei, Song Qingling (宋慶齡), and Academia Sinica secretary general Yang Quan (楊銓) to form the Chinese League for the Guarantee of Civil Rights (中國民權保障同盟) in reaction to Nanjing's 1932 "Emergency Law for the Prosecution of Threats to the Nation."[142] When Guomindang secret police assassinated Yang in 1933, after a series of the league's protest and rescue activities, Chinese intellectuals were profoundly shocked, and Li decided to name his most recently discovered fusulinida after Yang, to keep the memory of his friend alive.[143] Li repeatedly questioned if the state's interests were really synonymous with those of the nation, and whether naming species or hiring politically censored students at the Research Institute, he had few reservations about using his scientific position to express his political viewpoint.[144]

As the Japanese crisis deepened, academic opposition to the Nanjing regime became less vocal. Though intellectuals were hardly won over by the Guomindang's autocratic methods or its repeated attempts to interfere with their research, the foreign threat seemed to demand unity between state and academy, and organizations like the National Defense Planning Commission were demonstrating the efficacy of such cooperation. In 1934 Ding Wenjiang, who succeeded Yang as secretary general of the Academia Sinica, spearheaded the creation of the National Research Council (est. 1935) to coordinate domestic research policy and conduct research under the government's mandate.

This new organ of the central academy represented a general, though cautious, willingness of scholars and scientists to take up applied topics until the end of the national crisis, and Li encouraged Research Institute members to meet the practical needs of government bureaus and private industry as much as possible without sacrificing their own primary research. However, when war with Japan finally erupted after the Lugouqiao Incident of 1937, Li ignored Jiang Jieshi's invitation to consult on national affairs at his Lushan retreat, and he led the Research Institute to Guilin, in Guangxi instead of to the new capital of "Free China" at Chongqing.[145] Despite Academia Sinica acting president Zhu Jiahua's threat that institutes disobeying the summons to Sichuan would not receive further funding, Li argued that China needed to develop the scientific resources of the west, and convinced the Institutes of Physics and Psychology to go to Guilin as well.[146]

Conclusion

Throughout the Nanjing Decade, China's growing scientific corps grappled with the problem of "pure research" (sometimes "basic science") versus "applied science." Perhaps counterintuitively, this debate was rarely about disciplinary boundaries or research methodologies (which were of course contested, but not primarily in these terms). Rather, the tension between "pure" and "applied" reflected a set of underlying concerns about (1) status in the international community, (2) domestic professional identity, (3) the duties of the modern Chinese intellectual, and (4) the lessons of the Self-Strengthening movement. Each of these issues carried its own complex of historical and social baggage, but clarification of those intricacies is beyond the scope of this book. What is relevant here is that "pure" and "applied" were context-sensitive terms, and any attempt on the historian's part to define them categorically is likely to fail. Both words were as much weapons of discourse as indicators of concrete scientific commitments.

At issue was the meaning of "utility" (用, *yong*) that divided the "pure" from the "applied," and at stake were Republican scientists' sense of professional autonomy. Even in the Academia Sinica, where debates over "pure research" and "applied science" were quite vocal, few researchers opposed the idea of fostering science to help "stop the brutal and wild ambitions of foreign powers."[147] The fundamental problems were how this was to be done, on what basis it should be done, to what extent it should be done, and, crucially, who should decide such matters.

For Chinese geologists in the 1920s and 1930s, "utility" was not intrinsically threatening or disparaging. Both before and after the 1911 revolution, Chinese geologists came to their chosen field hoping to strengthen native industry and serve the nation. That they grasped a new sense of the materiality of the land and a new way of knowing it did not in any way dilute this intention. In fact, by blurring the previously sacred line between mental activity and physical labor, Chinese geologists expanded the "utility" of their science to include the training of useful men unafraid of hard work or self-sacrifice. In Republican China, surveying was as fundamental as it was utilitarian, and even theory building had practical applications.

Through mapping projects, collaborations, the Geological Society, and other international activities, Chinese geologists were more successful at projecting the image of a unified sovereign China than any Beijing administration had ever been. As a result, when the new Guomindang authority came to power, Chinese geologists were less worried about being branded as "useful" and more worried about being used. The debate between "pure" and "applied" science was not central to the Chinese geological community during the Nanjing Decade, but the definition of "utility" and how best to serve the national interest was just as pressing and just as tied to questions of autonomy.

These decisions hinged on previously unexpressed differences in geologists' understandings of "China" and nation. At one end of the spectrum Weng Wenhao and many members of the Geological Survey fought against the inefficiencies of the Nanjing regime but saw their hopes for the nation best aligned with a strong central state. Weng did not see participation in government as compromising scientific impartiality, he saw it as an opportunity to bring the values of the survey into the administration. At the other end of the spectrum, Li Siguang did not feel that the Guomindang had earned either his loyalty or the right to speak for Chinese science. He tacitly distinguished patriotism from (state-centered) nationalism, and asserted Chinese dignity through his scientific iconoclasm.

Ultimately, both Weng's stance and Li's were elite positions shaped not only by their leadership roles and the high quality of their geological work, but by the luxury of peacetime, however tense. Once the Japanese invaded China in 1937 and the War of Resistance began, Chinese interests and the interests of the state became harder to distinguish without threatening national survival, and the entire scientific community put the "pure" and "applied" sciences at the disposal of the wartime government.

FIVE

Resistance and Reconstruction: Emergence of a Domestic Community

Acting is easy, knowing is difficult.

—Sun Zhongshan[1]

As Johan Gunnar Andersson made clear in his 1939 book *China Fights for the World*, Chinese science was vital to one of the longest and most strategically important holding actions in military history. After the Lugouqiao Incident of July 7, 1937, the Guomindang government grudgingly joined with Chinese Communists in a Second United Front and resolved to resist Japanese aggression at all costs. From the outset, the Chinese knew that they could not match the superior training and firepower of the Japanese military, which projected conquest of the entire Chinese mainland in just three months. Instead the central government retreated inland, and its armies engaged the invaders in an eight-year war of attrition, never allowing Japanese forces to rest or consolidate their victories and compelling Japan to commit more and more resources to the China theater. In occupied areas, Chinese guerrillas (often organized by the CCP) constantly harassed Japanese troops and the puppet regimes they set up, forcing them to maintain large stationary forces at major cities and along rail lines. By 1939 Japan had penetrated almost a third of China, including the northern plains, the Yangzi valley, and major coastal cities, but it could not make significant advances in the Chinese hinterland, and the war had reached a costly stalemate. Once the European war erupted in 1939 and the United States joined the conflagration after Pearl Harbor, this stalemate became a key strategic element for the Allies. China's defensive strategy was used to deplete Japanese forces, preventing them from attacking Russia on a second front and keeping Japan tied up until the US could fully launch its Pacific campaign.[2]

Whether adapting its prewar policy of "trading space for time" (以空間換時間 *yi kongjian huan shijian*) to slow Japanese movements along the eastern seaboard, or employing "magnetic warfare" (磁鐵戰 *citie zhan*) to attract Japan's mechanized forces into ambushes on inhospitable terrain, Chinese tactics hinged on the ability of the government and population of the Interior (大後方, lit. "great rear area," often called "Free China") to survive a prolonged blockade while the army conducted its holding action against a technically superior enemy.[3] With more than three quarters of China's factories, commercial mines, railroads, and motorways in occupied territory and the Japanese cutting off Allied supply lines, scientists and technical personnel who followed the government inland devoted their energies to searching out natural resources, rebuilding communications and industry, and helping the population endure emergency conditions.[4] Chinese engineers rushed to establish power plants, refineries, foundries, arsenals, and mills, while Chinese scientists investigated problems such as vaccines against regional diseases, the nutritional and medicinal qualities of indigenous plants, building applications of local materials, soil analysis, irrigation systems, and alternative fuels.[5] For the first time, Chinese technical graduates were guaranteed employment in their chosen specialties and thousands more were put through accelerated training courses to fill the needs of the National Resources Commission, local industrial projects, and wartime research institutes.[6]

Despite the mobilization of personnel and resources for science during the War of Resistance, the state of the national economy and the nature of China's defensive war kept scientific efforts in a constant state of material scarcity. Government ministries were able to direct research toward practical problems but realities on the ground meant that within these parameters, specific projects were dictated by available supplies, local initiative, and transitory arrangements between neighboring organizations. In some cases, wartime conditions actually pushed scientists in the opposite direction, and in the absence of laboratory facilities, many Chinese physicists pursued computation- and theory-heavy studies in quantum mechanics or fluid dynamics rather than experimental work connected to the national war effort.[7] Whichever path Chinese scientists took, wartime relocation to the Interior and the Japanese blockade kept them almost completely out of reach of international advances and scientific exchange, and except for Chinese graduate students working in the United States and England, they did not participate in major Allied technological programs.

In his book, *Science Outpost*, Joseph Needham, recounted his experience with the Sino-British Science Co-operation Office (1942–46), which was a rare exception to this isolation. Needham recalled that

> at Chiating one discussed nuclear physics in a family temple within sight of the mountains of Tibet; in the caves of Kuangsi one found large power stations and engineers dying to talk to a technologist from the outside world; and among the aboriginal tribesfolk of Tali one helped a planktonologist to launch his boat on the lake of Erh-Hai below the Tower of the Five Glories.[8]

While this description gives a certain romance to China's plight, it also helps explain why Chinese efforts have largely been written out of general treatments of wartime science. The national government committed more specialists and more funding to research than ever before, but, practically, Chinese scientists were forced to work as individual cells conducting whatever research they could in response to immediate local demands. From the perspective of Western historians interested in the ways that World War II introduced fundamental shifts in the structure, scale, and style of scientific practice, Chinese wartime science was simply too makeshift, too isolated, and too ephemeral to have had a significant impact on global developments. Lying outside the lineage of the new forms of interdisciplinarity, industrial management, or government collaboration that emerged as postwar "big science," the Chinese experience has not been a compelling object of study for foreign scholars.[9]

The trying conditions and strange juxtapositions of science in the Interior have also been challenging for Chinese historians concerned with national modernization, and domestic scholars have been unable to fit wartime practices into a linear account of disciplinary progress. Most wartime scientists in China were proud of their ingenuity and perseverance but ambivalent about the quality of their scientific output, and as a result native historians have tended to view the War of Resistance as a setback or hiatus from what they see as China's real scientific growth. Attention to the heroism and self-sacrifice of patriotic scientists has only increased the perceived gulf between war work and the broader development of the sciences in modern China, and most studies of individual disciplines either end in 1937 with the basic establishment of domestic institutions or begin after the Communist takeover with the rise of science with "Chinese characteristics."

Historians of geology in China have followed this pattern and generally glossed over the War of Resistance as a "period of adversity" or "period of disruption" between the prestige of 1920s internationalism and the defiant successes of post-1949 geomechanics.[10] In fact, the war was an important transitional period for most of the modern sciences (including the laboratory disciplines for which they were most disruptive), and in many ways the dislocation of war was an opportunity in disguise for China's geologists. The "mountains of Tibet" and "caves of Kuangsi" (Guangxi) that formed such alien backdrops for physicists and engineers had long been frustratingly unavailable objects of fascination for coast-bound geologists. The rugged conditions of wartime research were familiar to the geological community, and though institutional and financial conditions were difficult, from a research standpoint the inland migration finally brought the frontier within reach.

Against this backdrop, this chapter presents the War of Resistance as a period of trial and opportunity, in which scarcity, isolation, and necessity both threatened the work of the previous decades and reinvigorated it. The first section examines the great migration of government and scientific institutions to the Interior and the subsequent adjustments and reorganizations entailed by the decision to pursue science in such alien circumstances. The next section traces the ways that the demands of the wartime economy diversified the pre-1937 geological community and stretched it beyond the reach of personal ties and direct interaction. Much as it did in the 1920s, the Geological Society of China bound the community together, but this time, instead of doing so through face-to-face meetings and the construction of a cosmopolitan forum, the Society midwived the birth of a new geological community by providing a "private" domestic space in the form of its Chinese-language journal, the *Dizhi lunping*. The chapter then concludes by examining the ways that Chinese geologists responded to Free China's petroleum shortage. I trace the origins of China's dependence upon foreign fuel and the ways that petroleum exploration represented a coming of age for the Chinese geological community.

Interior Designs

The tactical decision to retreat to the Chinese Interior in 1937 was not made lightly, by either politicians or private citizens. As early as 1932, in the wake of the Manchurian Incident, there were popular calls to empty Beiping of its cultural treasures for fear that irreplaceable relics would fall into the hands of invading Japanese. But Weng Wenhao and others of the *Duli pinglun* circle asserted that to "protect" Chinese culture in this way was to tacitly assent to foreign domination. Instead he insisted that "If Beiping falls, let it be taken, and not given up without a fight, artifacts or no."[11] Five years later, both Weng's position and the national situation had shifted. Under Weng's guidance the NRC had been systematically strengthening the production capacity of inland China and preparing for a drawn out struggle away from the vulnerable coastal areas. When the Japanese finally attacked Beiping at the end of July 1937, attempts were made to defend the ancient capital, but ultimately the sacrifice of Beiping was rationalized as part of China's emerging plan to regroup in the Interior and face the long-anticipated enemy from a more advantageous position.

Within days of the Lugouqiao Incident, Jiang Jieshi called high-level meetings with representatives from across the political spectrum to take the pulse of popular sentiment and rally support for resistance. The following week, academic and intellectual leaders gathered at Lushan to weigh in on national policy, and on July 17, Jiang issued a demand that Japan completely withdraw from Hebei and Chahar provinces. Jiang stressed that China had not courted conflict, but would accept it if Japan pressed forward. To the Chinese people he announced, "If the opening rounds of war do come, then there will be no division between northern and southern territories, no difference between the old and the young. Each and every person will bear the responsibility of defending the land and resisting [Japan]."[12]

Despite these strong words and Jiang's promise to join the Chinese Communists in a Second United Front against Japanese aggression, the Guomindang was fully conscious of Japan's military and economic superiority and painfully aware of China's inability to defend its most populous regions. The government desperately launched diplomatic missions to the international community in hopes of averting all-out war, but moral support from foreign powers fell far short of any real intervention.[13] Japanese troops extended their offensive southward to Shanghai, and on August 14, 1937, Jiang announced China's refusal to surrender in the "Declaration of Self-Defense and Resistance." The attack on Shanghai put the national capital at risk and, after securing its northern and western borders through a nonaggression pact with the Soviet Union in late August, the central government began planning its retreat to the Interior.

On October 30, the Guomindang named Chongqing as its provisional capital. Nestled in the fertile Red Basin of Sichuan but protected by surrounding mountains on all sides, Chongqing was an ideal refuge for the central government,[14] but many coastal research and academic institutions had already moved their books and equipment to the capital of Hunan at Changsha by the time official policy was announced. When Changsha came under enemy fire in 1938, most followed the government to Chongqing, but the process of relocation was difficult to coordinate and several institutions opted for other inland havens. Beida, Qinghua, and Nankai, which merged to form Lianda (Southwestern Associated University) during the war, led students and faculty several hundred miles on foot from Changsha to Kunming, Yunnan, in an effort to find a secure campus and avoid the glut of universities in Chongqing.[15] Smaller institutions, far from avoiding crowding, often accepted invitations to pool resources and share facilities, creating such unlikely pairings as the Liangguang (i.e., Guangdong and Guangxi joint provincial) Geological Survey's Yunnan offices. Others, including Li Siguang's Research Institute of Geology opted for the southwest in order to establish distance from central government control and help establish a new scientific center.

Regardless of where government and private institutions ended up, the mass migration inland was chaotic and unpredictable. The Luguoqiao Incident had caught Beiping by surprise and few institutions were prepared for Japan's rapid occupation of the city. Luckily much of the Geological Survey's equipment and books had been moved to Nanjing in 1935, and its Beiping offices were only sparsely provisioned by 1937. Members of the Beiping branch survey sought protection under the aegis of the Rockefeller-sponsored Peking Union Medical College through their connection to the Cenozoic Laboratory, but leading figures such as Yang Zhongjian and Xie Jiarong could not avoid Japanese notice and had to escape the city before being "asked" to collaborate with the occupying regime.[16] Stalwarts of the Beiping geological community, Zhang Hongzhao and Amadeus Grabau were too physically weak to evacuate, and spent the occupation under de facto house arrest for resisting Japanese service.[17] Pei Wenzhong, who had just received his doctorate from the University of Paris, actually rushed to occupied Beiping from Nanjing, to care for his family, and he stayed in the Cenozoic Laboratory until the early 1940s when he managed to rejoin the survey and conduct fieldwork in Mongolia, Gansu, and Qinghai.

Members of the Geological Survey in Nanjing had more time to prepare for relocation but less hope of international protection, especially with leading figures like Weng Wenhao, Huang Jiqing (黄汲清), and Pei Wenzhong on their way to Moscow to present at the 17th International Geological Congress when the Lugouqiao Incident erupted. Throughout late summer and early fall, the survey rushed research reports to publication in advance of anticipated disruptions in communication, and packed away important specimens, reference materials, and instruments while waiting for Beiping personnel to arrive from the occupied north. In November, just weeks before the "rape of Nanjing," twenty-odd survey members shepherded several boatloads of boxes up the Yangzi to Wuhan in Hubei province and onward about two hundred miles to Changsha, where temporary offices were set up on the premises of the Hunan Geological Survey.[18]

The survey took as much of a business-as-usual attitude as possible, sending members out to conduct fieldwork until its new buildings were completed in January 1938, and even keeping up publication of the *Bulletin of the Geological Survey of China, Memoirs of the Geological Survey of China, Paleontologia Sinica,* and the *Soil Bulletin* in spite of unstable circumstances.[19] Of the articles published in this transitional period, Xie Jiarong's estimate of China's petroleum reserves was the best marker of the survey's previous reach and its future scope for development. Based exclusively upon existing research on eastern China, known oil springs in Shaanxi and Sichuan, and secondhand information on untapped Xinjiang resources, Xie put China's petroleum base at 182 million tons.[20] This was far more than previous foreign estimates but far less than needed by the government in exile, and Xie's paper motivated the survey to quickly "face west" and explore the Interior.

By early 1938, the survey's temporary headquarters included research facilities, dormitories, and a small library. The chemical laboratories associated with fuel research arranged to use facilities at Hunan University and cooperate with the Hunan Institute of Mining Chemistry.[21] The soil laboratory also launched intensive studies of Hunan Province in conjunction with local authorities. Though facilities were far less sophisticated than at Nanjing, the survey was ready to settle down and focus on fieldwork. However, when Wuhan fell to the Japanese in late October, operations in Changsha were hastily moved to Chongqing, where the survey once again had to borrow rooms before constructing new facilities of its own.[22]

This repeated dislocation reflected the government's miscalculation of just how much territory it would have to forfeit in its strategy of trading space for the time to entrench itself in the interior. Since the 1911 revolution, efforts to develop China's economy and infrastructure had been concentrated near large cities and on the eastern seaboard, where domestic capital and foreign influence were strongest. Eager to build upon existing gains and bolster its urban power base, the Guomindang continued this trend of industrial development along the coast and in the lower Yangzi valley. Later, under the National Defense Planning Commission, several key projects were started in Hunan and Jiangxi, with the expectation that these inland areas could be defended against foreign attack. Once these provinces were lost in 1938, lynchpins of the NDPC resistance strategy like the Central Steel Works had to be abandoned or hastily rebuilt in the southwest. The newly formed Industrial and Mining Adjustment Administration (工礦調整処) scrambled to remove over six hundred public and private iron works, arsenals, textile factories, and chemical industries far enough inland to be out of Japanese reach but still available to the new power center at Chongqing.[23]

The Industrial and Mining Adjustment Administration was only able to salvage a fraction of China's prewar production capacity, but its efforts gave the wartime economy a starting point. Infrastructure investments like railroads and motorways could not be transplanted in the same way, and by early 1939 sixteen major rail lines were in enemy control or had been destroyed by retreating Chinese troops.[24] Free China possessed less than 10 percent of China's fifteen thousand kilometers of railroads, and over the course of the war, supply shortages and mountainous terrain kept inland construction down to just over one thousand kilometers of new track.[25] The Japanese also captured most of China's paved roads and blocked Chinese links to the sea.[26] This not only shut down access to foreign manufactured goods and income from international trade, but it also created a serious fuel shortage that threatened to cripple the Chongqing regime.

Over the course of the conflict, the Chongqing administration managed to construct new motorways to connect Free China internally but they were still unable to link up with the international community, and after 1942, when Japan captured protected land routes through Indochina and Burma, China depended almost exclusively on US air shipments "over the [Himalayan] Hump" for critical supplies.[27] Loss of working mines in the north and east, where the bulk of China's known coal and iron reserves were located, exacerbated the problems created by the loss of imported fuel and complicated China's plans for building a wartime base in the Interior. The retreat also left China's fertile lower Yangzi valley and other traditionally cultivated regions along the coast in occupied territory, which placed the burden of supporting the large migrant population squarely on Sichuan, Yunnan, and other host provinces.

The migration of the central government to Chongqing produced tensions between national leaders and provincial power holders, who had retained a great deal of local autonomy throughout the Nanjing Decade. Wartime development projects were therefore both economically and politically necessary for the Guomindang to legitimize its new inland base, and the movement of academic and research institutions to the Interior was critical to most of these government initiatives. Of the 108 universities and technical colleges operating in China before 1937, over 75 percent were located in areas eventually occupied by the Japanese, and sixty-two relocated one or more times during the war.[28] Research institutes and professional societies, which were clustered even more tightly in Beiping, Shanghai,

Nanjing, and Guangzhou before the war, established new inland centers at Chongqing, Kunming, and Guiyang, and provided the technical base for "scientific resistance."[29]

Intellectuals across the academic spectrum applied their specialized knowledge to problems related to the war effort. Linguists surveyed local dialects in Guizhou, Guangxi, Yunnan, and Sichuan; historians studied the development of army organization in the late Qing; economists conducted fieldwork on the fluctuation of commodity prices during wartime; and playwrights used theater to spread patriotic messages to frontier populations coping with the influx of east coast Han refugees.[30] So many engineers were trained to run the factories, machine works, public utilities, and large-scale construction projects needed to keep the Interior functioning, that foreign observers complained that they were actually "cluttering up" government projects.[31]

Scientists who made it inland often found themselves investigating surprising problems unrelated to their prewar research but vital to the survival of the besieged inland population. Beyond the problems of industrial growth and economic development lay practical issues of public health, nutrition, irrigation, and energy. Chemists like Zhu Ruhua of the Academia Sinica Institute of Chemistry applied her skills to the production of antimalarial drugs, and university professors like Gu Yuzhen joined the National Bureau of Industrial Research to work on the cracking of vegetable oils for fuel.[32] In Kunming, members of Qinghua's Botany Department formed a new Institute of Agricultural Research, and, as Tang Peisong, the biochemist who previously pioneered Chinese research on cellular respiration, described his altered wartime role, "I am known in southwest China as the Castor Oil Man now, whether you believe it or not, and in spite of myself."[33]

Toward the end of the war, a few microfilms trickled in through British and American efforts, but from 1941 to 1945, Chinese scientists were effectively disconnected from the external world. They did not receive foreign journals, could not consult with international experts, and often lacked basic laboratory or field equipment. In James Reardon-Anderson's view, wartime "education and research had to compete with the business of staying alive," but for China's displaced intelligentsia, science and survival often went hand in hand.[34]

Scientific organizations in exile were not only research units, but they were also residential communities responsible for all aspects of their members' lives, and research was often directed at the provision of daily necessities. Physicists in Kunming made water heaters using cast-off parts from the Institute of Meteorology's broken trucks, and yeast from alcohol fermentation plants was turned into meat substitute by the physiological division of a local hospital.[35] The books, paper, pencils, clothes, food, and physical plant needed to keep students and scholars working were similarly produced through ingenuity and an exchange of technical skill across neighboring scientific organizations. Joseph Needham observed that "when microscope slides could not be had, window-panes broken by air raids were cut up, and the unobtainable coverslips were replaced by local mica."[36] In other cases, the Geological Survey produced the first locally designed seismograph from salvaged parts, and chemical reagents were replaced by indigenous plant dyes.[37]

Though displaced intellectuals complained of "vermin-infested beds" and the "academic wasteland" of western China, they were also extremely proud of their resourcefulness and the impact that their small victories had on the quality of life in the Interior.[38] Isolation from foreign models and equipment created opportunities for incorporating indigenous methods into modern processes, and inland salt and soda refineries, for instance, successfully employed native pumping and evaporation methods.[39] Chinese scientists often wrote to one another to share their successes and exchange ideas for creatively exploiting available materials, and as one physiologist recalled of his efforts to establish a tuberculosis treatment facility with just a homemade pneumothorax machine and hand-carved benches, "those months were the happiest of my life."[40]

The central government strongly emphasized a utilitarian approach to science through education, funding, and propaganda programs, yet scientists themselves initiated many of the applied research projects conducted during the war. Major objectives such as alternative energy sources, increased food production, and substitution of native building materials for scarce imports were set by the NRC and other national agencies, but specific lines of investigation were inevitably dictated by local needs and the resources or personnel at hand rather than top-down planning. In his analysis of wartime Chinese science, Reardon-Anderson emphasizes researchers' turn to pure or basic science to defend their intellectual autonomy from the state. This suggests that pursuit of practical research agendas amounted to tacit acknowledgement of Guomindang control over scientific priorities, but many scientists who voluntarily shifted to applied topics during the war still actively resisted state intervention.

Li Siguang defied Guomindang threats to terminate funding if his Research Institute of Geology did not follow the government to Chongqing, and he led several scientific organizations to Guilin instead. The Research Institute was well-known for its independent-mindedness and its emphasis on theoretical studies, but by headquartering in the far southwest rather than the busy new capital, Li ensured that even primarily economic surveys would make fundamental contributions to the understanding of a geologically unknown region. This strategy paid off, and, while studying the structure of China's mountainous southwest, the institute managed to produce so much economic data on industrial ores (e.g., tin, tungsten, mercury, and antimony) that the Guomindang could not afford to rescind its funding and leave the institute in the hands of powerful Guangxi warlords.[41] Li also maintained an explicit policy of placing the institute "at the disposal" of any public or private groups that required geological expertise for the war effort, and thus framed much of the institute's applied research as service to immediate wartime needs rather than as acceptance of central government directives.[42]

In contrast, the Geological Survey of China was uniquely positioned to execute state policy without compromising its sense of self-determination. As soon as the War of Resistance reached Shanghai in August 1937, Weng Wenhao became director of the Industrial and Mining Adjustment Administration and took charge of relocating China's industrial capacity to the Interior.[43] A month later plans were set in motion to incorporate the Ministry of Industry, the National Resources Commission, the Industrial and Mining Adjustment Administration, and technical divisions of the National Economic Council and Junweihui into a new Ministry of Economic Affairs headed by Weng.[44] Weng made sure that the Geological Survey was directly under Ministerial control and not subordinate to the ever-expanding NRC, which allowed him to keep NRC and survey activities coordinated but distinct throughout the war.

Formally, Weng's move into officialdom left the directorship of the Geological Survey open and allowed him to pass the torch to China's second generation of native geologists, who had received undergraduate training at Beida before completing advanced degrees overseas. As his first replacement, Weng chose Huang Jiqing for his combination of technical credentials, administrative skill, and international experience. Huang Jiqing graduated from Peking University Department of Geology in 1928 with several geological papers in print. He immediately entered the Geological Survey and worked closely with Weng, Zhao Yazeng, and others before being hand-picked for a China Education and Culture Foundation scholarship to the University of Neuchâtel. He completed his PhD under Émile Argand in 1935, spent several months doing fieldwork in Europe and North America, and returned to take over the Geological Department of the Survey. Weng personally groomed Huang for leadership during the transition of power from the Ministry of Industry to the Ministry of Economic Affairs, and remained in close communication with Huang throughout the survey's chaotic adjustment to wartime operations. After several years of distinguished service, from November 1937 to June 1940, Weng felt that Huang had vindicated his trust. However, Huang was severely reprimanded by central authorities for failing to devote survey resources to one of president of the Executive Yuan Kong Xiangxi's (孔祥熙) private mining ventures, and he resigned his post as survey director in order to return to active research.[45]

Kong Xiangxi and Weng Wenhao were often at odds politically and Huang's reprimand was clearly an attempt to check Weng's influence over wartime economic development. Geologist Yin Zanxun (尹贊勛), who received his doctorate from the University of Lyon and headed the Jiangxi Provincial Survey, was named acting director of the survey after Huang's resignation but never won Weng's complete confidence, and the survey did not have an official director until Li Chunyu (李春昱) took over in August 1942. Li, like Huang and Yin, had been an outstanding graduate of the Peking University Department of Geology, and after finishing a doctorate at Berlin University under Hans Stille in 1937, he accepted Weng's invitation to lead the Sichuan Provincial Survey before taking over the Geological Survey of China. Li's experience at the provincial level not only demonstrated his ability to gain the respect of China's younger geologists, but it also kept him in close contact with Weng's ministry in Chongqing, which was critical to the success at the national survey.

According to Yang Tsui-hua, who concludes her study of Republican geology at the outset of the war in 1937,

> the National Resources Commission and the National Economic Council [NEC], rather than the Geological Survey, played the leading role during the war in exploring for mineral resources, conducting engineering projects, and developing the frontier.[46]

While it is certainly true that the NRC and the NEC, like their American counterparts, the Office of Scientific Research and Development and the War Production Board, oversaw Chinese scientific efforts during the war, the Geological Survey of China remained the hub of actual geological research, and its adaptation to wartime circumstances had far-reaching effects on later Communist-era geology.

Weng Wenhao maintained a special interest in the Geological Survey long after leaving active scientific duty and consulted with survey directors throughout the war. The strength of his teacher-student relationships kept him involved in everyday survey affairs, and through weekly correspondence with both Huang Jiqing and Li Chunyu, Weng advised on everything from fieldwork assignments and funding allocations to personnel decisions and office politics.[47] As a result, the survey not only remained intact during the War of Resistance, but it also actually managed to expand, and many of the new geological organs that sprung up during the war were either operated in conjunction with the survey or staffed by its alumni.[48] Central economic policy was as often informed by the Geological Survey's needs and concerns as vice versa, and government orders rarely felt like impositions from above.

A Truly Chinese Geological Community

In the early years of the Republic, the Chinese geological community coalesced around a small group of "returned student" pioneers who established China's first geological training and research institutions in Beijing. The tightly knit cadre of Geological School and Peking University graduates that they produced dominated the first two decades of Chinese geology and established an international reputation for both the Geological Survey and the Geological Society of China. Those few who did not remain in the survey taught at Beijing-area schools and later founded departments of geology in other major cities. Wherever these early members of the geological community worked, they were vertically bound to Beijing by student-teacher and subordinate-leader ties, and horizontally linked as former schoolmates and colleagues. They all honed their geological expertise in eastern China, engaged directly with foreign research through the Geological Society or advanced study overseas, and shared an identity as part of China's scientific elite. Once the Research Institute of Geology became active in south China and provincial surveys cropped up in the wake of the Northern Expedition, the Chinese geological community began to splinter, but the founding generation of geologists was influential enough to manage this new diversity and sustain a sense of camaraderie through personal contact.

After the outbreak of hostilities in 1937, the geological community swelled in response to wartime needs and rapidly outgrew these informal bonds. Zhang Hongzhao was isolated in occupied Beiping, Li Siguang was maneuvering to protect his research institute from Guomindang interference, Weng Wenhao could only maintain personal contact with survey administrators, and Ding Wenjiang had died of carbon monoxide poisoning while examining coal mines in Hunan in 1936. Younger luminaries found themselves scattered across the Interior as they were called upon to establish new geological organizations and lead critical lines of research. Xie Jiarong moved from Beida and the survey's Beiping branch to the Jianghua Mining Bureau in Hunan in 1937 before relocating to Yunnan to head the NRC's new Mineral Prospecting Office in 1940.[49] Yang Zhongjian stayed with the survey after escaping occupied Beiping, but was assigned to critical posts in Kunming, Xinjiang, and Gansu before being sent abroad to work with the American Museum of Natural History.[50]

During the war, the survey both reorganized its internal departments and expanded its regional coverage. As the military situation near Changsha deteriorated in 1938, the survey established field offices in Guilin and Kunming as a way of hedging bets in case Chongqing proved as indefensible as Changsha. Wang Hengsheng set up the Guilin office in August 1938, but once Chongqing's position seemed secure, the strong presence of Li Siguang's Research Institute made it wasteful to maintain separate geological operations there, and the office closed in April 1939.[51] In contrast, the Kunming office under Yang Zhongjian was bolstered by the proximity of Lianda and the discovery of exciting paleontological finds, and it lasted until October 1940 when the constant threat of a Japanese drive into Yunnan brought fieldwork to a standstill.[52]

Though short-lived, these field offices demonstrated the utility of local research stations, and helped the survey realize Ding Wenjiang's longstanding dream of comprehensively exploring the geology of the southwest in relation to the rest of China. When the emphasis on economic development moved to the undeveloped northwestern frontier in the early 1940s,[53] the survey built on its experience with field offices in the southwest and created a five-man Geology and Mining Team to anchor a new scientific community in Lanzhou, Gansu. A year later, with preliminary studies and planning completed, the Survey and the Gansu Construction Bureau jointly established the Northwest Branch of the National Geological Survey under Wang Yuelun.[54] The Northwest Branch operated on dedicated budget allocations rather than fieldwork expenses like the older Guilin and Kunming offices, but in contrast to the former Beiping branch survey, it was truly regional in focus.[55] It was responsible for survey research and mapping in Gansu, Qinghai, Xinjiang, Shaanxi, and Ningxia provinces and represented the survey in collaborations with the NRC's Lanzhou Petroleum Prospecting Station, the Ministry of Economic Affairs' Central Industrial Laboratory, the China Petroleum Company, and the Gansu Mining Company.[56] The Northwest Branch also hosted visiting geologists from survey headquarters and served as a base of operations for several large-scale expeditions to Xinjang and Qinghai.

Research at the Northwest Branch continued from 1943 until 1949 and set a precedent for postwar survey branches in Taiwan and Changchun (Manchuria) that reconfigured the Geological Survey as a hub and spoke organization rather than a central dispatching agency.[57] This new model allowed the survey to extend its work in a coordinated fashion to what had previously been inaccessible frontier territory. In the years preceding the war, provincial surveys had proven disappointing in this regard and Xie Jiarong had publicly complained that provincial surveys were often just "meal tickets" for local officials rather than productive research units. He hoped that the national survey could someday open regional branches to replace the least effective of these organizations, but understood that in the near term the best he could hope for was better communication with the Geological Survey.[58]

The cataclysm of war offered an opportunity to improve relations, but results were mixed. Existing provincial surveys that relocated inland were largely cut off from regular funding, and, lacking resources to replace lost libraries and facilities, many turned to the Geological Survey for practical assistance. The Hunan Survey, which had played host in Changsha at the start of the war, relied on the Geological Survey to process all field data after Japanese bombs forced its laboratory and publication activities to close.[59] Similarly, the Jiangxi Survey made arrangements to publish in the Geological Survey *Bulletin* in order to focus exclusively on fieldwork, and it managed to continue scientific mapping as well as urgent wartime prospecting.[60] In contrast, the Guangdong and Guangxi (Liangguang) Survey, which followed Zhongshan University to Yunnan and lost contact with the Geological Survey, terminated all activity until the end of the war.

Provincial surveys established during the war were even more variable. Except for the Sichuan Survey, which was established on Weng Wenhao's personal recommendation in 1938 to study the area around the new capital, local wartime surveys were largely initiated by provincial authorities interested in regional economic development. Some, like the Fujian Geological and Soil Survey of the Fujian Construction Bureau, could not attract personnel qualified to carry out serious geological work and simply identified promising mineral deposits.[61] Others, such as the Xinjiang Survey, were locally instigated but staffed by geologists from the Geological Survey who insisted on careful geological mapping and close contact with survey administrators.[62]

Besides the Geological Survey, the Research Institute of Geology, and provincial surveys, geologists were also spread across several displaced academic departments during the war. The most famous of these, the Department of Geology, Geography and Meteorology of Lianda, was formed after the amalgamation of Peking University, Qinghua University, and Nankai University. These universities relocated to Changsha in 1937 before making the trek to Kunming in 1938 and anchoring that city's vibrant intellectual community. Faculty members from Lianda's parent institutions connected to the Geological Survey scattered across the Interior but several well-known professors such as Sun Yunzhu and Wang Lie upheld the Beida geological tradition, and at different points Yang Zhongjian, Wang Hengsheng and Peter Misch joined the faculty. Laboratory and reference facilities were almost nonexistent, but the Yunnan countryside provided considerable scope for investigation. Geological students participated in the work of the Kunming branch survey, the Yunnan Provincial Survey (est. 1942), and several interdisciplinary university projects on the natural history of southwest China. Economic hardships, military service, and Japanese bombings led to a high attrition rate at the university, but over 150 students graduated from the Department of Geology, and twenty-one of them later became academicians of the Chinese Academy of Sciences.[63]

Not all geology departments were as effective as that of Lianda. National Central University, whose geology program was second only to Beida's before the war, followed the government to Chongqing and boasted an impressive roster of professors from the Geological Survey and the Research Institute of Geology. However, wartime conditions in Chongqing limited student fieldwork, and graduates received only basic research training. National Zhongshan University, which was forced to relocate repeatedly during the war, was barely able to maintain a full course load in geology and did not graduate any students after 1941.

During the war, mining companies, petroleum prospecting groups, and large-scale construction projects came and went in a desperate attempt to sustain China's besieged economy, and this raised demand for basic geological expertise far above the number of available college graduates or geologists. While the most successful of these organizations turned to the Geological Survey for assistance, many simply advertised for technically minded individuals and offered on-the-job training in required techniques.[64] The proliferation of geological institutions, private companies, and ad hoc educational settings diluted the former unity of the Chinese geological community, and the geographical separation of these organizations magnified the problems of coordination and information exchange that were already beginning to emerge before 1937. Moreover, individuals at the edges of the geological community, who had the least connection to core members, were also most in need of support to sustain their activities.

The Geological Society of China was a natural forum for bringing these diverse new elements together but mechanisms such as the *Bulletin of the Geological Society of China* and occasional meetings, which were designed to help build community in the face of foreign pressure proved ineffective for building domestic ties. During the War of Resistance, the international presence declined rapidly and after 1941, only Amadeus Grabau and Peter Misch remained. This shift rendered many of the Geological Society's linguistic practices obsolete, and in the absence of an active foreign audience, the use of Western languages stratified the Chinese ability to participate without adding any real cosmopolitanism.

As the Geological Society's sole organ, the *Bulletin* was caught between its established role of representing the best of Chinese geology to the world and its duty to accurately reflect the changing composition of the geological community at home. At the 10th Annual Meeting in 1933, Weng Wenhao repeated the Society leadership's stance:

> Our President J.S. Lee [Li Siguang] has already emphasized the importance of continuing the high standard of our *Bulletin* and its representative nature for the whole country. It is sincerely hoped that his appeal for more scientific papers from different places outside Peiping [Beiping] will meet with quick response from our members, and the board of editors will receive and publish with great pleasure any communication of scientific signification.[65]

This call was repeated several times, with the editorial board insisting that they "warmly welcomed" contributions "from all the institutions working on Chinese geology."[66]

The problem was not that geologists outside of Beiping were shy about submitting their work, or even that they saw the *Bulletin* as the intimidating preserve of older, more established geologists in China's academic center. The problem was that the limited resources, overtaxed personnel, and explicitly practical mandates of China's local geological institutions kept "scientific significance" out of reach for the majority of geological workers. When Weng Wenhao wrote, "it seems very desirable from several points of view to maintain this *Bulletin* for publishing the main discoveries or investigations of scientific significance in this country while more detailed and elaborated reports are being printed by different government and private institutions,"[67] he was hoping the *Bulletin*'s relationship to China's regional geological journals would reprise its division of labor with the survey's publications in the early 1920s. Unfortunately by the 1930s, the distinction of detailed and utilitarian from broad and theoretical effectively divided the expanding Chinese geological community into those with advanced research capacity and those without.

Finally, at the 12th Annual Meeting in 1936, Xie Jiarong proposed that the Society establish a Chinese-language bimonthly, called the *Dizhi lunping* (hereafter, *Lunping*). The motion was carried unanimously, and the *Lunping*'s first issue appeared before the end of the year, with Xie as chief editor and a roster of China's younger stars heading nine subdisciplinary departments.[68] According to the Society's description of its publications, the *Lunping* would "introduce the newest geological concepts," "summarize research results," and offer book reviews, news, and notes.[69] The formation of the *Lunping* enabled the *Bulletin* to remain "the only important journal for the study of Chinese geology" by once again freeing it to focus on advanced scientific contributions without worrying about questions of geographical representation.[70] While the *Lunping* served the Society's native majority, the reinvigorated *Bulletin* continued to publish the work of China's best and brightest in internationally accessible form, and it kept foreign work on China available to domestic audiences long after the Geological Society's hosting days had waned.[71]

In his introduction to the 1942 *Overview of the Geological Society of China*, Weng Wenhao explained that the Society "published the *Bulletin* first in order to reach out to the world, and put out the *Lunping* second, in order to present [geology] to the people of the nation."[72] He described the two journals as natural complements, grouping them both under the Society's ideal of "not asking where a contribution comes from" in the interest of "spreading achievements to all colleagues everywhere."[73] This approach was indeed the keystone of China's cosmopolitan reputation, but Weng's retrospective depiction glosses over the *Bulletin*'s struggles for geographical diversity within China, and typifies the pioneering generation's preoccupation with bringing China into the scientific community of nations.

For China's second generation of geologists, who were trained natively before going abroad and who promoted geology against imperialism during the May Fourth era, the story of the *Bulletin* and the *Dizhi lunping* was less about orderly progress than about unfortunate necessity. In his introduction to the *Lunping*'s inaugural issue, Xie Jiarong is apologetic that the Society's first publication officially adopted Western languages only, explaining that, at the time, "there was no alternative," and hoping the choice was not mistaken for "a purposeful attempt to be unconventional."[74] He continues that it has "truly been a regret" that the Society has not had a Chinese journal, for despite the great success of the *Bulletin*, "how can the academic publications of a country depend exclusively on foreign languages?"

The *Lunping* was in no way the handmaiden of the Society *Bulletin*. Nor did it simply satisfy a point of Chinese pride. By publishing in Chinese, the *Lunping* not only made geological knowledge accessible to new audiences and a wider pool of domestic contributors, but it also created a "private" space where Chinese geologists could communicate on a national scale without the pressures of international scrutiny. No serious Chinese geologist could avoid learning at least one foreign language, but foreign geologists had no need to learn the Chinese language to study Chinese geology, and, with the exception of certain Japanese researchers, few of those living and working in China ever learned more than a few conversational phrases. Even Amadeus Grabau, who died in Beijing after almost thirty years at the heart of the Chinese geological community, worked exclusively in Western languages and depended on Chinese collaborators for local data. In scientific circles, Chinese was a medium that brought knowledge inward but did not export it outwards. Though every Society member was entitled to a copy of both the *Bulletin* and the *Lunping*, the *Lunping* did not even have an English table of contents and its substance was thus completely opaque to most foreigners.

Relieved of the need to either demonstrate China's scientific prowess to the outside world or hold foreign interest, the *Lunping* devoted itself to the practical problems of the native community, chief among which were isolation and scarcity. By publishing rougher, more preliminary research, it allowed geologists in the periphery to get feedback from distant colleagues and subdisciplinary specialists on scientific questions. It also put regional mineral surveys and applied research in wider context, aggregating local data and facilitating collaboration. The *Lunping* circulated firsthand accounts of international conferences and foreign geological organizations, reports of new instruments and methods, and professional information, such as job announcements and funding opportunities. Its book reviews were substantive digests of important articles and reports on all aspects of geology. They provided Chinese-language summaries of research by Chinese and foreign geologists alike, on both Chinese and non-Chinese topics. These reviews covered rare and recently acquired materials as often as newly published work, making important references available to geologists in remote and underfunded areas.[75] Similarly, students studying overseas frequently contributed reviews of literature located abroad, alerting compatriots to critical new citations that would otherwise be unavailable. Through its features and reviews, the *Lunping* concentrated and redistributed scarce resources, and by lowering the bar to include such prosaic items, it actually raised the general standard of Chinese research.

The *Lunping* traded refinement for functionality in terms of production values as well. At 40 cents, its cover price for nonmembers was five times lower than that of the *Bulletin*, making it readily affordable for students. During the War of Resistance, when the *Bulletin* opted to print in Hong Kong and accept delays rather than compromise quality, the *Lunping* switched to newsprint and cheaper ink.[76] Though wartime stoppages were inevitable, the *Lunping*'s commitment to timeliness made it a vital source of community news.[77] Besides summaries of research activities at China's major geological institutions, the news section announced upcoming publications, scholarships and awards, student field trips, and open events like talks and exhibitions. It also communicated important initiatives, like the national map project and donation drives for money, maps, or instruments. Personal items such as marriage and sickness were also covered, as were updates on geologists thought to be missing or dead. As people and institutions were reassigned and relocated, the *Lunping* helped geologists keep track of each other despite the rapid changes and uncertainty of war. In each issue Chinese geologists read about who was in the field, which colleagues were studying abroad, and what areas were safe for research.

In addition to building community through mutual aid and reportage, the *Lunping* also encouraged scientific discussion. The *Bulletin* did not shy away from controversial theories, like eustasy or continental drift, but as a publication designed to accommodate a transient foreign population and represent the breadth of Chinese geology, it rarely had enough continuity to support extended debate. The *Bulletin* was, moreover, explicitly intended to provide a congenial atmosphere for scientific cooperation.[78] The *Lunping*, on the other hand, was not bound by the constraints of hosting, and it catered to an extremely invested readership with complex personal loyalties. Within this Chinese-language context, even rank and file native geologists felt comfortable directly engaging one another on the quality and interpretation of data, and the editors of the *Lunping* welcomed "scientific discussion by correspondence." A few of these exchanges, such as those regarding Quaternary glaciation, were indecorously ad hominem, but more often than not they were nuanced and constructive, establishing, for example, the basis for standardization of terminology, mapping practices, and periodization.[79]

Many other things that could not easily be expressed in the *Bulletin*, fit quite naturally in the *Lunping*, and leaders of the Geological Society often used the Chinese journal to bring their personal influence to bear on difficult issues. The war with Japan, which dominated all aspects of Chinese life from 1937 to 1945, appeared in the *Bulletin* only as occasional references to "unpleasant conditions . . . caused by the present day situation"[80] or "frequent bombardment."[81] But while the *Bulletin* maintained as much of a business as usual air as possible, the *Lunping* featured war-related articles quite often, and did not mince words. The "private" space of the *Lunping* allowed Weng Wenhao to address the sensitive issue of loyalty and collaboration with the enemy head on, and in one of two open letters meant to provide a "compass," or guide to action, for the entire Chinese geological community, Weng wrote with great emotion:

> Scientific truths do not have national borders, but scientific personnel, scientific data, and scientific workspaces all do. We absolutely cannot use science as a pretext to abandon the nation, and cannot use the excuse of protecting our institution's data to forget the Republic of China. Under present circumstances, we would rather sacrifice everything that our institutions have to win back our character and preserve our national dignity. My saying this is not without cause; these are reactions to actual circumstances, so it grieves me even more to speak.[82]

For those workers trapped in occupied regions, this was often reinforced by offers of aid upon reaching the Interior. Matters of scientific conduct and research ethics were treated with similar candor, and Yang Zhongjian and others frequently offered heartfelt opinion pieces to Chinese geologists at large.[83]

These personal messages from the geological elite to the native community helped them extend their influence over colleagues outside of their direct pedagogical and professional networks. With this authority, leaders set research agendas and tried to discipline the increasingly disparate Chinese geological community on everything from the labeling of samples to the etiquette of scientific criticism.[84] Though the *Lunping* was not the standard-bearer for Chinese geology in international circles, it was a working mechanism of communication among native researchers. By realistically reflecting the increasingly complex nature of the domestic geological community, its mixture of original research and community features made coordination of wartime mobilization and the maintenance of scientific standards possible when direct contact was not.

Petroleum Geology and Shifting Foreign Influence

While the bulk of new geological personnel in China were struggling to keep up with the high expectations of their older colleagues, China's geological elite were desperately seeking petroleum to meet the needs of the war effort. The search defied conventional wisdom, and by war's end Chinese geologists had only identified a few viable deposits. Still, the effort gave Chinese investigators a chance to explore previously remote regions and read the landscape in a new way.

Modern petroleum exploration in China began with foreign attempts to find local fuel reserves for their own, rapidly growing treaty-port consumption. Several reports by interested amateurs were published in foreign-language journals in China, and reprinted abroad.[85] This preliminary work attracted the attention of both international corporations and Qing officials, who initiated a half-hearted attempt to locate oil on Taiwan island in 1878. Two foreign "experts" were hired to conduct preliminary surveys, but after considerable outlays for foreign drilling equipment, no oil was discovered and the Qing court soon lost interest in the project.[86] Several decades later, in 1903, German business interests spurred by hints of natural gas in Ferdinand von Richthofen's China reports contracted with the Qing court for exclusive exploration and drilling rights at Yanchang in Shaanxi Province. Local Chinese recovered control of these fields in 1907 and hired a Japanese petroleum engineer to continue surveying the area. This work continued until the fall of the dynasty but none of the test wells produced oil and the republican revolution interrupted further study.[87]

In 1913 the American Oriental Mine Company sent two of its geologists to Sichuan to explore Chinese mineral resources. They were focused on iron reserves but published several incidental reports suggesting that petroleum production was unlikely to ever be commercially viable in China. Despite this grim outlook, the Standard Oil Company of New York approached the Beiyang government in late 1913 with a plan for resuming work at the Yanchang fields, and the two parties formed the Chinese-American Petroleum Affairs Office. The Petroleum Affairs Office was staffed by a team of six American geologists under M. L. Fuller and F. G. Clapp and nine Chinese mining students from the Beiyang Academy in Tianjin. Six teams were sent out to survey Hebei, Henan, Shandong, Shaanxi, and parts of Manchuria and Mongolia on horseback. One year and 3 million yuan later, several oil seepages were recorded but test wells were not producing enough oil to justify transportation costs. Exploratory work ceased in 1915.[88]

According to M. L. Fuller who was the scientific leader of the Standard Oil project, China's lack of petroleum could be attributed to two main causes. Either sandstone underlying potential oil producing strata had allowed oil to seep away rather than segregate into pools, or oil-producing strata were so disturbed by folding and faulting that pockets of oil were released into the surrounding matrix.[89] Anticlinal structures characteristic of petroleum deposits were also absent or insufficient to trap oil in exploitable quantities. In 1922, Eliot Blackwelder of Stanford University echoed Fuller's assessment and added even more basic reasons for the failure to discover workable wells in China. The Paleozoic rocks dominating Chinese geology could not bear oil, and China's Mesozoic and Cenozoic strata were almost exclusively of continental type rather than the marine type generally associated with petroleum deposits.[90] These arguments were so conclusive that in pedagogical contexts the Chinese case was often used to illustrate the non-oil-bearing nature of continental formations.[91]

Though estimates varied considerably, China's oil reserve was generally thought to be between 1.3 to 4.4 billion barrels, putting it at about the same level as that of Japan, whose territory was twenty-five times smaller. Later surveys by the British Royal Academy and Texaco Corporation supported these conclusions. The overwhelming agreement by foreign specialists that China proper was "oil poor" not only quelled foreign interest, it retarded the domestic petroleum industry for several decades, despite positive indications from native scientists.

Oil seepages had been recorded in the Yumen area for centuries and Weng Wenhao and Xie Jiarong conducted the first native Chinese petroleum reconnaissance there in 1921.[92] Weng and Xie found favorable conditions in the area and, having identified a broad anticline of porous sandstone both capped and underlain by less permeable shale, they concluded that Yumen was likely to have significant commercial value. Several years later, Zhang Renjian, a US-trained Chinese geologist interested in the development of China's northwestern provinces, circulated a report on petroleum samples from Yumen that urged Gansu officials to invest in modern extraction equipment and harvest the field, but the local government considered the project too risky.[93] The mid-1930s saw the first of several "develop the northwest" movements in national political circles, and in 1936 a Northwest Geological and Mineral Reconnaissance Team was formed to reexamine the economic viability of developing the Yumen fields. Despite previous geological work and the positive opinion of the team's Chinese geologist, Sun Jianchu (孫健初), lead scientists J. Marvin Weller and Frederick A.

Sutton concluded that prospects for finding oil in Gansu and neighboring Qinghai were slim. Based on the prestige of these two American petroleum geologists, development plans were halted indefinitely.[94]

Pessimism about oil prospecting in China proper did not, however, deter foreign activity in more remote frontier regions. Russian and later Soviet geologists were extremely interested in historical reports of oil in Xinjiang, for instance, and Vladimir A. Obruchev surveyed the area for petroleum deposits in 1905.[95] By 1909 informal arrangements were made between Russia and local authorities to extract small quantities of oil using traditional methods. Because Xinjiang was only loosely under central control, this activity continued until the War of Resistance. In 1924 the Sinclair Corporation attempted to stake a claim as well. The American oil company conducted its own petroleum surveys, and results were promising, but transportation issues and difficulty negotiating with the local warlord, Sheng Shicai, made further investment economically unfavorable.[96] Because the Sinclair Corporation did not publish its findings, and the Geological Survey could not yet extend its fieldwork as far west as Xinjiang, official Chinese reports like the survey's periodical *Summary of Mineral Resources in China* relied exclusively upon Russian (and later, Soviet) research and informal travel accounts until the early 1930s.[97]

Guizhou, unlike Xinjiang, was not a historically oil-producing region, but in 1918 seepages were found and the Standard Oil Company of New York, undeterred by its losses in Shaanxi, mounted a two-year survey of petroleum prospects in southwestern China.[98] Results of this research were never released and Chinese geologists were unable to penetrate the area until after the central government moved to Nanjing. Petroleum deposits on Taiwan island, ceded to Japan after the Sino-Japanese War, were even less accessible to Chinese scientists. Frontier areas did not alleviate any of China's reliance on foreign oil, and Xie Jiarong's 1937 estimate of China's regional petroleum reserves explicitly omitted northern and southwestern provinces. Because of foreign secrecy, too little firsthand data was available, and none of these resources contributed to China's domestic output anyway.[99]

Foreigners found commercial development of frontier petroleum reserves difficult because modern infrastructure was so lacking in the north and southwest. But China's frontiers were at least available to foreign explorers, whose money and military backing often softened local warlords. Before 1928, China's best-trained geologists were completely unable to enter Xinjiang, Xikang, Tibet, Guizhou, and parts of Yunnan, let alone Japanese-controlled Manchuria and Taiwan. The national government's move to

Nanjing opened up exploration in the southwest, but regional powerholders were still covetous of natural resources and resistant to central interference. In the northwest, Chinese geologists found their personal safety at risk as local leaders only nominally loyal to the Guomindang "warned" survey workers that they could not be protected against "uncontrollable bandits."[100] In this way, the incompleteness of national political integration directly affected Chinese geologists' ability to assess petroleum reserves, and it was not until the War of Resistance, when frontier leaders were closer to central authority and bound by issues of survival, that the scientific climate changed.

Because there were so many near misses in China's early petroleum exploration, Chinese historians in the 1980s accused foreign geologists of conspiring to keep China dependent upon imported oil. These scholars labeled the "oil-poor theory" an instrument of imperialism used to retard Chinese development, and the Chinese who accepted the theory as weak or reactionary.[101] However, foreign petroleum interests had many other well-documented methods of monopolizing market share. It is unlikely they would have willingly ignored new oil fields when precedents in coal mining so clearly illustrated the political and economic advantages of developing China's natural resources. There was even less reason for Chinese geologists to help foreigners suppress domestic oil production, and Chinese underestimation of local petroleum reserves is better analyzed in terms of available prospecting methods and lack of access to frontier regions.

The paucity of oil reserves discovered in China was not for lack of trying. Chinese survey geologists in the Nanjing Decade went to great pains to resurvey areas that foreign petroleum geologists had dismissed. From 1931 to 1934, Wang Zhuquan and Pan Zhongxiang (潘鍾祥) led exploratory teams to Shaanxi that retraced Fuller and Clapp's footsteps.[102] They identified local anticlinal structures, used paleontological methods to revise the age determinations of several key strata, and found evidence that by finely differentiating potential oil-bearing layers around Yanchang, it might be possible to harvest small pockets of oil at shallow depth. During the same period, Tan Xichou and Li Chunyu reexamined the structure of Sichuan and took issue with American geologist George D. Louderback's pessimistic view of the province's oil prospects.[103] Tan and Li delineated eight promising anticlinal regions and six oil-bearing strata, but in both Shaanxi and Sichuan test wells were too shallow and these favorable opinions did not seem to bear out.[104] Petroleum prospecting in the 1920s and 1930s was largely limited to tracing known seepages and locating anticlines suitable for pooling of existing oil. This theoretical simplicity placed the burden of proof on drilling and extraction equipment that was largely beyond the means of Chinese geologists.

During the War of Resistance, members of the Geological Survey's Fuel Laboratory studied extraction of oil from low-density shales, and others produced fuel from corn and alcohol to keep the war effort going. Under these emergency circumstances, Chinese geologists could neither afford better equipment nor afford to wait for better equipment; they had to act immediately with the resources available. In a letter to his mentor, Weng Wenhao, Huang Jiqing described his feelings about China's petroleum prospects:

> Exploratory drilling in the Sichuan basin is an inherently risky move. Success and failure are both possible, and perhaps failure is more likely. But considering that hope of finding oil within China's borders is not great anyway and that Sichuan is the most suitable base of operations during the War of Resistance, we propose to take a chance on exploration. If it works, it will cement the nation's fortunes; if it fails, a conclusively negative result will still qualify as an important gain. Western European nations have very little hope of finding domestic oil fields, but to this day they work vigorously and tirelessly, expending their nations' money to continue exploration; it is enough to show that citizens planning for their nation's future must find a way even when none exists.[105]

Huang's words captured the feelings of many Chinese geologists, who hoped against hope that petroleum reserves could be found where none had been found before. The Sichuan basin was both tantalizing and frustrating, with several shallow layers of promising age and structure but clearly continental origin. Deeper marine layers were inaccessible and though teams from the Geological Survey, the Sichuan Survey and the NRC joined forces to explore the basin, test wells produced no oil until after the war.[106]

Huang led Sichuan petroleum exploration from 1939 to 1941, and then personally responded when the Gansu Provincial Development Bureau requested help evaluating the petroleum potential of the Gansu Corridor (河西走廊 *hexi zoulang*). Huang and the rest of the survey geologists working out of Lanzhou not only concluded that the Yongchang region was likely to be petroleum-rich, but they also set a precedent for central-provincial cooperation that eventually led to the 1943 establishment of the Northwest Branch Survey.[107] The following fall, after pledging his loyalty to the nation and the Guomindang, de facto "king of Xinjiang," Sheng Shicai, launched a "go to Xinjiang" (到新疆去 *dao Xinjiang qu*) policy to help bring the region into the national economy. Huang Jiqing, Yang Zhongjian, Cheng Yuqi, Zhou Songjun, Bian Meinian, and Weng Wenbo launched the first major all-Chinese expedition to Xinjiang in October 1942. They investigated mineral resources, geology, and oil fields both north and south of the Tianshan mountains under joint provincial and NRC auspices.[108] The expedition produced a detailed study of the Dushanzi oil and natural gas field and laid the groundwork for technological exchange agreements with the Soviet Union that made large-scale extraction possible by 1944.[109] More importantly the Xinjiang project marked the growing integration of Chinese territories previously aloof from national politics.

The single greatest wartime triumph in Chinese petroleum geology, however, was still the successful opening of the Yumen oil field in northwestern Gansu under the direction of Sun Jianchu. Though the discouraging assessment of American geologists Weller and Sutton shut down research and investment in 1936, the area was a traditional source of oil for native consumption, and Sun was convinced that it was commercially viable. With reports from Shaanxi and Sichuan optimistic but uncorroborated by actual production, the NRC decided to reopen prospecting at Yumen and sent Sun to investigate in October 1938. Within a year he identified an asymmetrical dome-shaped anticline of Cretaceous age at Laojunmiao as the most likely petroleum deposit and convinced Guomindang authorities to borrow specialized drilling equipment from Yanchang in Communist-controlled Shaanxi.[110] This unprecedented bit of cooperation paid off, and all of ten test wells struck oil, with two tapping large pools suitable for modern extraction on a large scale.[111] The NRC formed a Gansu Petroleum Bureau to continue exploration, and development around Yumen, and Sun's success attracted enough investment capital for the purchase of up-to-date equipment from the United States. Delivery was slowed by wartime conditions, but by mid-1941 China's first commercial oil field was in full production.

Sun's work was not only a practical economic achievement, but it was also the first of a series of Chinese challenges to conventional wisdom concerning the depositional conditions required for petroleum formation. Sun's structural and stratigraphical analysis of the Yumen region did not differ significantly from that outlined by Weller and Sutton after their preliminary studies, and Sun maintained contact with Weller throughout his Yumen research. The key difference between Sun's optimism and Weller's naysaying was that Weller, as a geologist familiar with the major oil fields of North and South America, did not believe that continental deposits could produce oil.

In fact, for many leading foreign geologists, petroleum was by definition the thermal maturation product of marine life from the geologic past.

The crisis of China's wartime oil crunch forced the Chinese to make the most of the territory and conditions available and gave them incentive to look for oil where foreign geologists would not. In essence foreign geologists were looking for sites likely to bear oil whereas Chinese geologists in the late 1930s and early 1940s were looking for oil at sites where they needed to find it. The success of the Yumen oil field demonstrated that this was possible, and gradually Chinese geologists began to theorize the nature of continental oil deposition from their empirical research. In 1941, while studying petroleum geology at Kansas State University, Pan Zhongxiang published, "Non-Marine Origin of Petroleum in North Shensi, and the Cretaceous of Szechuan, China," in the *Bulletin of the American Association of Petroleum Geologists*. He concluded that if, rather than studying the structure of these areas to locate suitable marine beds for test drilling, geologists traced known oil seepages and oil sands to their origin beds, they would find them consistently associated with fresh-water (i.e., continental) sediments.[112]

In 1943, also while studying petroleum geology in the United States, Sun Jianchu published "Outline of a Plan to Develop China's Petroleum Resources," which included a speculative map of the distribution of oil in China, and "Discussion of Petroleum Distribution Based on China's Geography and Geology" in Chinese journals.[113] These works applied his experiences at Yumen to future prospecting in China and showed the new horizons opened by the possibility of continental oil deposits. Huang Jiqing pushed these ideas further in his 1943 (Chinese) and 1947 (English) reports on oil in Xinjiang. Huang took the possibility of nonmarine petroleum deposits as given, and he proposed a "polygenetic" theory that took into account the more complex processes required for continental sediments to form workable petroleum reservoirs.[114] Later Chinese geologists proposed theories for understanding the conditions under which organic matter in freshwater sediments could be transformed into petroleum, providing both causal mechanisms and new structural cues for prospecting work.

The new theory of nonmarine petroleum deposits was so empowering to Chinese geologists that even when John L. Rich of the American War Production Mission at Chongqing concluded in 1945 that oil fields were "non-existent" in the Sichuan Basin, Chinese steadfastly continued their search. In fact, according to Xie Jiarong, the basin's alternating system of steep and gentle anticlines indicated "without a doubt" the existence of petroleum deposits. Once drilling strategies improved in 1947 and multiple wells were sited for each test location, this faith was finally vindicated, and

Chinese geologists continued to improve their theories of nonmarine petroleum production through the postwar era into the Communist period.

Conclusion

The tragedy of war uprooted millions of Chinese and transported east-coast Han intellectuals to the unfamiliar Interior, throwing the national economy into disarray and challenging scientists to make use of their new, often impoverished surroundings. But the great migration inland was also an opportunity for the geological community to reinvent itself by accommodating new members, reorganizing its institutions, integrating previously inaccessible territories, and finding a scientific voice independent of international convention.

The necessities of the wartime economy and the burden of patriotic service emphasized action rather than contemplative knowing. Scientists did not forsake "pure" or "basic" research, but they could not waste energy on anything that might slow down the production of tangible results. Whether theoretical studies in physics, practically oriented research on alternative industrial materials, or economically significant surveys of previously inaccessible frontier regions, wartime science was about what investigators could do, not what they could imagine doing.

For Chinese petroleum geologists this meant reconsidering previously dismissed observations, such as natural oil seepages and local lore, and following up on these exhaustively in hopes of exploiting even limited fuel resources. By trying to understand the geology of natural seepages and differentiate between productive and nonproductive antisynclines, Chinese gained confidence in their empirical instincts and opened themselves up to the theoretical possibility of oil-bearing continental deposition. Their small successes in the early 1940s defied conventional wisdom, but were less a challenge to foreign scientific authority than a demonstration of how much lack of access and poverty had previously stifled the economic and scientific potential of China's frontiers.

When the War of Resistance ended in August 1945, Chinese geologists were eager to push forward with petroleum prospecting and other geological investigations, and they looked forward to rejoining international circles as an expanded and self-sufficient scientific community. However, the euphoria of victory faded quickly as the demobilization (復員 fuyuan) process began. The domestic situation was so chaotic that government institutions could not reach coastal areas to reclaim their occupied property from the Japanese. When they did, they often discovered that enterprising individuals masquerading as Chinese officials had beaten them to it, and that buildings, equipment, and data that had managed to survive the Japanese occupation did not make it through the first year of "peacetime."

Weng Wenhao was put in charge of demobilization and reconstruction, but unlike the 1937 retreat to the Interior, demobilization required Weng to organize a mass movement into territory that was largely out of Guomindang control, and no amount of peacetime planning could quiet the brewing GMD-CCP power struggle. Large swathes of northern and eastern China that had been occupied by the Japanese were "liberated" by Communist forces who refused to surrender control to their domestic rivals, and, despite several high-level truces, the end of war with Japan marked the beginning of a decisive civil conflict.

CONCLUSION

Grounded Visions

> Friends, we should . . . remain determined in our loyalty to the nation. If today one method does not work, then tomorrow use another and continue the effort.
>
> —Chu Anping[1]

After their decisive victory at the Battle of Huai-Hai in the winter of 1948–49,[2] Communist forces consolidated control of northern and central China and took Beiping without a fight. They celebrated by renaming the former capital "Beijing," and then set their sights on the south. With CCP success increasingly likely, the Guomindang began secretly planning a retreat to the island of Taiwan, where they hoped to continue the Republic of China in exile and regroup for an eventual attack on the mainland.

Though the civil war continued unabated, the GMD quietly transported cultural artifacts, archives, and bullion to Taiwan, and ordered important institutions, like the Academia Sinica, to relocate to the island before fighting intensified in the south. Realistically, however, the regime was too weak to organize or enforce such a retreat, and several academic leaders questioned the wisdom of following a government that was so willing to give up the mainland in order to preserve itself.[3]

In January 1949, eleven members of the Research Institute of Geology wrote to Li Siguang, who was in England at the time, vowing to stay at their posts and resist relocation. In their own words, no one who broke this promise would ever be accepted into the geological community again.[4] Survey director Li Chunyu reported a similar resolve to Weng Wenhao, and laid in rations of rice and preserved vegetables to tide the survey over during its break with the Guomindang.[5] Having learned from the demobilization experience, survey members even organized a security brigade to protect their property from possible looting. When the Communists reached Nanjing in late April 1949, they found geological facilities intact and Chinese geologists from several institutions all patiently waiting to be "liberated" by the new, decisive revolution.[6]

Some post-1949 revolutionary accounts have tried to frame geologists' decision to stay on the mainland as a "progressive" (進步 *jinbu*)—and therefore political or ideological—choice, but I would argue that it was the last *nonpolitical* choice that mainland geologists would make for several decades. At the Geological Survey, for instance, Li Chunyu was so committed to maintaining a nonpolitical work environment that when the weakening GMD tried to impose political surveillance on the geologists in late 1948, Li and two other longtime GMD members volunteered for the party's spy organization, Zhongtong (中統), just to preempt the government from sending in outsiders. By all accounts this neutralized Guomindang political interference and also forced the Survey's only underground CCP member to remain inactive for fear of compromising Li and others.[7]

As we have seen throughout this study, Republican geologists were always politically engaged, but they consistently put their scientific and corporate identities first, and they formulated their views of politics and nation through the lens of geological practice. So what were geologists really choosing when they refused to move to Taiwan in 1949? How did geologists' multiple visions of Chinese modernity collapse onto the Communist state? And what effect did this have on the geological community? In this conclusion I use geologists' choice to defy the Guomindang and stay on the Chinese mainland to revisit the relationship between nationalism and geology that evolved over the Republican period, and raise questions about the meaning of geology in postrevolutionary China.

Landed Interests

In order to be effective, Chinese geologists had to establish their own scientific authority in ways that were fungible in both domestic and international contexts. But as relative newcomers to modern science, they also had to find creative methods of insuring that these exchanges took place on fair and equitable terms, whether at home or abroad. Put simply, if they wanted geology to serve China, Republican geologists had to devise ways of doing independent research and making their contributions known without the benefit of an internationally recognized scientific tradition. As a result, practices of fieldwork and professional communication became as much expressions of patriotic mission as tools of scientific investigation. When postwar China devolved into civil war and threatened these sources of authority and action, Chinese geologists bet the future of the nation on the outlook for science and their prospects for studying the land.

Fieldwork

Fieldwork allowed the first generation of geological pioneers to establish a native community by "indigenizing" geological concepts and practices. References in foreign-language textbooks to the Appalachian mountains or the Scottish highlands that made little sense to Chinese students on paper were translated into observable local realities that illuminated comparable processes or phenomena. In this way, students learned geology through contact with the Chinese land, and an unfamiliar science became immediately relevant to domestic problems. Technical skills such as sketching, surveying, and notetaking then brought this firsthand knowledge back into the classroom for further study, and made Chinese examples a part of future teaching.[8] A few places, such as the Western Hills around Beijing (Beiping), were so much a part of native geological training that they became not just shared points of reference, but rites of passage and even sites of scientific pilgrimage.

The social experience of fieldwork also cemented personal bonds, transferring the traditional student-teacher relationship into a new active context and tying peers together through a unique form of shared labor. Instead of associating geological exploration with adventure and heroism as in the West, Chinese teachers used fieldwork to instill values of nationalistic sacrifice and commitment, and members of the community credited the physicality of fieldwork with everything from incorruptibility to love of truth, loyalty to the nation, and longevity.[9] The strenuous, against-the-elements aspects of fieldwork helped geologists inject a new dynamism into the role of the scholar in public and intellectual life, and these pedagogical traditions did not end with the Geological School or the early years of the Peking University Department of Geology. As the geological enterprise expanded in China, students of these original training programs initiated younger generations into the geological community by bringing a similar perspective on fieldwork to their own teaching.[10]

By putting Chinese research on the same empirical footing as that of foreign investigators, fieldwork not only facilitated the growth of geology within China, but it also made the evaluation and exchange of Chinese findings meaningful in international circles. Chinese insistence on rigorous field methods and documentation helped them gain the respect of outsiders and gave them a sound basis on which to discuss matters of fact and interpretation. This slowed the activity of foreign explorers and enabled Chinese to weigh in on matters of general geological significance.

In a sense, fieldwork spun the land into scientific gold, both producing economically important results and placing Chinese territory within a global history of development. Even when Chinese views were controversial, as with Li Siguang's Pleistocene glaciation theory, geologists from around the world who came to gather evidence firsthand were able to rely on the guidance of Li's prior studies. Chinese geologists' persistent faith in fieldwork paid off even more in situations where foreign theories were ill-adapted to local conditions. During the War of Resistance this approach not only led to the discovery of oil reserves where foreign experts least expected, but it also made important contributions to the emerging recognition of nonmarine petroleum deposits.

The patriotic feelings that Chinese geologists attached to fieldwork and the land after decades of scientific development are perhaps best captured by the lyrics of the Geological Society song that was introduced in 1941.

> Vast is my China! Vast is my China!
> Eastern water, western mountains
> Southern hills and northern loess
> Truly are worthy of boasting
> . . .
> Once our hammers' target is found,
> Investigating and discussing together
> Brings us boundless pleasure
> Vast is my China! Vast is my China![11]

Like the Society emblem adopted four years earlier, the full lyrics of the song emphasized characteristic geological features of different regions of China and several of the more significant findings of the previous two decades. But the anthem injected Chinese researchers into the imagery, and was intended to be sung as they worked "to encourage boldness while climbing mountains and cure the exhaustion of laboring in the field."[12] Even during the dark days of the war, geologists were excited about exploring China's distant frontiers and cementing national unity by helping recapture the occupied east.

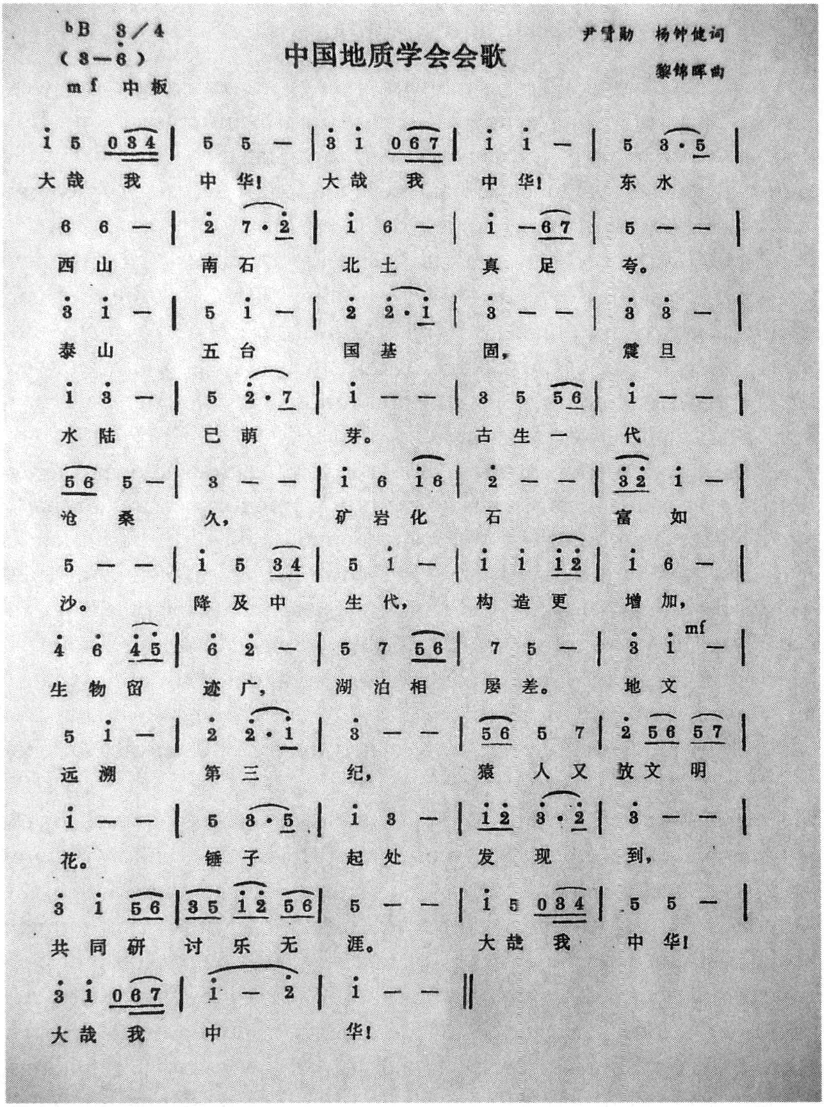

3. Lyrics for the anthem of the Geological Society of China written by Yin Zanxun and Yang Zhongjian. Music provided by Li Jinhui, taken from "Dizhijie xiaoxi" 地質界消息 (Geological community news), *Dizhi lunping* 地質論評 (Geological Review) 5, no. 6 (1940), between pp. 573 and 574.

Scientific Communication

The pre-Republican origins of Chinese interest in modern geology were rooted in fears of foreign encroachment and the growing sense that the Chinese were losing touch with their own territory. Statements like that of the French geologist Emmanuel de Margerie at the 1913 International Geological Congress—where he recommended that a colonial power take responsibility for filling in the "gap" of China's eighteen provinces—fueled Chinese geologists' dedication to fieldwork as a way to reconnect with the land, harness its resources, and challenge foreign knowledge claims.[13] However, for the first several years of Chinese geological activity, foreign researchers were largely unaware of native advances, and continued to think of China as an open field for foreign exploration.

Chinese geologists hoped to make their investigations known internationally, but as members of a young scientific community without the backing of a strong, unified government, they found it difficult to participate as equals in international fora. They were not, for example, invited to the 1920 Pan-Pacific Congress, where the director of Japan's Geological Survey reported on Chinese territory. In 1926 China was actually rejected from the Pacific Science Council for lack of a recognized national scientific organ.[14] Just as worrisome was the possibility that Chinese geologists who did take part in international conferences would become mere "data collectors" for more established geological powers.

The solution that Chinese geologists devised was a geological society of their own that would be international in character and allow them to tap into broader scientific networks. This strategy hinged on the lure of the Chinese land, which brought researchers from around the world through Beijing en route to their own fieldwork. Chinese made themselves indispensable as hosts to these foreign scientists, and the meetings of the Geological Society of China became popular social events in the capital. Through the Society, the Chinese not only broadcast their own geological findings, but they also gained access to foreign data that would otherwise have left the country, and the English-language based *Bulletin of the Geological Society of China* then captured this cosmopolitan mix on paper. As the *Bulletin* gained prominence, it publicized the Society's activities around the world, and Chinese researchers used it as their calling card at major international congresses and events.

The Geological Society experiment brought home the lesson that dissemination of research was as critical to scientific progress as fieldwork, and as the population of native geological workers grew in the late 1930s and 1940s, the Society established a Chinese-language journal, the *Dizhi lunping,* to serve the special needs of the domestic community. The *Dizhi lunping* allowed native investigators at isolated institutions across the country to stay in touch with China's geological elite and exchange information about new findings, professional opportunities, and research methods. This was particularly valuable during the War of Resistance, when the journal became a major tool for coordinating research efforts, and an outlet for the community's most practical and patriotic sentiments.

It was no accident that the Geological Society song was announced exclusively in the *Dizhi lunping* or that the Society emblem received only minimal coverage in the *Bulletin*. Chinese Society members understood that national and international understandings of Chinese geology were not always aligned, and that scientific access to Chinese territory had very different meanings for groups within and without the local geological community. Nevertheless, Chinese geologists insisted on making as much research on Chinese territory available to as many investigators as possible, and they kept both the *Bulletin of the Geological Society of China* and the *Dizhi lunping* in print throughout the many crises of the late Republican period.

Choice

This kind of persistence in the face of adversity led former Geological Survey director Huang Jiqing to high expectations for postwar geology, and in an article written just before war's end, he looked forward to an energetic revival of geological research in China followed by a new golden age. China would require at least five times the number of existing personnel and several new geological specialties to handle the research needs of its reclaimed territory. But if the political situation was stabilized, Huang anticipated that geological education would expand, and the Chinese geological community would once again host foreign researchers.[15] The winning combination of fieldwork and scientific exchanges would continue where it had left off.

When the war finally did end, the Japanese threat was over and Chinese geologists took possession of Japanese research materials in Manchuria, Shanghai, and Taiwan. Branch surveys established during the war in Gansu and Yunnan offered a model for Geological Survey expansion, and new branches were planned in Manchuria and Taiwan.[16] The debilitating treaty system was also terminated, and Chinese geologists seemed poised to complete their exploration of all of the former Qing territories.

But contrary to wartime fantasy, there was no real peace after the Japanese surrender. Jiang Jieshi and Mao Zedong met in Chongqing in late August 1945 with promises of conciliation and support, but the agreement collapsed in less than three months. United States General George Marshall managed to negotiate several brief truces in early 1946, but by June full-scale civil war was raging in north China and the fighting did not stop until Communists controlled the whole of the mainland in 1950.

Though Chinese geologists, like other scientists, had willingly endured hardship and privation during the war against Japan, they found themselves angry and disillusioned with post-war realities. Inflation was so rampant that the cost of living in Shanghai was five hundred times higher on March 1948 than on VJ Day. By August 1948 prices for household staples had increased tenfold again, and a forty-nine-pound bag of flour cost 21.8 million yuan.[17] Geologists were not only on fixed salaries, but they were also rarely paid on time, and daily necessities were often hard to come by.[18]

Under these circumstances all promises of expansion that the government made to the survey and the Academia Sinica's Research Institute near war's end went up in smoke, and funding did not even cover basic operational expenses. Geologists could rarely work in the field, and when they tried to, security issues sent them right back indoors.[19] In 1947, Yang Zhongjian wrote:

> Demobilization is still continuing in stop and go fashion, and in both politics and economics there is absolutely no way out. Our international reputation has suffered a disastrous decline, and the fighting in this civil conflict is even more vigorous than before. The little world that is the Geological Survey is just as chaotic and cannot move forward. Welcome to post-victory China![20]

Years later, Yang would recall the period leading up to 1949 as the worst time of his life, filled with anger, depression, and hopelessness about his inability to do research. Beiping, where Yang's paleontological research was mostly based, was in the thick of GMD-CCP fighting, and fieldwork was out of the question. Government funds did not even allow indoor work, like specimen processing, to continue, and political conditions were too unstable for foreign funding agencies to offer any assistance. Yang kept busy doing whatever tasks he could find, but described the Beiping Survey branch as barely functioning and severely demoralized.[21]

Research conditions were little better in Nanjing, where the bulk of China's geological specialists were concentrated in the late 1940s. Neither the

Geological Survey nor the Research Institute could relocate to Nanjing before late 1946, and cleanup work continued well into 1947, as geologists tried to inventory losses and assess their research options. Li Chunyu spent almost all of his time trying to glean extra funding from other organizations, but most of them were in the same straits as the survey and could not offer much help.[22] The state-owned Sino-Petroleum Corporation did commission some geological work, but even petroleum geology, with its obvious economic potential and promising wartime successes, could not overcome the disorganization and financial mismanagement of the Guomindang. Several petroleum prospecting teams were disbanded, and the geological crew at Yumen oil field was reduced to maintenance levels. From 1945 to 1949 only a handful of petroleum surveys were launched and even fewer completed because of funding or security problems.[23] Average yearly oil production in China actually dropped slightly from 1945, despite the addition of fields in Taiwan.[24]

As the ongoing civil conflict ground fieldwork at Nanjing institutions to a halt, geologists occupied themselves indoors with tasks such as writing up past fieldwork and indexing or translating material captured from the Japanese.[25] The most significant accomplishment was Huang Jiqing's 1:1,000,000 scale map project, which had been started in 1936 but delayed by the War of Resistance. The project, which engaged a significant portion of the survey's sidelined field geologists and brought together material from all of China's geological institutions, added data from recent studies of the northwest and southwest to earlier geological maps and was a major synthesis of over three decades of research.[26] Even so, several local areas were left blank because there was no near-term possibility of conducting fieldwork in those places, and Huang and his team of geologists "took the view that it was better to go without than to produce something shoddy."[27]

Maddeningly, geologists in areas more removed from actual fighting found themselves in an opposite but equally difficult position. Researchers at the Northwest Branch Survey, for instance, had unobstructed access to a region (including Xinjiang and Tibet) that was the "hinge between Europe and Asia" and whose varied and unfamiliar features could "lead the traveler to the heights of fascination." But their numbers had been reduced when the War of Resistance ended, and their offices were given to the newly founded Lanzhou University by local authorities, so the branch had to function without a library, specimen rooms, drafting facilities, laboratory space, or even a place to assemble. These geologists could do nothing *but* fieldwork with their limited funding, and this compromised their ability to draw firm conclusions. From 1946 until the China Textile Corporation donated facilities in late 1948, indoor work was so impossible that none of the branch's three-year backlog of field notes could be worked up for publication, and prewar Swedish, Soviet, and British reports on China's northwest remained the state of the field in international circles.[28]

Geological Society activities continued as best they could during the civil war, and the 1947 annual meeting was even held in Taiwan to introduce Chinese geologists to the newly recovered province and support the local survey branch.[29] However, Communist guerrilla tactics and the Guomindang's inability to provide assurances of safety kept foreign researchers out of China, and the Geological Society was never the international organization that it had been in the 1920s and early 1930s. The domestic conflict also dashed Chinese hopes that foreign geologists' desire to "get near the [Chinese] land and appreciate it directly" might allow them to host an international geological congress.[30] China did manage to send a few representatives to foreign geological and paleontology conferences, but not enough, in Yang Zhongjian's view, to represent the nation's physical size.[31]

Through all the trials of the Republican period—from warlords to the War of Resistance—the small community of Chinese geologists consistently found ways to defy the odds and pursue an ambitious program of research. Funding never came easily, the nation was never effectively unified, and outsiders were always as much a threat as an inspiration, but geologists were willing to endure significant hardships in order to continue their work. However, they were unwilling to suffer not working at all, and by the end of 1948, even Guomindang supporters like Li Chunyu were angry. According to Li, everyone talked about "developing the northwest" and "building the nation" but they sat in the big cities, with bright lights and song, consuming imported luxuries, and doing nothing to maintain control of the countryside to make that development possible.[32] As the civil war dragged on, the Guomindang isolated themselves in urban oases, while Communists won over the areas in between and cut the cities off from their surrounding territory. Though geologists had originally emerged as an urban elite—trained in universities, familiar with the West, and connected to elite culture—the mechanisms of fieldwork and exchange that they developed to establish scientific authority and serve the nation rooted them in the land. When 1949 came, and they were forced to make a choice between a Chinese government in exile and a Chinese government on the mainland, geologists chose the one element of the nation that had always been reliable, and equated the object of their loyalty with their object of study. They chose the land, and waited for what would come.

Improvisation

The Chinese geological community's decision to stay on the mainland in 1949 was neither the result of direct contact with the CCP nor clear assurances of what was in store for them under the new government. But it did reflect their long-standing tradition of bold improvisation in the face of domestic uncertainties. This flexibility allowed Chinese geologists to draw on diverse sources of support, shield their research from sudden political changes, and maintain a united front despite internal differences. By dedicating their efforts to the shared goal of China's well-being without limiting themselves to any narrow definition of Chinese identity, the geological community also experimented with multiple aspects of Chinese modernity, and they were confident that they could accommodate the Communist vision without compromising their own.

Though China's pioneering geologists had a great deal to offer, they were beggars not choosers throughout the entire Republican period. Both the Beiyang and Guomindang governments were too politically and economically troubled to cultivate the native geological enterprise, so most of its growth depended on geologists' willingness to improvise from available resources and respond creatively to unexpected opportunities. Born under the Qing, they nevertheless grasped the chance to make geology part of the new republic as soon as the revolution succeeded. Strapped for staff and equipment, they took over the Imperial College's abandoned facilities and trained their own personnel. Unable to finance fieldwork in one place, they would simply explore another that was of interest to local coal mining concerns or railway companies. Chinese geology was an art of the possible, not a strictly unfolding plan.

By training their sights on the overall development of geology in China and remaining flexible about details and timing, Chinese geologists achieved many of their own goals while catering to the interests of both native philanthropists and foreign funding agencies. When the remains of Peking Man were first announced in 1926, for instance, the Chinese geological community quickly turned its attention to paleoanthropology. Though it had no experience in this field, the Geological Survey convinced the Rockefeller Foundation to fund a Cenozoic Research Laboratory to study both Peking Man and the "tertiary and quaternary deposits of northern China" more broadly.[33] Similarly, when Chinese industrialist Jin Shuchu expressed interest in funding a fuel laboratory to address China's energy needs, geologists accepted his offer and expanded the laboratory to include chemical analysis, a mineralogical and petrographic lab, paleobotany, and even a photography department.[34] In this way, Chinese geologists capitalized on private interests to strengthen their own basic research capacity, which allowed them to redirect their tiny government budget toward vital tasks like fieldwork and publication.

Over time, the geological community used a similar model to establish new academic departments, specialized journals, provincial surveys, and other technical facilities like the soil laboratory and seismological station. However, one of its boldest improvisations was in taking advantage of the foreign presence in China to frame its national Geological Society as an outward-facing, multilingual organization that encompassed study of all the former Qing territories. This move turned the threat of foreign exploration on its head, and made international interest an asset to the national geological enterprise. Within the structure of the Geological Society, outsiders linked the native community to global science and vouchsafed Chinese expertise, while local hosts harvested information from foreign expeditions and tapped into the transnational knowledge-making process.[35] By creatively acknowledging their status as a peripheral scientific community, Chinese geologists established a unique intellectual niche that allowed them to accomplish more with less.

This ability to work with prevailing conditions and still move toward their own goals became especially important for geologists during the Nanjing Decade and afterwards. When Weng Wenhao agreed to enter government service, he used his position as head of the National Defense Planning Commission (and its successor the National Resources Commission) to fold many of the geological community's nation-building objectives into official state policy. In turn, the Geological Survey coordinated its scientific agenda with the government's new economic program, and this created room for newer institutions like Li Siguang's Research Institute of Geology to focus on theoretical problems without too much government interference. The model of engagement developed during the Nanjing Decade allowed the geological community to work even more closely with the state during the War of Resistance and still remain aloof from party politics.

The key to Chinese geologists' ability to respond to changing circumstances with such patriotic opportunism was their conviction that scientific authority was more dependable as a basis for nationalism than political authority. Governments might come and go, but geological knowledge would always benefit the nation, so the development of the geological enterprise was inherently patriotic. Their reliance on the needs and standards of geology to make choices in China's name allowed geologists to act decisively without ever articulating what they actually meant by "China," and several different visions of Chinese modernity lay hidden behind their elision of geology and nationalism.

Often these visions would come in and out of focus as the community shifted from one improvisation to the next and accentuated different priorities. The Geological School of the 1910s, for instance, emphasized "Chinese students in a Chinese school under Chinese teachers studying Chinese geology,"[36] while the Geological Society of the 1920s and 1930s projected an image of cosmopolitanism that welcomed colleagues from all over the world to participate as equals on Chinese soil. Fieldwork promoted individual geologists as physically active and morally independent thinkers, while the institutional developments of the Nanjing Decade stressed the role of policy, collective planning, and coordination with the state. Because the community was constantly responding to outside pressures, Chinese geologists did not see these contrasting positions as exclusive or final, and they did not cause insuperable conflicts or divisions within the community.

Rifts and Faults

For Chinese geologists, their refusal to move to Taiwan was simply the latest in a long line of improvisations aimed at keeping geological research alive, and they saw no danger in highlighting the affinities between their own concerns and those of the new Communist regime. They found common ground on long-term planning for scientific and economic development, expansion of the geological workforce, exploration of the northwest, and geological education for the masses. But the founding of the Communist state marked a radical break from the politics of the Republican period, and geologists underestimated the ideological distance between their prerevolutionary views of labor and education and those of the new regime.

Within a few months after the official establishment of the People's Republic of China on October 1, 1949, the Geological Society of China was recommending that, in order to perfect their science, "geological workers should first learn to grasp the Marxist-Leninist standpoint, perspective and method."[37] Later, geologists like Cheng Yuqi (formerly a member of both the Geological Survey and the Academia Sinica) would look back on Republican geology and claim that "even though a certain number of geological organs were founded, their effect was minor, and the development of this scientific discipline was extremely sluggish." Republican geology, in this view, "was merely used by reactionary power-holders to simulate peace and prosperity," and though Chinese geologists had chosen the mainland to protect their hard won successes, they soon denied that such successes ever existed.[38]

This radical revisionism makes it extremely difficult to evaluate geologists' transition to Communism.[39] On one hand, the new regime's emphasis on rapid geological development was a welcome relief from the Republican era. Within a few years the People's Republic had disbanded the Geological Survey and established a Ministry of Geology at the national level. By the late 1950s, almost fifty thousand newly minted "geological workers" were scouring the country from Manchuria to Tibet, and, according to Joseph Needham, there was "no other nation where the sciences of the earth [were] so highly regarded."[40] But, as "China" became synonymous with the party-state, geologists became increasingly suspect if they strayed from the state-sanctioned orthodoxies of geomechanics and Pleistocene glaciation, or questioned the credentials of the masses to produce science.[41] The revolution required geologists to denounce their former selves for "feudal, comprador mentality," including collaboration with the capitalist West, privileging of technical expertise, and isolation in scientific "fiefdoms" like the Geological Survey.[42] Those markers of success which had once distinguished geologists' patriotic service to the nation now fueled struggles and debates about who had or had not been an imperialist tool or a capitalist roader.

However we appraise the denunciations and recantations of individual scientists, "China" as an object of loyalty had taken a radical turn, and the relationship between science and nation was fundamentally altered. Republican geology had evolved around a fluid, transhistorical understanding of "China" that commanded allegiance by encompassing—even demanding—significant diversity.[43] Geologists positioned their science to serve the nation in whatever form it ultimately took, and they welcomed a wide variety of political and scientific perspectives to strengthen their cause. But once the state and its revolutionary ideology claimed exclusive rights to "China" as the People's Republic, decades of creative opportunism in the name of an a priori attachment were suddenly measured against the teleology of the Communist state. Instead of supporting Chinese modernity as an ongoing, open-ended process, science was expected to speed the nation forward on its predetermined path, and differences within the community that had once been enriching and empowering became divisive and disloyal.

The period under Mao represented a total reconceptualization of geological nationalism that the counterrevisionism of the post-1978 period has only complicated and not undone. But with so many of the same personnel leading the geological enterprise, Communist-era geology cannot be divorced from its Republican past. Li Siguang helped transform Communist geology into a political and scientific enterprise; Weng Wenhao returned to the mainland in 1951 and lived under the shadow of his service to the

Guomindang; Zhang Hongzhao died in 1953, a Qing *xiucai* turned comrade; and Ding Wenjiang's memory was alternately vilified and lionized in succeeding waves of party struggle. Yang Zhongjian, Xie Jiarong, Huang Jiqing, Li Chunyu, Yin Zanxun, and other Republican giants continued to lead China's geological research, and members of the Geological Survey and Research Institute seeded new institutions around the country. Yet, we must remain skeptical of recent claims that Communist geology simply took up where Republican geology left off, for the meaning and design of the geological endeavor was changed in ways that even surviving members of the community cannot fully articulate. What is clear, however, is that whether or not science and politics were linked for Chinese geologists, science and nation always were, and the legacy of the Republican experience was that, when all else failed, geologists looked to the land for both the past and their future.

ACKNOWLEDGMENTS

When I began this project, it took a lot of explaining to convince people that the history of Chinese geology needed to be told. Now there are several related studies on natural history, paleoanthropology, geography, and oceanography in China in print or in the works, with more to come as archives open up and interest in Chinese science increases. In the course of writing my own contribution to this emerging field, I have accrued many debts, and I cannot hope to fully express the gratitude I have for the generosity, wisdom, and good will of all my teachers, friends, and family. I write with the hope that I will be able to pay my debts in person and in kind, rather than in words, but I would like to mention a few of those who understood what all the fuss was about from the outset.

At Harvard, Peter Galison, as well as Philip Kuhn and Everett Mendelsohn, encouraged me from the earliest attempts to tell this story, and guided me through the process of revision and reimagination with persistence and patience. I am continually grateful for their inspiration and advice.

I was fortunate enough to have received support for this project from the Social Science Research Council, the Mellon Foundation, the Dibner Institute, and the Harvard Asia Center, and this has made all the difference. I am especially grateful to the Needham Institute for providing me with a home away from home and a wonderful space for research and writing. My year as a Fulbright fellow shaped this book in every way.

From the beginning, I found an intellectual home at the Harvard-Yenching Library, and I am greatly indebted to the librarians and archivists there. Over the long years of research and writing, I have also benefited from the librarians and archivists at the Second Historical Archives in Nanjing, the Academia Sinica Institute of Modem History Archives, the Institute of Vertebrate Paleontology and Paleoanthropology (IVPP) Archives, the American

Museum of Natural History, the Columbia University Columbiana Collection, the Hoover Institute, the National Library of China, the Nanjing University Library, the Harvard Archives, the Peking University Library, the University of Wisconsin–Madison Library, the Georgetown University archives, the Stanford University Department of Geology, the University of Illinois at Urbana-Champaign Archives, and the New York Public Library. I would like to particularly acknowledge 朱文通 of the National Geological Library of China whose kindness was as valuable as his assistance locating documents. Over the course of writing, the librarians at Vassar College also provided me with a helpful and congenial environment that made much of this work possible. While I was in China, the Institute for the History of Natural Sciences (IHNS), Chinese Academy of Sciences (CAS) offered me a welcoming research community and Peking University's Geology Department provided an invaluable link to the past.

In addition I am grateful for the kindness of many individuals, including 张九辰、李学通、马胜云、吴凤鸣、李扬、崔克信、张立君、潘云唐、陈宝国、楊翠華, and countless others. The families of Li Siguang, Weng Wenhao, Huang Jiqing, Li Chunyu, and Yang Zhongjian were particularly generous with irreplaceable materials and reminiscences that gave me a truer sense of my historical actors as real people. Allan Mazur, members of Amadeus Grabau's extended family, Robert Dott, Jr., Markes Johnson, and Vin Morgan all shared precious documents and expertise that shaped my thinking and spurred me in my research.

Bridie Andrews was an invaluable sounding board at the very start of this project, and I could not have completed this work without the feedback and camaraderie of the Eccentric Circle and the writing group at Harvard University. Fa-ti Fan, Danian Hu, Sigrid Schmalzer, and Zuoyue Wang read sections of the manuscript and have been wonderful models of how the history of modern Chinese science should be done. The historians of science and technology at York University, Toronto, gave me many opportunities to discuss my work, and I hope this book has captured some of the spirit of that dynamic community.

In the late stages of writing, I had the privilege of working at the Max Planck Institute for the History of Science in Berlin in Dagmar Schaefer's research group. The text bears the indelible imprint of our many conversations and of the scholars that made the group and the institute their home away from home. The finishing touches on this book were completed at Fordham University, which has welcomed me with open arms and broadened my intellectual horizons. Indexing was supported by the Ames Fund for Junior Faculty.

Several sections of this book were presented at conferences and on each occasion, the questions and suggests of participants and colleagues helped me refine my understanding of Republican China and the history of science. I hope that this book reflects in some small measure the insights of all of those who have read or heard part of the manuscript, especially my anonymous referees. Though their impact can be seen throughout, all errors and omissions remain my own.

As this book neared completion, I had the great fortune of working with an outstanding team at the University of Chicago Press, headed by Karen Merikangas Darling. Karen, Susan Karani, and Micah Fehrenbacher all but carried me across the finish line, and I thank them for their boundless patience and enthusiasm. Parts of chapter 2 are adapted from my essay "Taking to the Field: Geological Fieldwork and National Identity in Republican China," published in *Osiris: Science, Technology, and National Identity* (2009), and I am also grateful to Chicago for permission to use this material.

Beyond the world of libraries, publishing, and ivy-covered walls, however, my debts are even greater. I would not have survived these years of grappling with Chinese geology if not for my friends and family. J Clifford Dyer deserves particular thanks, and I am grateful for the unending support of my husband and son. Through thick and thin, they have made everything worthwhile. Above all, I wish to thank my parents, who mean the world to me and are good people in ways I can never hope to match. They have given me a model of China which will guide me through all of my days, and I dedicate this book to them.

NOTES

INTRODUCTION

1. Wang Yangming 王陽明, Chuan xi lu 傳習錄 (Record of teaching and practicing), Xu Ai lu 徐愛錄 (Dialogues with Xu Ai), Juan shang 卷上 pt. 1, Diwutiao 第五條 chap. 5.
2. "Proceedings of the Thirteenth Annual Meeting," *Bulletin of the Geological Society of China* (hereafter *BGSC*) 17, no. 1 (March 1937), xiii–v.
3. "Dizhijie xiaoxi" 地質界消息 (Geological community news), *Dizhi lunping* 地質論評 (Geological review) 2, no. 1 (1937), 89–90. The emblem was designed by geologists Zhang Hongzhao (章鴻釗), Xie Jiarong (謝家榮), Yang Zhongjian (楊鍾健), and Amadeus Grabau (usually transliterated as 葛利普, "Gelipu"), and executed by the famous calligrapher Zhang Hairuo (張海若). Grabau, a German-American geologist who arrived in China in 1920, devoted the rest of his life to Chinese geology. He was the only foreigner to have achieved undisputed insider status within the local geological community, and a memorial in his honor still graces the campus of Peking University. See chaps. 2 and 3 for more on Grabau.
4. The *Dizhi lunping* description explicitly avoided the term 山水 (*shanshui*), which means "landscape" in general usage. Here, mountain alludes to the "internal forces" involved in orogeny and volcanic activity, while water references processes such as erosion and marine transgression and regression. This last process was central to the earth theories of Amadeus Grabau, as expressed in later works such as *The Rhythm of the Ages* (1940).
5. China's eastern region is riddled with lakes and rivers, and bounded by the sea; the west is dominated by the mountains of the Himalayan and Kunlun ranges; the south is characterized by the rocky terrain of both its karst mountains and the Fujian foothills; and the north is noted for its remarkable loess plains.
6. The issue of representation and inclusiveness is even more interesting once we note that Amadeus Grabau was one of the emblem's four designers.
7. The platform at the Altar of the Gods of Land and Grain, which represented the state, can still be found in Zhongshan Park, Beijing. According to Weng Wenhao, this arrangement of soils is a relatively accurate representation even by modern standards. China's far northeast has black virgin soil; the south and southwest have red lateritic soil; the west has deserts and calcium-rich, gray-white soil; the eastern seaboard has water-logged, green-blue soil; and the central loess plain is yellow. Weng Wenhao 翁文灝, "Zhongguo de turang yuqi xiangguan de rensheng wenti" 中國的土壤與其

相關的人生問題 (Chinese soil and related life problems), in *Weng Wenhao xiansheng yanlun ji* 翁文灝先生言論集 (Collection of Mr. Weng Wenhao's speeches) (Shanghai: Bingzi xuehui yinxing, 1936), 168–76.

8. On sanitation in discourses of modernization, see in particular Ruth Rogaski, *Hygienic Modernity: Meanings of Health and Disease in Treaty-Port China* (Berkeley: University of California Press, 2004).

9. For a wonderfully nuanced analysis of how both British naturalists and native participants made sense of their encounter through natural history, see Fa-ti Fan, *British Naturalists in Qing China: Science, Empire, and Cultural Encounter* (Cambridge, MA: Harvard University Press, 2004).

10. Several notable works touch on ideas of social Darwinism in modern China, but for a good introduction to the topic, see James Reeve Pusey, *China and Charles Darwin* (Cambridge, MA: Harvard University Press, 1983).

11. For an incisive take on this phenomenon, see Wang Hui (Howard Y. F. Choy, trans.), "The Fate of 'Mr. Science' in China": The Concept of Science and Its Application in Modern Chinese Thought," *positions* 3, no. 1 (Spring 1995): 1–68.

12. In his broad overview of the environmental sciences, Peter Bowler notes that "geological theorizing took on a distinctly nationalistic flavour, as the scientists of each country sought to explain the phenomena that were most obvious in their local territory." While the reference is to theorizing that asserts a unique national identity through regional features, the conflation of science and nationalism is regularly accomplished through the assumption that the territory of a nation is in some senses the substance of the nation itself. See Peter Bowler, *The Norton History of the Environmental Sciences* (New York: W.W. Norton & Co., 1993), 231. For an example of the idea of the land *qua* nation interpreted through an explicit body metaphor (i.e., "geo-body"), see Thongchai Winichakul, *Siam Mapped: A History of the Geo-Body of a Nation* (Honolulu: University of Hawaii Press, 1994). Though many groups make rhetorical use of body metaphors, few have deployed them as persistently, or with as much self-conscious force, as power holders and nationalists in Siam/Thailand.

13. The history of geology is a thriving subfield of the history of science in China, and it is pursued by both professional scholars and a broad range of individuals associated with contemporary geology in China. I will only list a representative few as either points of entry or useful syntheses. For insight into institutional achievements, see Yang Xiaoming 杨小明 and Li Qiang 李强, "1912–1937 nian Zhongguo dizhi kexue zhishi zengzhang de jianzhihua yanjiu" 1912–1937 年中国地质科学知识增长的建制化研究 (Study of the structural organization of the proliferation of Chinese geological knowledge), *Zhongguo kejishi zazhi* 中国科技史杂志 (The Chinese journal for the history of science and technology) 31, no. 2 (2010): 165–75; Wang Yangzhi 王仰之, *Zhongguo dizhi diaochasuo shi* 中国地质调查所史 (History of the Geological Survey of China) (Beijing: Shiyou gongye chubanshe, 1996); Xia Xiangrong 夏湘蓉 and Wang Genyuan 王根元, *Zhongguo dizhi xuehui shi* 中国地质学会史 (History of the Geological Society of China) (Beijing: Dizhi chubanshe, 1982). For an analysis of the development of the geological community, see Zhang Jiuchen 张九辰, *Dizhixue yu minguo shehui: 1916–1950* 地质学与民国社会: 1916–1950 (Geology and Republican society: 1916–1950) (Jinan, China: Shandong jiaoyu chubanshe, 2005). For an example of Republican geology understood as the early stage of geology in the People's Republic, see Wang Hongzhen et al., "Development of Geological Science in China in Twentieth Century," *Journal of China University of Geosciences* 11, no. 3 (September 2000): 197–203. Two

important works by nonmainland authors are Tsui-hua Yang Lee's "Geological Sciences in Republican China, 1912–1937" (PhD diss., State University of New York at Buffalo, 1985), which ends at the War of Resistance; and Liu Shaomin 劉昭民, *Zhonghua dizhixue shi* 中華地質學史 (History of Chinese geology) (Taibei: Taiwan shangwu yinshu guan, 1985), which locates modern Chinese geology in both native and foreign geological traditions.

14. For an explicit example that uses George Basalla's model of diffusion, see Yang Jingyi 杨静一, "Pangpeili yu jindai dizhixue zai Zhongguo de chuanru" 庞佩利与近代地质学在中国的传入 (Pumpelly and the introduction of modern geology to China) *Zhongguo keji shiliao* 中国科技史料 (Chinese historical materials of science and technology) 17, no. 3 (1996): 18–27.
15. See Henrika Kuklick and Robert Kohler, "Science in the Field," *Osiris*, 2nd ser., vol. 11 (1996) for an excellent introduction to particular challenges of outdoor scientific environments.
16. Mott Greene, *Geology in the Nineteenth Century* (Ithaca, NY: Cornell University Press, 1982), 290.
17. Dajian Zhu and Homer Le Grand, "Plates, Politics and Localism: Geological Theory in China," *History and Anthropology* 11, nos. 2–3 (1999): 292.
18. For some of the most influential examples, see Martin J. S. Rudwick, *The Great Devonian Controversy: The Shaping of Scientific Knowledge among Gentlemanly Specialists* (Chicago: University of Chicago Press, 1985); James A. Secord, *Controversy in Victorian Geology: The Cambrian-Silurian Debate* (Princeton, NJ: Princeton University Press, 1986); A. Hallam, *Great Geological Controversies* (Oxford: Oxford University Press, 1989); David Oldroyd, *The Highlands Controversy: Constructing Geological Knowledge through Fieldwork in Nineteenth-Century Britain* (Chicago: University of Chicago Press, 1990).
19. For an interesting treatment of continental drift theories and plate tectonics in China, see Jing Yi Yang and David Oldroyd, "The Introduction and Development of Continental Drift Theory and Plate Tectonics in China: A Case Study in the Transference of Scientific Ideas from West to East," *Annals of Science* 46 (1989): 21–43.
20. For the classic model of science diffusion in colonial contexts, see George Basalla, "The Spread of Western Science," *Science* 156 (May 5, 1987): 611–22. Most modern discussions of colonial science have emphasized the interactions between native and foreign agents as critical to the movement of science, but many implicitly retain the idea of transmission.
21. See Benjamin Elman, "China and the World History of Science, 1450–1770," *Education about Asia* 12, no. 1 (Spring 2007): 40–44.
22. Naomi Oreskes, *The Rejection of Continental Drift* (New York: Oxford University Press, 1999), 314.
23. This tension is apparent in Ding Wenjiang's article on "The Responsibility of Chinese Geologists," where he talks of science as borderless and nationless, and yet stresses the proud and weighty responsibility of Chinese geologists to contribute research on the peculiar features of their own territory. See Ding Wenjiang 丁文江, "Zhongguo dizhi xuezhe de zeren" 中國地質學者的責任 (The responsibility of Chinese geologists), *Guoli Beijing daxue dizhi xuehui huikan* 國立北京大學地質學會會刊 (Bulletin of the Geological Society of the National University, Peking), no. 5 (April 1931), 1–13. In "The Integrative Revolution: Primordial Sentiments and Civil Politics in the New States," Clifford Geertz describes a related phenomenon ("the desire to be recognized as responsible agents whose wishes, acts, hopes, and opinions 'matter,' and

the desire to build an efficient, dynamic, modern state") as "powerful, thoroughly interdependent, yet distinct and actually opposed motives." See *Old Societies and New States* (New York: Free Press, 1963), 108.

24. For more on this aspect of the international dynamic, see chap. 3.
25. See for example, Anthony D. Smith, *Nationalism and Modernism: A Critical Survey of Recent Theories in Nations and Nationalism* (New York: Routledge, 1998), 191.
26. When the problem has been perceived by theorists of Western nationalism, it has been dealt with as "proto-national," "primordial," or simply an unexplained exception.
27. Ernest Gellner, *Nations and Nationalism* (Ithaca, NY: Cornell University Press, 1983).
28. A great deal of research has been done on the relative successes (and overall sincerity) of this project, which used to be understood under the rubric of "Sinicization." However, the persistence of this long-standing state cosmology is not generally contested, and it was the explicit basis of both the domestic governance and tribute relations for over two millennia.
29. Benedict Anderson, *Imagined Communities: Reflections on the Origins and Spread of Nationalism* (London: Verso, 1991).
30. See Kai-Wing Chow, *Publishing, Culture, and Power in Early Modern China* (Stanford, CA: Stanford University Press, 2004); and Benjamin Elman, *A Cultural History of Civil Examinations in Late Imperial China* (Berkeley: University of California Press, 2000) for more on the examination system since the Tang dynasty.
31. For more, see R. Bin Wong, "Two Kinds of Nation, What Kind of State?," in *Nation Work: Asian Elites and National Identities*, ed. Timothy Brook and Andre Schmid (Ann Arbor: University of Michigan Press, 2000), 109–23.
32. In 1984, Paul Cohen led an important critique of the "impact-response" approach centered on its assumption of China as passive and reactive, and its corollary that the only important stories are those that concern adjustment to the Western "impact." He championed a "China-centered" approach that has since been very popular but has recently come under fire for still assuming a clear-cut China-West dichotomy. See Paul Cohen, *Discovering History in China: American Historical Writing on the Recent Chinese Past* (New York: Columbia University Press, 1984). Without going into the arguments for both sides (or the recent reaction), my point here is simply that there are crucial ways in which Chinese historical actors framed their experiences in terms of impact, shock, crisis, and threat that cannot be swept under the table for fear of Eurocentrism. The key is in the handling of the question.
33. Joshua Fogel does well to remind us that "China" and "nation" can never be uncritically read into history, even when the terms present themselves in primary texts. However, most of the individuals involved in the development of modern Chinese geology were participants in the nationalizing project even before the founding of the republic. As I will elaborate later, my usage of these two terms intentionally allows my actors to slip across multiple meanings, so that we can follow them as they move between choices about science and nation. Elsewhere I have opted to use "China" and "Chinese" to describe historical, territorial, and ethnic subjects without encumbering my prose with unwieldy disclaimers. References in passing should be understood as conventional expedients (tacitly prefaced with "what is now known as" or "which corresponds to the modern"), rather than reality claims about the primordial nature of the modern nation-state. Joshua Fogel, "Introduction," in *The Teleology of the Modern Nation-State* (Philadelphia: University of Pennsylvania Press, 2005), 1–7.

34. William Kirby, "When Did China Become China? Thoughts on the Twentieth Century," in *The Teleology of the Modern Nation-State*, ed. Joshua Fogel (Philadelphia: University of Pennsylvania Press, 2005), 105–14. For a rich perspective on this process, see Lydia H. Liu, *The Clash of Empires: The Invention of China in Modern World Making* (Cambridge, MA: Harvard University Press, 2004).
35. John Fitzgerald, "The Nationless State: The Search for a Nation in Modern Chinese Nationalism," in *Chinese Nationalism*, ed. Jonathan Unger (Armonk, NY: M. E. Sharpe, 1996), 57.
36. In Fitzgerald's view, this "floating referent" masked "the ideal of a unitary state" and therefore suggests that the Chinese nation cannot exist outside of the state framework ("Nationless State," 57, 58).
37. Prasenjit Duara, *Rescuing History from the Nation: Questioning Narratives of Modern China* (Chicago: University of Chicago Press, 1995). See pp. 7 and 27–28 for direct explanations of these points, which are amplified throughout the book and its companion essays, "Historicizing National Identity, or Who Imagines What and When," in *Becoming National*, ed. Geoff Eley and Ronald Grigor Suny (New York: Oxford University Press, 1996), 151–70; and "De-Constructing the Chinese Nation," in *Chinese Nationalism*, ed. Jonathan Unger (Armonk, NY: M. E. Sharpe, 1996), 31–55.
38. Emily Honig, *Creating Chinese Ethnicity: Subei People in Shanghai, 1850–1980* (New Haven, CT: Yale University Press, 1992).
39. Lucian Pye, *The Spirit of Chinese Politics* (Cambridge, MA: MIT Press, 1968), 5.
40. In fact, James Townsend refers to a "long wrenching 'identity crisis' [that] makes contemporary Chinese nationalism unusually intense." The idea of China's struggles with identity is far more common than Pye's notion that "the Chinese have been generally spared the crises of identity common to most other transitional systems." However, Townsend also states that "the resolution of the [identity] crisis [became] something like the religion of modern China," which points, in its own way, to the strength of "China" as an object of loyalty. Townsend, "Chinese Nationalism," in Unger, ed., *Chinese Nationalism*, 5; and Pye, *Spirit of Chinese Politics*, 5.
41. Pye, *Spirit of Chinese Politics*, 5.
42. For examples of studies that express these ideas in other national contexts, see Christer Nordlund, "'On Going Up in the World': Nation, Region and the Land Elevation Debate in Sweden," *Annals of Science* 58 (2001): 17–50; and Oliver Zimmer, "In Search of Natural Identity: Alpine Landscape and the Reconstruction of the Swiss Nation," *Comparative Studies in Society and History* 40, no. 4 (October 1998): 637–65. A third form of geological nationalism that involves the authority to name and interpret as a function of political dynamics between center and periphery can been seen in William Eagan, "The Debate over the Canadian Shield, 1880–1905," *Isis* 80, no. 2 (June 1989): 232–53.
43. The significance of Republican geologists outside of scientific circles is quite remarkable and underscores the special character of geology among the various sciences that developed in early twentieth-century China. Ding Wenjiang was not only a famous polemicist, military strategist, and eugenicist, but he was also a member of China's unofficial delegation to the Versailles Conference and briefly the director-general of Shanghai. Li Siguang was the youngest member of Sun Zhongshan's Revolutionary Alliance and later, under the Communist regime, became a household name as one of the key figures recognized as both "Red" and "expert." Weng Wenhao was the architect behind Guomindang economic preparations for the War of Resistance and served as premier (or president of the Executive Yuan) of the Republic of China in

1948. See Charlotte Furth, *Ting Wen-chiang: Science and China's New Culture* (Cambridge, MA: Harvard University Press, 1970); Gu Xiaoshui 谷小水, *"Shaoshu ren" de zeren*『少數人』的責任 ("Responsibility of 'the few'") (Tianjin, China: Tianjin guji chubanshe, 2005); Li Siguang 李四光, *Chuanguo dipingxian* 穿过地平线 (Crossing the horizon) (Beijing: Baihuawen chubanshe, 1998); Li Xuetong 李学通, *Huanmie de meng—Weng Wenhao yu Zhongguo zaoqi gongyehua* 幻灭的梦—翁文灏与中国早期工业化 (Vanishing dream: Weng Wenhao and China's early industrialization) (Tianjin, China: Tianjin guji chubanshe, 2005).

44. Peter Perdue, "Where Do Incorrect Political Ideas Come From? Writing the History of the Qing Empire and the Chinese Nation," in Fogel, ed., *Teleology of the Modern Nation-State*, 185.

CHAPTER ONE

1. Jennings Mason Gentzler, *Changing China: Readings in the History of China from the Opium War to the Present* (New York: Praeger Publishers, 1977), 24–25.
2. Knowledge of mineral resources was frequently a focal point of Chinese struggles against Western mining interests, but as several careful studies have shown, what was appropriated was the "fact" of coal or iron, rather than an interest in the geological basis for this assessment. Other methods, involving international law, notions of citizenship and ownership, boycotts, and media campaigns, were also appropriated to the Chinese context, but despite the active participation of students with modern schooling, there is little record that individual protestors looked to mastering geology as a workable solution. For more on popular Chinese strategies for dealing with foreign resource extraction, see En-Han Lee, "China's Response to Foreign Investment in Her Mining Industry (1902–1911)," *Journal of Asian Studies* 28, no. 1 (November 1, 1968): 55–76; and Roger Thompson, "'If Shanxi's Coal Is Lost, Then Shanxi Is Lost!': Shanxi's Coal and an Emerging National Movement in Provincial China, 1898–1908," *Modern Asian Studies* 45, no. 5 (2011): 1261–88.
3. This understanding of China proper not as a spatially delineated region, but "the space of reference for Chinese culture and history" can also be found in Pamela Kyle Crossley, *The Wobbling Pivot, China Since 1800: An Interpretive History* (Malden, MA: John Wiley & Sons, 2010), 66.
4. The "new Qing History" developed in the 1990s and began with a reevaluation of the ethnic and cultural relations between Han Chinese, their Manchu conquerors, and other frontier peoples. It has highlighted the importance of Manchu archives, and spurred new approaches to Qing economic development, urban studies, popular uprisings, and rituals, among others. For a starting point, see Joanna Waley-Cohen, "The New Qing History," *Radical History Review* 88 (Winter 2004): 193–206. Recent responses from mainland China can be found in Ding Yizhuang, "Reflections on the 'New Qing History' School in the United States," *Chinese Studies in History* 43, no. 2 (Winter 2009/2010): 92–96; and Dang Wei 党为, *Jin sanshinian lai de meiguo qingshi yanjiu: yi xinqingshi wei xiansuo* 近三十年來的美国清史研究: 以新清史为线索 (Thirty years of American Qing history research: Taking the new Qing history as a thread) (PhD diss., Peking University, 2010).
5. Waley-Cohen, "New Qing History," 200.
6. Michel de Certeau, *The Practice of Everyday Life* (Berkeley: University of California Press, 1988). See in particular pt. 2, "Spatial Practices."
7. For those who did move across the empire, native place associations (*huiguan*, some-

times rendered as native place guilds or lodges) reinforced many of these ties. The classic work on these native place associations is He Bingdi 何炳棣, *Zhongguo huiguan shilun* 中國會館史論 (Historical essays on Chinese native place associations] (Taibei: Taiwan xuesheng shuju, 1966). Subsequently, an extensive literature has developed on the ways that native place associations not only reified difference, but also ultimately supported a new national sensibility in the late Qing and early Republican periods. See, for example, Bryna Goodman, *Native Place, City, and Nation: Regional Networks and Identities in Shanghai, 1853–1937* (Los Angeles: University of California Press, 1995); and Richard Belsky, *Localities at the Center: Native Place, Space, and Power in Late Imperial Beijing* (Cambridge, MA: Harvard University Asia Center, 2005). Though much of the work on *huiguan* is aimed at urban history and economic history, it highlights the same kind of reconceptualization of territorial thinking and self-other identification with place that I see in the emergence of Chinese interest in geology as a distinctive scientific discipline.

8. *Neidi* is often taken by Western authors to mean "China proper," but while this usage was common, *neidi* could also refer to areas inland from the coast and was therefore very contextually variable.
9. For more, see Honig, *Creating Chinese Ethnicity*.
10. Even for those elites who had access to maps as representations of the empire, the emotional resonance of cultural markers played a significant role in their understanding of available maps, and this reading from the familiar to the unfamiliar often privileged long-established spatial concepts over (sometimes unpleasant) political realities. For an analysis of this phenomenon from the Song dynasty, see Hilde De Weerdt, "Maps and Memory: Readings of Cartography in Twelfth- and Thirteenth-Century Song China," *Imago Mundi* 61, no. 2 (June 2009): 145–67.
11. See Richard Strassberg, *Inscribed Landscapes: Travel Writing from Imperial China* (Berkeley: University of California Press, 1994) for some notable examples.
12. Xu Xiake, who wrote during the late Ming, is best known for his travel diaries, which have appeared in several modern editions. For more information, see Julian Ward, *Xu Xiake (1587–1641): The Art of Travel Writing* (Richmond, Surrey: Curzon, 2001). See Xu Hongzu 徐宏祖, *Xu Xiake youji* 徐霞客遊記 (Travel diaries of Xu Xiake), ed. Ding Wenjiang 丁文江 (Shanghai: Shangwu yinshuju, 1928), for modern interpretations of his writings in geological and geographical terms.
13. Because of the relationship between gazetteers to Chinese administrative units and local Han elites, they were not produced for Inner Asian frontier regions. Geographical materials on these areas were kept in the Qing court and often held off-limits to even imperial historians, though general information can be found in official histories and several independent works by merchants and travelers are available.
14. There are several references on gazetteers in China. See, for example, Timothy Brook, *Geographical Sources of Ming-Qing History* (Ann Arbor: Center for Chinese Studies, University of Michigan, 2002); and Fu Zhenlun 傅振伦. *Zhongguo fangzhi xue* 中国方志学 (The study of Chinese local gazetteers) (Fuzhou: Fujiansheng difangzhi bianzuan weiyuanhui bangongshi, 1984).
15. Of approximately 8,000 extant pre-1911 gazetteers, over 5,600 are from the Qing.
16. There are many collections of Gu's writings, which include philosophy, geography, philology, economics, and statecraft, among other scholarly topics. For a quick reference to his life and times, see Willard J. Peterson, "The Life of Ku Yen-wu (1613–1682)," pts. 1 and 2, *Harvard Journal of Asiatic Studies* 28 (1968) and 29 (1969):

114–56, 201–47. A good introduction to some of the other Ming loyalist thinkers is Lynn Struve, "Huang Zongxi in Context: A Reappraisal of His Major Writings," *Journal of Asian Studies* 47, no. 3 (August 1988): 474–502.

17. Laura Hostetler, *Qing Colonial Enterprise: Ethnography and Cartography in Early Modern China* (Chicago: University of Chicago Press, 2001), 70.
18. In 1672 and 1729 each province was ordered to compile an official gazetteer. Endymion Wilkinson, *Chinese History: A Manual* (Cambridge, MA: Harvard University Press, 2000), 156. Work on the *Da Qing yitongzhi* commenced under the direction of Xu Qianxue in 1687. Joseph Needham, *Science and Civilisation in China* 3, sec. 22.6 (Cambridge: Cambridge University Press, 1959), 521.
19. These surveys involved eight Jesuit fathers, including Jean-Baptiste Régis, Joachim Bouvet, Dominique Parrenin, and Ehrenberg Xavier Fridelli. Cordell Yee points out that tests of the new (mostly French) Jesuit methods were conducted on small regions surrounding Beijing (including studies of the Great Wall and neighboring waterways) before the large surveys were commissioned.
20. These trigonometric surveys relied on determinations of latitude and longitude for thousands of points in China but ignored altitude and topography in the interests of speedy completion. In fact, the Jesuit survey was finished before the great mapping of France commissioned by Louis XIV under Colbert, and given the vast size of the Qing Empire, this is a reflection of the quantity of geographic data already available in gazetteers and local maps. The longevity of Jesuit maps in gazetteers is later a subject of criticism for Ding Wenjiang.
21. In her book *Qing Colonial Enterprise,* Laura Hostetler points out interesting subtleties to the influence of eighteenth-century Jesuit maps on Western understandings of China. In the most common European versions of the Jesuit maps, place names are indicated in Chinese throughout, suggesting total continuity across the empire and leading to an elision of Qing territory and China. An alternate and less widely available version of the map uses Chinese characters for place names in China proper and employs Manchu script to label all other territories, suggesting that China was perhaps merely one portion of the greater Qing Empire. The public functions of these maps are also evident in the restriction of Jesuit surveyors from extensive travel in regions bordering Russia (from whom great tracts of land in Siberia were gained in the 1689 Treaty of Nerchinsk). Since the maps were not only designed for distribution as gifts to other ruling houses but were also produced by foreigners, the mapping project sometimes traded detail for national security. Separate mapping projects for military use were conducted, and much of this data was kept from the Jesuit surveyors (74–79).
22. Mary Gertrude Mason, *Western Concepts of China and the Chinese, 1840–1876* (New York: [s.n.], 1939), 5–6.
23. An even earlier representation of China as part of Abraham Ortelius's *Theatrum Orbis Terrarum* was presented by Matteo Ricci to several Ming scholars and officials, and was ultimately used by the Wanli emperor as an imperial gift. Notably, Ricci recentered the map on Chinese territory, without changing the scale. In 1623 Giolio Aleni modified this map to include information from Chinese gazetteers. As Chinese geographic knowledge increased, Ferdinand Verbiest and later Michel Benoit made similar updates to this picture of the world. Joanna Waley-Cohen, *The Sextants of Beijing: Global Currents in Chinese History* (New York: W.W. Norton, 1999), 112–13.
24. The original *Atlas Sinensis* is essentially the Latin version of Luo Hongxian's 16th cen-

tury *Guang yu tu*. It was quickly was translated from Latin into Dutch, German, English, French, and Spanish as part of Joan Blaeu's *Atlas Major*, and it included both maps and description. It also greatly clarified the physical relation between Korea, Japan, and the Chinese mainland. See Benjamin Schmidt, "Mapping an Exotic World: The Global Project of Dutch Geography, circa 1700," in *The Global Eighteenth Century*, ed. Felicity Nussbaum (Baltimore, MD: Johns Hopkins University Press, 2003), 20–37.

25. Grueber and D'Orville were both trained as geographers. Michael Gorman argues that *China Illustrata* was part of Kircher's attempt at a "Consilium Geographicum," or Geographical Plan of the world using Jesuit networks and a "magnetic" approach to solving the problem of longitude calculations at sea.
26. Donald Lach, *Asia in the Making of Europe* (Chicago: University of Chicago, 1965), 1571–1619, 1634–54.
27. Obviously, praise was not unanimous (e.g., Montesquieu on despotism and decay), but it was easily the dominant position and there was in no sense any disagreement on the age and sophistication of China's civilization, even on the part of those who doubted that Confucianism created a society of true reason.
28. David Jones, *The Image of China in Western Social and Political Thought* (New York: Palgrave, 2001), 14–66.
29. For an interesting take on the ways that the notion of Chinese decadence took hold in Europe, see Larissa Heinrich's work on smallpox and visual representation *The Afterlife of Images: Translating the Pathological Body between China and the West* (Durham, NC: Duke University Press, 2008). Anthony Clark's *Beating Devils and Burning Their Books: Views of China, Japan, and the West* (Ann Arbor, MI: Association for Asian Studies, 2010) offers further insight by considering mutual constructions of difference that arose in the wake of eighteenth- and twentieth-century contacts.
30. Non-Westerners involved in the junk trade were given considerably more latitude, and the Portuguese maintained Macao as their base of operations. For more on the everyday lives of traders and administrators in the Canton system, see Paul Van Dyke, *The Canton Trade: Life and Enterprise on the China Coast, 1700–1845* (Hong Kong: Hong Kong University Press, 2007).
31. In the interests of space this treatment leaves out and thus unfortunately sanitizes, many of the unsavory details by which the treaty system was established in China. The system itself was ever-changing, involving multiple nations and negotiators, whose entangled agreements and strong-arm tactics affected the rights of all others through the most favored nation clause and independent European alliances. I omit details of many famous iniquities (such as the burning of the Summer Palace), the subtleties of Qing strategic assumptions, and also some of the main economic provisions (tariff, internal *likin*, foreign shipping rights, etc.), because my primary concern is territorial, but even on that count, there are many provisions (including Russian rights in Central Asia, etc.), which I have left out to streamline my point. However, study of the treaty system was the cornerstone for the development of China studies in the West and many works are available on all aspects of the subject, including the emerging role of Japan at the end of the nineteenth century. A classic example that remains a good starting point for understanding both the system and subsequent historiographical shifts is John King Fairbank, *Trade and Diplomacy on the China Coast: The Opening of the Treaty Ports, 1842–1854* (Cambridge, MA: Harvard University Press, 1953). For insight into the ways that perceptions of the treaty

system have continued to shape Chinese domestic and international politics, see Zheng Yangwen, "The 'Peaceful Rise of China' after the 'Century of Unequal Treaties': Will History Matter?" in *Negotiating Asymmetry: China's Place in Asia*, ed. Anthony Reid and Zheng Yangwen (Honolulu: University of Hawaii Press, 2009), 159–91.

32. Evariste Régis Huc, *Souvenirs d'un voyage dans la Tartarie, le Thibet, et la Chine pendant les années 1844, 1845 et 1846* (Paris: Librairie d'Adrien le Clerc, 1850). In particular, Huc's account, which was widely translated, lured the geologist Raphael Pumpelly to China in 1863.
33. For excellent analyses of travelers' accounts of China in this period, see Sybille Fritzsche, "Narrating China: Western Travelers in the Middle Kingdom After the Opium War" (PhD diss., University of Chicago, 1995); and Nicholas Clifford, *"A Truthful Impression of the Country": British and American Travel Writing in China 1880–1949* (Ann Arbor, MI: University of Michigan Press, 2001.
34. For example, see the accounts of Isabella Bird (Bishop) or William Edgar Geil in Frances Markley Roberts, *Western Travelers to China* (Hong Kong: Kelly & Walsh Ltd, 1932). According to Herbert Hoover, foreigners eager to get in on the China boom had about "ten days" to write their China story if they did not want to be scooped. *The Memoirs of Herbert Hoover: Years of Adventure, 1874–1920*, vol. 1 (New York: Macmillan, 1951).
35. Clifford, *Truthful Impression of the Country*, 35.
36. Two useful examples of geological writings by missionaries are Rev. J. Edney, "The Bituminous Coal Mines West of Peking," *Journal of the North China Branch of the Royal Asiatic Society* (Shanghai) 5 (1867): 243–50; and W. A. P. Martin, "Notes on China, Geological and Geographical," *American Journal of Science* 47 (1869): 98–103. It is interesting to note the ways in which Sinology and natural studies were thus intertwined in the writings of missionaries, businessmen, and travelers. For more, see Fa-ti Fan, "Hybrid Discourse and Textual Practice: Sinology and Natural History in the Nineteenth Century," *Historia Scientiarum* 38, pt. 1 (March 2000), 25–56.
37. This encouraged the rapid growth of paleontology and helped build the geohistorical approach detailed in Martin Rudwick's *Bursting the Limits of Time: The Reconstruction of Geohistory in the Age of Revolution* (Chicago: University of Chicago Press, 2005) and *Worlds before Adam: The Reconstruction of Geohistory in the Age of Reform* (Chicago: University of Chicago Press, 2008).
38. Raphael Pumpelly, *Across America and Asia* (New York: Leopold & Holt, 1870), 194. For a good overview of Pumpelly's work in Japan, see Kenneth R. Aalto, "American Contributions to the Geological Mapping of Hokkaido, Late Nineteenth Century," *Earth Sciences History* 30, no. 1 (2011): 1–19.
39. Badong is now in Hubei Province, but its Chinese name indicates its historical connection to Sichuan.
40. See notes cited in David Oldroyd and Yang Jingyi, "On Being the First Western Geologist in China: The Work of Raphael Pumpelly," *Annals of Science* 53 (1996): 107–36.
41. Pumpelly correctly observed what is now known as the Huang Ling anticline in the Three Gorges region but presently the limestones he dated as Devonian are understood to be Cambrian and the coal measures he dated as Triassic are now attributed to the Jurassic.
42. Tao Shilong 陶世龙, "Cong Pangpeile dao Weilishi" 从庞培勒到维里士 (From Pumpelly to Willis) *Dizhixue shi luncong* 地质学史论丛 (Compendium of the history of geology) 3 (1995), 16. For a balanced study of the Lay-Osborn flotilla episode, see

Jack Gershon, *Horatio Nelson Lay and Sino-British Relations, 1854–1864* (Cambridge, MA: Harvard University Press, 1972).

43. Raphael Pumpelly, "Report of the Chinese Government on a Preliminary Examination of the Coal Districts of the Si Shan, in the Province of Chili," *Papers Relating to Foreign Affairs, Accompanying the Annual Message of the President of the Second Session Thirty-Eighth Congress. 1864 Part III* (Washington, 1865): 363–68.
44. Bailey Willis, "Memorial of Raphael Pumpelly," *Bulletin of the Geological Society of America* 36 (1925), 63. Pumpelly was also fond of going to Chinese lapidary shops in search of mineral or fossil specimens.
45. Pumpelly returned to Central Asia in 1903–4 with his son and the geographer Ellsworth Huntington. While this trip did produce several papers on geological topics, the Carnegie Institution funded it primarily to explore the hypothesis that human civilization developed in Asia. For more, see Raphael Pumpelly, *Explorations in Turkestan; Expedition of 1904* (Washington, DC: Carnegie Institution of Washington, 1908).
46. Raphael Pumpelly, *Geological Researches in China, Mongolia, and Japan* (Washington, DC: Smithsonian Institution, 1866), 67–69.
47. See Oldroyd and Yang for an informative analysis of weaknesses in Pumpelly's fieldwork and interpretation, but note the sheer joy with which Pumpelly described the impact of his brief Yangzi tour: "This great anticlinal mountain axis had given me the key to the structure not only of China but of eastern Asia as well." Raphael Pumpelly, *My Reminiscences*, vol. 1 (New York: H. Holt, 1918), 381. As for his willingness to revise this ideas, by 1879 Pumpelly was so fully convinced of Richthofen's aeolian loess theory he was publishing papers to support it. "Richthofen's 'China,'" *The Nation*, April 4, 1878, pp. 231–32; Raphael Pumpelly, "The Relation of Secular Rock Disintegration to Loess, Glacial Drift and Rock Basins," *American Journal of Science* 17 (1879): 133–44.
48. Peggy Champlin, *Raphael Pumpelly: Gentleman Geologist of the Gilded Age* (Tuscaloosa, AL: University of Alabama Press, 1994), 72–73. Ferdinand von Richthofen, *Ferdinand Freiherr von Richthofen: Tagebücher aus China* (Berlin: Dietrich Reimer [Ernst Vohnsen], 1907), 1:1.
49. Bailey Willis, "Preface," in *Research in China*, vol. 1, pt. 1, (Washington, DC: Carnegie Institution, 1907), xi. This view was commonly expressed by Republican-era Chinese geologists as well, albeit with more consideration of Richthofen's geopolitical impact.
50. John Schrecker, *Imperialism and Chinese Nationalism: Germany in Shantung* (Cambridge, MA: Harvard University Press, 1971), 4–5.
51. E. G. Ravenstein, "Obituary: Ferdinand Freiherr von Richthofen," *Geographical Journal* 26, no. 6 (December 1905): 679–80.
52. For example, see Ferdinand von Richthofen, "The Natural System of Volcanic Rocks," *Memoir California Academy Sciences* 1, pt. 2 (1868): 1–94.
53. Earlier Richthofen had used his mining connections to lobby the California Legislature to retain the Office of the State Geologist, which remained an insecure post until 1880, when a State Mining Bureau was finally established. Champlin, *Raphael Pumpelly*, 72–73.
54. Almost immediately, Richthofen found fossil evidence that contradicted Pumpelly's assumption that China's limestone formations were largely Triassic and consistently associated with overlying coal measures. "First Preliminary Notes of Geological Explorations in China (letter addressed to Prof Whitney, Shanghai, March 1, 1969)."

Proceedings of the Academy of Arts and Sciences 8 (1873), 12; Ferdinand Richthofen, "On the Existence of the Nummulitic Formation in China," *American Journal of Science* 3rd ser. 1 (1871): 110–13. Throughout his explorations in China, Richthofen sent short reports of his initial findings to German geological and geographical journals.

55. Richthofen, *Tagebücher*, 1:29.
56. Richthofen was the first to systematically study China clay, and he named "kaolin" after the Chinese "kaoliang." Ferdinand von Richthofen, "On the Porcelain Rocks of China," *American Journal of Science* 1, 3rd ser. (1871): 179–81.
57. Richthofen, *Tagebücher*, 1:141, translation after Jürgen Osterhammel, "Forschungsreise und Kolonialprogramm: Ferdinand von Richthofen und die Erschießung Chinas im 19. Jahrhundert," *Archiv für Kulturgeschichte* 69 (1987), 171.
58. Ferdinand von Richthofen, *Baron Richthofen's Letters: 1870–1872* (Shanghai: North China Herald, n.d.), 5, 29. These letters, written as they were in English for an audience with direct business interests, remained the most widely referenced of Richthofen's works in China, even after the first volumes of his grand compendium *China* began to appear in 1877. For more, see Guo Shuanglin 郭双林 and Dong Xi 董习, "Lixihuofen yu *Lixihuofen nanjue shuxin ji*" "李希霍芬与《李希霍芬男爵书信集》" (Richthofen and *Baron von Richthofen's Letters*) *Shixue yuekan* 史学月刊 (Journal of historical science) 11 (2009): 52–60.
59. The metalliferous deposits of Yunnan, especially tin and copper, were quite well known through small-scale native mining, and the Gansu Corridor was the main link between China proper and Central Asia. Besides the novelty of exploring such remote regions to the north- and southwest of China, Richthofen hoped to trace part of Marco Polo's famous path through these two provinces. Instead, he communicated reports of these areas (and Guizhou) that he gathered from informants in Shaanxi and Sichuan.
60. Richthofen received funding from both the Ministry of Commerce and the Ministry of Culture. *China: Ergebnisse eigener Reisen und darauf gegründeter Studien* (Berlin: Reimer, 1877), 1:xi.
61. Ravenstein, "Obituary," 681. Richthofen moved to the University of Berlin in 1886 and remained there until his death in 1905.
62. Richthofen, *China*, 1:xxx.
63. *China* also had a marked influence on Eduard Suess's *Das Antlitz der Erde* (Prague: F. Tempsky, 1885–1909), who used Richthofen's ideas about the Kunlun mountains to support his tectonic theories.
64. Richthofen remarked in his *Letters* and his notes on Henry Yule's edition of *The Travels of Marco Polo* that it was remarkable that Marco Polo never commented on the loess of north China on his journey and some scholars have added this to their evidence that Marco Polo never actually reached "Cathay" himself.
65. For discussion of loess as a characteristic feature of Chinese geology, see Ding Wenjiang, "Zhongguo dizhi xuezhe de zeren" 中國地質學者的責任 (The responsibility of Chinese geologists), *Guoli Beijing daxue dizhi xuehui huikan* 5 (1931), 1–13.
66. Richthofen includes a discussion of some early objections to his theory of loess in the appendix of his *Letters*, and several other detractors remained until the early twentieth century. Though distinctions have been made among different loess formations, a few of which are pseudoloess or regenerated loess and may be aqueous sediment, the massive loess formations of China are now generally recognized as aeolian in origin.

67. Ferdinand Richthofen, *China*, 1:22, trans. follows Amadeus Grabau, "The Sinian System," *BGSC* 1, nos. 1–4 (1922), 66.
68. Ferdinand Richthofen, *China*, 2:73, for a succinct definition, but see also vol. 3, footnote, p. 148 (dated 1905) and scattered references for further elaborations. The definition of the Sinian, as a "system" and as a "period" changed markedly from Pumpelly and Richthofen's early usage and is still contested today, though it is generally accepted as Precambrian, approximately 1,000 to 500 million years old.
69. In 1933, director of the Geological Survey of China, Weng Wenhao, stated that "Richthofen's work has saved ten years of labour for Chinese geologists." Wong Wen-hao, "Richthofen and Geological Work in China," *BGSC* 12 (1933), 313. Indeed, the first Chinese students of geology in 1916 had to learn German in order to understand Richthofen's *China*, and they referenced it throughout their field reports.
70. For interesting details on arguments against Li's Kaiping venture because of their proximity to imperial tombs, see Ellsworth Carlson, *The Kaiping Mines (1877–1912)* (Cambridge, MA: Harvard University Press, 1957), 16.
71. George Steinmetz, "'The Devil's Handwriting': Precolonial Discourse, Ethnographic Acuity, and Cross-Identification in German Colonialism," *Comparative Studies in Society and History* 45, no. 1 (May 2003): 76. See also Jürgen Osterhammel and Ute Wardenga, "Ferdinand von Richthofen Als Erforscher Chinas: Ein Beitrag Zur Entstehung Und Verarbeitung Von Reisebeobachtungen Im Zeitalter Des Imperialismus," *Berichte zur Wissenschaftsgeschichte* 13, no. 3 (1990): 141–55.
72. *German Diplomatic Documents, 1871–1914*, trans. E. T. S. Dugdale, vol. 3, "The Growing Antagonism, 1898–1910," (New York: Harper & Brothers, 1930), 6; Helmuth Stoecker, *Deutschland und China um 19. Jahrhundert* (Berlin: Rutten & Loening, 1958), p. 72, as cited in Schrecker, *Imperialism and Chinese Nationalism*, 5.
73. Ibid., 16–17.
74. Ibid., 17–18.
75. See Benjamin Elman, "Naval Warfare and the Refraction of China's Self-Strengthening Reforms into Scientific and Technological Failure," *Modern Asian Studies* 38, no. 2 (2003): 283–326, for a discussion of the historicization of China's preparation and defeat.
76. Russia had originally enlisted the help of Great Britain, but Russian maneuvering in China in this period is complicated by its dual interests in northeastern China (Manchuria) and in far western Central Asia where it was engaged in the shadowy "Great Game" against Britain through the late nineteenth and early twentieth centuries.
77. Germany's failure to immediately gain Jiaozhou was partially due to China's post-Sino-Japanese War policy of courting Russian interests, but the two concessions granted Germany were still sizable. William Kirby, *Germany and Republican China* (Stanford, CA: Stanford University Press, 1984), 10; Schrecker, *Imperialism and Chinese Nationalism*, 48–49.
78. As with all of China's foreign relations from the Opium War until the War of Resistance, both the tripartite intervention and Germany's occupation of Jiaozhou involved complex power relations between multiple European nations and Japan, which are too involved to detail here. Suffice it to say that the global balance of power established in China was often more critical to foreign governments than any specific gains vis-à-vis China itself.
79. Alfred von Tirpitz, *My Memoirs*, vol. 2 (New York: Dodd, Mead & Co., 1919), 92. Richthofen's influence was not only felt through his published research but also his active participation in the *Kolonialrat*, where he not only argued for Jiaozhou but

against earlier candidates such as Formosa, the Pescadores, Mirs Bay (near Hong Kong), and the islands off of Amoy (Xiamen). Steinmetz, "'Devil's Handwriting,'" 45; Dugdale, trans., *German Diplomatic Documents*, 6.

80. The concessions also gave Germany the right to operate mines within ten miles of the railways they built.

81. In fact, more detailed surveys revealed that many of these concessions were less productive than originally assumed, especially when extraction and transportation costs were considered. Shellen Wu has a illuminating discussion of the ways that the development of foreign mining concessions, right recovery efforts, and international law were linked together. See especially chap. 5, "Nations, Empires and Mining Rights (1895–1911)," in Shellen Xiao Wu, "Underground Empires: German Imperialism and the Introduction of Geology in China, 1860–1919" (PhD diss., Princeton University, 2010).

82. One consequence of this narrow-mindedness is the legacy of foreign-run rail lines in China. Many of these were built for strategic military purposes or for commercial gain within a single sphere of influence, and do not make any sense from the perspective of national topography, resource distribution, or trading centers. For details of railroad and mining development after the Sino-Japanese War, see Cheng Lin, *The Chinese Railways, Past and Present* (Shanghai: United China Press, 1937); Chi-ming Hou, *Foreign Investment and Economic Development in China, 1840–1937* (Cambridge, MA: Harvard University Press, 1965); Tim Wright, *Coal Mining in China's Economy and Society, 1895–1937* (Cambridge: Cambridge University Press, 1984); Clarence Davis, "Railway Imperialism in China, 1895–1939," in *Railway Imperialism*, ed. Clarence B. Davis, Kenneth E. Wilburn, and Ronald E. Robinson (New York: Greenwood, 1991), 155–73. For Qing policies, a good overview is Wang Shanshan 王珊珊, "Wanqing zhengfu tielu sixiang bianqian yanjiu" 晚清政府铁路思想变迁研究 (Research on the development of official late Qing railway thinking) (MA thesis, Shandong University, 2008).

83. The Carnegie Expedition modified its route plan to avoid both German and British military reconnaissance teams, neither of which made their findings public. However, Willis was able to consult some topographical military maps while in Beijing.

84. The German ambassador to China was particularly suspicious of Willis, whose route passed through Shandong Province. Andrew Carnegie did not, however, have any personal business interests in China and very little in Bailey Willis's research indicates an interest in mining. A. W. Grabau, "The First Year of the Third Asiatic Expedition's Activities: Report of Last Friday's Meeting," *Peking Leader*, Tuesday, October 3, 1922, p. 2; David Oldroyd and Yang Jing-yi, "Bailey Willis (1857–1949): Geological Theorizing and Chinese Geology," *Annals of Science* 60 (2003): 9.

85. The sudden appearance of trilobites in the evolutionary record was an important puzzle and then Secretary of the Carnegie Institution, Charles Walcott, was America's leading authority on trilobites and Cambrian fauna. Bailey Willis et al., *Research in China*, vol. 1, pt. 1, *Descriptive Topography and Geology* (Washington, DC: Carnegie Institute of Washington, 1907), xi; Oldroyd and Yang, "Bailey Willis," 4; Ellis Yochelson, *Charles Doolittle Walcott, Paleontologist* (Kent, OH: Kent State University Press, 1998).

86. Bailey Willis, *Friendly China: Two Thousand Miles Afoot Among the Chinese* (Stanford, CA: Stanford University Press, 1949), 21.

87. Willis uses the term "Adam Trilobite" to describe his interest in Precambrian fauna in

his popular travel account *Friendly China*. There he also describes his traveling companions, who included an interpreter, Li San, and several Chinese coolies and servants. Distances were generally traversed by cart and boat where possible, but actual investigations were made on foot. Sargeant notes that one of his "native assistants" quickly learned to set up surveying equipment and take rudimentary measurements. These assistants accompanied the American research staff on their field excursions, while the rest of the crew prepared camp or traveled by cart.

88. Willis, *Friendly China*, 127.

89. However, vol. 3 of *Research in China* was devoted to the rich paleontological material that they did find. The expedition's negative result on trilobite ancestry contributed to Walcott's notion of a worldwide Lipalian Interval (i.e., "the era of unknown marine sedimentation between the adjustment of pelagic life to littoral conditions and the appearance of the Lower Cambrian fauna." Charles Walcott ("Cambrian Geology and Paleontology," *Smithsonian Miscellaneous Collections* 57 [1914]: 14) to account for the apparent "explosion" of fauna at the Cambrian boundary.

90. Oldroyd and Yang provide a detailed analysis of the problems with Willis's tectonic theory but point out that he was fond of using "multiple working hypotheses" and very comfortable proposing and discarding ideas as he considered new data. Oldroyd and Yang, "Bailey Willis," 16–23.

91. A good discussion of the term can be found in Gao Zhenxi 高振西, "'Sinian' zhi yiyi zai Zhongguo dizhixue shang zhi bianqian" 'Sinian' 之意義在中國地質學上之變遷 ("Variations of the Term 'Sinian' as Used in Chinese Geology"), *Guoli Beijing daxue dizhi xuehui huikan* 4 (1930): 59–67.

92. This text was provided by the Chinese minister to the United States. Willis, "Preface," in *Research in China* 1, pt. 1, xiv. In *Friendly China* Willis also provides several examples of his attempts to speak "pidgin Chinese" to his Chinese assistants.

93. Joseph Barrell, "Review of *Research in China*" *Science* 29, no. 737 (February 12, 1909): 257.

94. Wang Lixin (王立新) refers to this trend as *kaiyan kan shijie* (開眼看世界, "opening one's eyes to observe the world") in "Meiguo chuanjiaoshi yu yapian zhanzheng hou de 'kaiyan kan shijie' sichao" 美国传教士与鸦片战争后的'开眼看世界'思潮 ("American Missionaries and Post-Opium War 'Kaiyan kan shijie' Thought"), *Meiguo yanjiu* 美国研究 (American studies quarterly), 11, no. 2 (1997): 27–51.

95. *Haiguo tuzhi* was officially published in 1844 with 50 fascicles, and subsequent printings expanded substantially, reaching 100 fascicles by 1852. The complete 10 fascicle version of *Yinghuan zhilue* was finished in 1848, published in 1849, and reprinted several times in China and Japan. See chap. 5 of Wataru Masuda, *Japan and China: Mutual Representations in the Modern Era* (New York: St Martin's Press, 2000) for more. For more information on these two gazetteers, see Jane Kate Leonard, *Wei Yuan and China's Rediscovery of the Maritime World* (Cambridge, MA: Harvard University Press, 1984); and Fred Drake, *China Charts the World: Hsü Chi-yü and his Geography of 1848* (Cambridge, MA: Harvard University Press, 1975). Leonard makes the important point that Wei Yuan's interest in foreign nations does not have to be viewed within a modernization or Westernization framework and can be understood within established traditions of statecraft and Chinese intellectual discourse. However, he was still bringing foreign information to bear on these issues and reinterpreting China's relationship to the rest of the world.

96. Wang Lixin, "Meiguo chuanjiaoshi yu yapian," 28–29, 33–34. Wei showed particular

interest in the political and social systems of foreign countries and traced the ways commercial interests vying for power in China had previously dealt with nations in the South Pacific.

97. Wei Yuan 魏源, *Haiguo tuzhi* 海国圖志 (Illustrated gazetteer of the maritime states), bk. 4, leaf 43.

98. Xu Jiyu 徐繼畬, "Xu Jiyu shi" 徐繼畬識 (Xu Jiyu), *Yinghuan zhilue* 瀛環志略 (A brief account of the maritime circuit) ([n.p.]: Shan yun lou, 1866), 8. David Abeel was his primary informant, but he also consulted prominent figures such as Rutherford Alcock and Robert Morison. For more see Drake, *China Charts the World*; and Yao Zhenli 姚振黎, "*Yinghuan zhilue* yanjiu" 《瀛環志略》研究 (Research on *Yinghuan zhilue*) in *Diqijie qingdai xueshu yantaohui lunwenji* 第七屆清代學術研討會論文集 (Collected papers of the Seventh Qing Dynasty Academic Symposium) (Gaoxiong, Taiwan: National Zhongshan University Chinese Literature Department, 2002), 71–100.

99. For example, Rebecca Karl points out that though ideas such as *yaxiya* (亞細亞, Asia) crop up in Wei and Xu's geographies, they did not yet fit within the larger epistemological framework of elite readers and had little impact at the time. See Rebecca E. Karl, *Staging the World: Chinese Nationalism at the Turn of the Twentieth Century* (Durham, NC: Duke University Press, 2002), 155–56.

100. Though the Self-Strengthening movement was significantly more nuanced in philosophy and application, its general outlook is usually associated with Zhang Zidong's dictum *zhongxue weiti xixue weiyong* (中學為体，西學為用), or "Chinese learning as substance, Western learning for utility." Later Self-Strengtheners increasingly supported nontechnical reforms in education, bureaucratic organization, and economic policy, but all did so in defense of some idealized Chinese essence that they felt was irreducible. A wealth of literature exists on various aspects of the Self-Strengthening movement (or Foreign Affairs movement). For a few starting points, see Kim Kwan Ho, *Japanese Perspectives on China's Early Modernization: The Self-Strengthening Movement, 1860–1895* (Ann Arbor, MI: Center for Chinese Studies, University of Michigan, 1974); Lai Chi-kong, *Recent Historiography on the Self-Strengthening Movement* (Davis, CA: Institute of Governmental Affairs, UC Davis, 1989); and Zhou Jianbo 周建波, *Yangwu yundong yu Zhongguo zaoqi xiandaihua sixiang* 洋務運動與中國早期現代化思想 (The Foreign Affairs movement and early modernization thought in China) (Jinan, China: Shandong Renmin Chubanshe, 2001).

101. For more information on Self-Strengthening translation projects, see Adrian Arthur Bennett, *John Fryer: The Introduction of Western Science and Technology into Nineteenth-Century China* (Cambridge, MA: Harvard University Press, 1967); Tsuen-hsuin Tsien, "Western Impact on China through Translation," *Far Eastern Quarterly* 13, no. 305 (1997): 305–27; David Wright, *Translating Science: The Transmission of Western Chemistry into Late Imperial China, 1840–1900* (Boston: Brill, 2000).

102. David Wright points out that this was modeled on the method used to the translate Buddhist sutra into Chinese, but that in contrast to Buddhist transmitters, John Fryer and his Western colleagues often put sole blame for inaccuracies on the shoulders of their Chinese collaborators. David Wright, "The Translation of Modern Western Science in Nineteenth-Century China, 1840–1895," *Isis* 89, no. 4 (1998): 662, 664.

103. Hua Hengfang 華衡芳, *Jinshi shibie* 金石識別 (Classification of metals and rocks), (Shanghai: Jiangnan zhizaoju, 1872), 1. These translations were also abridgements and the original sources are not always listed. Several sources claim that *Jinshi shibie* is a translation of Dana's *Manual of Mineralogy*, but I follow influential modern

Chinese geologist, Huang Jiqing's attribution. Huang examined multiple editions of both of Dana's works and concluded that Hua most likely was working from an early edition of *System*. Huang Jiqing 黃汲清, "Xinhai geming qian dizhi kexue de zhongguo xianqu" 辛亥革命前地質科學的中國先驅 (China's Pre-Xinhai Revolution geological pioneers), *Zhongguo keji shiliao* 中國科技史料 (China historical materials of science and technology) 1 (1982): 2–13.

104. Hua Hengfang, "Zi xu" 自序 (Author's preface), *Dixue qianshi* 地學淺釋 (Shanghai: Jiangnan zhizaoju, 1873), 1; Weng Wenhao, "Lixihuofen yu Zhongguo zhi dizhi gongzuo" 李希霍芬與中國之地質工作 (Richthofen and Chinese geological work), *Fangzhi yuekan* 方志月刊 (Local Records Monthly) 6, no. 12 (1933): 37–38.

105. Huang Jiqing, "Xinhai geming qian dizhi kexue de Zhongguo xianqu," 2–3.

106. Wright,"Translation of Modern Western Science," 672.

107. For more on this idea, see chap. 3, "Making Modern Economies," in R. Bin Wong's *China Transformed: Hisotrical Change and the Limits of European Experience* (Ithaca, NY: Cornell University Press, 1997).

108. Carlson, *Kaiping Mines*, 4.

109. Numerous examples of the class issues that shaded perceptions of mining can be found in "Zhongguo renmin daxue Qingshi yanjiusuo" 中国人民大学清史研究所 (People's University Institute of Qing History), *Qingdai de kuangye* 清代的矿业 (Qing period mining) (Beijing: Zhonghua shuju, 1983).

110. Pumpelly, *My Reminiscences*, 411–30.

111. Huo Youguang 霍有光, "Waiguo shili jinru Zhongguo jindai dizhi kuangchan lingyu ji yingxiang" 外国势力进入中国近代地质矿产领域及影响 (The reach and influence of foreign involvement in modern Chinese geology and mining), *Zhongguo keji shiliao* 中国科技史料 (Chinese historical materials of science and technology) 15, no. 4 (1994): 3–20. On the other hand, Zhang Zhidong frequently complained that foreign teachers and experts would take much longer than necessary or focus on minor problems just to extend their posts, so deficiencies in advising existed on both sides of the Sino-foreign interaction.

112. Sun Yutang 孫毓棠, *Zhongguo jindai gongye shiliao, 1840–1895* 中國近代工業史料 *1840–1895* (Historical materials of modern Chinese industry) (Beijing: Kexue chubanshe, 1957), 613, 618, 622, 637.

113. For a striking example, see chap. 4 of Wu, "Underground Empires," on the fascinating role played by German engineers who were introduced as "translation interns" and positioned to both serve the technical needs of the Qing court and "win influence" (147) for German industry.

114. Wright, *Coal Mining in China's Economy*, 152–53; Carlson, *Kaiping Mines*, 30–31. These mining reforms (revised successively in 1902, 1904, 1907, and 1914) were criticized by both foreign and Chinese interests for different reasons, and in the political climate of the times, they often ground disputes to a stalemate between the government and foreign mining concerns. However, they marked an innovative application of international law to address Chinese rights issues, and they formed the basis of many of the successful reforms enacted during the Republican period. See V. K. Ting, "Mining Legislation and Development in China" (Appendix 2) in *Mineral Enterprise in China* by William W. Colling (New York: Macmillan, 1918), 212–21; and Zhou Yan 周彦 and Li Dan 李丹, "Qingmo xinzheng yu Zhongguo jindaihua" 清末新政与中国近代化 (New policies of the late Qing and Chinese modernization), *Lishi dang'an* 历史档案 Historical Archives, no. 2 (2010): 62–67.

115. Wright, *Coal Mining in China's Economy*, 142. See also Ian Phimister, "Foreign Devils,

Finance and Informal Empire: Britain and China, c. 1900–1912," *Modern Asian Studies* 40, no. 3 (July 1, 2006): 737–59.

116. Carlson, *Kaiping Mines*, 54–74. The actual circumstances were very complex and hotly contested in both China and Britain. In fact the scandal over the Kaiping Mines became a campaign issue for Herbert Hoover, who was involved in setting up the Chinese Engineering and Mining Company. See chap. 6 of Hoover, *Memoirs of Herbert Hoover*, 35–72, for a very different version of the Kaiping transfer.

117. For more on the railway Rights Recovery movement, see Mary Backus Rankin, "Nationalistic Contestation and Mobilization Politics: Practice and Rhetoric of Railway-Rights Recovery at the End of the Qing," *Modern China* 28, no. 3 (July 1, 2002): 315–61.

118. Gu Lang's real name was Rui Tiqian (芮體乾), and he was enrolled in the School of Mines and Railways as such, but he changed his name to Gu Lang upon arrival in Japan, and this is the only name under which he published from that point on.

119. In addition to geological and geographical works, Zhou read philosophy, poetry, psychology, and many other subjects in foreign languages. Many of these are referenced in his writings, but we also have a list of his German-language book orders while in Japan. Lu Xun 鲁迅, *Lu Xun Yiwen quanji* 鲁迅佚文全集 (Complete collection of Lu Xun's lost works), vol. 1 (Beijing: Qunyan chubanshe, 2001), 14–24.

120. Previously, topics related to modern geology were included under the heading *dixue*, which encompassed a wide variety of geographical, geomantic, and topographical matters. The term *dizhi* can be found in classical Chinese texts, where it refers to the quality or composition of the earth rather than any anachronistic analog of modern geology. Zhou's adoption of *dizhi* followed contemporary Japanese usage, and its intention was less to define the field of geology than draw attention to it as a distinct body of knowledge with a unique contribution to make toward China's future. Li Erong 李鄂荣, "'Dizhi' yici heshi chuxian yu woguo wenxian" '地质'一词何时出现于我国文献 (When did the term *dizhi* first appear in Chinese documents), *Zhongguo keji shiliao* 中国科技史料 (Chinese historical materials of science and technology 5, no. 3 [1984]: 53–57) traces *dizhi*'s classical usage and claims a long history for the term, but I believe that the examples in the article actually support the position that *dizhi*'s connection to geology is extremely recent, and comes into Chinese usage through Japan. If, however, Wu Rulun's proposed syllabus for the newly established Imperial College of Beijing in 1903 antedates Zhou's article in *Zhejiang Tide*, Wu would be the first person to bring the term *dizhi* from Japan to China. Wu Zhaoqian 吴昭谦, "Qingchao monian Wu Rulun dui chuanbo xifang dixue zhi gongxian" 清朝末年吴汝纶对传播西方地学之贡献 (The contribution of Wu Rulun to the dissemination of Western geoscience in the late Qing dynasty), *Anhui dizhi* 安徽地质 (Anhui geology) 8, no. 2 (1998): 72–75. For an illuminating discussion of how Chinese terms for geoscientific subjects developed, see Zhang Jiuchen 张九辰, "Zhongguo jindai dixue zhuyao xueke mingcheng de xingcheng yu yanhua chutan" 中国近代地学主要学科名称的形成与演化初探 (Preliminary investigation of the formation and evolution of common terms for geoscientific subjects in modern China), *Zhongguo keji shiliao*中国科技史料 (Chinese historical materials of science and technology) 22, no. 1 (January 1, 2001): 26–36.

121. Zhou Shuren 周树人, "Zhongguo dishi luelun" 中國地質略論 (Brief outline of Chinese geology), *Zhejiang chao* 浙江胡 Zhejiang Tide 8 (1903): 60.

122. Zhou Shuren and Gu Lang, "Zhongguo kuangchan zhi" 中国矿产志 (Record of Chinese mineral resources) in Lu Xun, *Lu Xun yiwen quanji*, 1:29. *Record* was reprinted

several times in Shanghai, Nanjing, and Tokyo and it remained popular through 1912.

123. Lu Xun , *Lu Xun yiwen quanji,* 1:37.
124. Zhou Shuren, "Zhongguo dishi luelun," 75.
125. Ibid., 61.
126. Zhou Shuren and Gu Lang, "Zhongguo kuangchan zhi," 32. For an interesting discussion of the ways in which self-criticism functions to help transform a loss (here, the loss of credible territorial knowledge) into the possibility of redemption (the active pursuit of sovereign territorial awareness), see William Callahan, "National Insecurities: Humiliation, Salvation, and Chinese Nationalism," *Alternatives* 29 (2004): 207–8.
127. Zhou Shuren and Gu Lang, "Zhongguo kuangchan zhi," 31.
128. Zhou Shuren, "Zhongguo dishi luelun," 6.
129. Zhou Shuren 周樹人, "*Zhongguo kuangchanzhi* guanggao" 中国矿产志 广告 (Advertisements for *Record of Chinese Mineral Resources*) in Lu Xun , *Lu Xun yiwen quanji,* 1:25.
130. Zhou Shuren, "*Zhongguo kuangchanzhi* zhengqiu ziliao guanggao" 国矿产志 征求资料广告 (Solicitation of data for *Record of Chinese Mineral Resources*), in Lu Xun , *Lu Xun yiwen quanji,* 1:26.
131. Zhou Shuren and Gu Lang, "Zhongguo kuangchan zhi," 32.
132. In 1909 Zhang Xiangwen founded the Earth Studies Society in order to arm the modern Chinese citizen with geography, geology, and other territorial knowledge in the lead-up to constitutional representation. Later Zhang and his fellow Society members were active participants in the revolutionary movement. Their efforts, though radical, emphasized geography (especially of foreign places) over geology in the tradition of Wei Yuan and Xu Jiyu, but the Society's journal provided an important forum for geological pioneers to set out their scientific agenda (see chap. 3). Gu Lang, by then dean of studies at an industrial college in Tianjin, helped the journal review submissions, and the journal published the first modern geological and mining maps produced by a Chinese investigator.

CHAPTER TWO

1. Ding Wenjiang, "Chongyin Xu Xiake youji ji xinzhu nianpu xu" 重印*徐霞客遊記*及新著年譜序 (Introduction to the reprint of *Xu Xiake youji* with newly written chronology), in *Xu Xiake youji* 徐霞客遊記 (Travel diaries of Xu Xiake), 1:3.
2. See Roy Porter, "Gentlemen and Geology: The Emergence of a Scientific Career, 1660–1920," *Historical Journal* 21 (1978): 809–36, for an account of this phenomenon over a longer period.
3. For more, see Rudwick, *Great Devonian Controversy*, 41; Secord, *Controversy in Victorian Geology*; Oldroyd, *Highlands Controversy*; and Bruce Hevly, "The Heroic Science of Glacier Motion," *Osiris* 11 (1996): 66–86.
4. For an in-depth study of Republican China's supposed disregard for empirical evidence and the epistemic politics of the ensuing "culture of fact" that developed as a response in the late Qing and early Republic, see Tong Lam, *A Passion for Facts: Social Surveys and the Construction of the Chinese Nation-State, 1900–1949* (Berkeley: University of California Press, 2011). As I argue elsewhere, Chinese self-criticisms and humiliations in the face of foreign knowledge regimes were at once a "trauma" and also a self-assertive act with ties to longstanding tropes of self-cultivation through introspection (*fanxing* 反省).

5. The more common contemporary term for fieldwork is *tianye diaocha* (田野调查), which can be rendered more literally as "field" or "open land" investigation, but this is largely a post-1949 usage. Other contemporary translations, such as *waiqin* (外勤), or "outdoor effort," still emphasize the labor involved in fieldwork, while the simple *shixi* (实习) refers to the "real practice" aspect of both fieldwork and internships. These terms are widely used in many fields, not just in geology.
6. Zhang Hongzhao 章鸿钊, *Liuliu zishu* 六六自述 (Self-narrative at sixty-six) (Wuhan, China: Wuhan dizhi xueyuan chubanshe, 1987), 21.
7. Ibid., 22.
8. Ibid., 22.
9. Ibid., 24. Ding Wenjiang goes into great detail on the miserable state of both Chinese inns and local maps in the first few installments of his 1932 series of "Manyou sanji" 漫遊散記 (Random notes from my wanderings) articles in *Duli pinglun* 獨立評論 (Independent review).
10. Zhang Hongzhao 章鴻釗, *Zhongguo dizhixue fazhan xiaoshi* 中國地質學發展小史 (Short history of the development of geology in China) (Shanghai: Shangwu yinshuguan, 1940), 12.
11. Zhang Hongzhao, *Liuliu zishu*, 25.
12. Ibid.
13. V. K. Ting, "Chinese Students." *Westminster Review* 169 (January 1908): 48–55.
14. Furth, *Ting Wen-chiang*, 24–25.
15. For more on the financial dimensions of Ding's experience in England, see Li Xuetong 李学通, "Ding Wenjiang liuying xuefei kao" 丁文江留英学费考 (Study of Ding Wenjiang's overseas study costs), *Lishi yanjiu* 历史研究 (Historical research), no. 6 (2009): 181–86.
16. G. W. Tyrrell, "Dr. V. K. Ting," *BGSC* 20, no. 3–4 (1940): 369; *Glasgow University Calendar* (Glasgow: n.p., 1910–11), 547, 550; *Glasgow University Calendar* (Glasgow: n.p., 1911–1912), 581. For more honors and an image of Ding at Glasgow, see http://www.universitystory.gla.ac.uk/biography/?id=WH22363&type=P.
17. See Hu Shi's biography of Ding for the most concise description of these travels. Hu Shi 胡适, *Ding Wenjiang zhuan* 丁文江传 (Biography of Ding Wenjiang) (Haikou, China: Haikou chubanshe, 1993).
18. Ding Wenjiang, "Manyou sanji", *Duli pinglun* (July 10, 1932), 21–22. Ding mentions keeping a field notebook, collecting specimens, examining outcrops, sketching and carrying equipment he brought from England. He also explains how he corrected local gazetteers on imperial trade routes and studied waterways.
19. Li Xuetong 李学通, *Weng Wenhao nianpu* 翁文灏年谱 (Chronicle of Dr. Weng Wenhao) (Jinan, China: Shandong jiaoyu chubanshe, 2005), 11.
20. "Liu Ou guanfeisheng guiyue" 留歐官費生規約 (Stipulations for government-sponsored students studying in Europe), *Jiaoyu zazhi* 教育雜誌 (Education journal) 5, no. 11 (1913), 45, item 18. See chap. 4 for more on Li Siguang's background and training, including his previous study of shipbuilding in Japan and his motivations for going abroad after the republican revolution.
21. Li Xuetong 李学通, *Shusheng congzheng: Weng Wenhao* 书生从政: 翁文灏 (A scholar in politics: Weng Wenhao) (Lanzhou: Lanzhou daxue chubanshe, 1996), 13.
22. Wong Wen-hao, "Contribution à l'étude de la porphyrite quartzifère de Lessines," *Mémoires de l'Institut géologique de l'Université de Louvain*, t.1 (1913), 298–325.
23. Ma Shengyun 马胜云, *Li Siguang nianpu* 李四光年谱 (Chronicle of Li Siguang) (Beijing: Dizhi chubanshe, 1999), 33.

24. J. S. Lee, "The Geology of China" (MS thesis, University of Birmingham, 1918).
25. Li Xuetong, *Shusheng congzheng*, 13, 15.
26. Ma Shengyun, *Li Siguang nianpu*, photos (1916), 32, 35.
27. Ibid., 37. Li took many photographs on his excursions in the Alps.
28. Weng was offered the position of chief mining engineer in a Sino-British joint venture mine in Hubei Province (Li Xuetong, *Shusheng congzheng*, 25), and William Boulton recommended Li as mining engineer to a British-run mine in India (Ma Shengyun, *Li Siguang nianpu*, 35). In Weng's case this was a particularly difficult decision because the Reiss Bradley Company, which had extended the offer to him, had long-standing business ties to the Weng family. However, Weng thought that becoming a mining engineer would be a waste of his specialized scientific education and did not want to work for a foreign corporation. Li Xuetong, *Weng Wenhao nianpu*, 15.
29. In the late Qing, this problem is quite complex and can be attributed in part to the lack of existing technical projects that sought Chinese staff. Many government and private efforts were too entangled with foreign money, or with the desire to catch up quickly by using foreign experts, to provide employment opportunities to any but the best returned students. Having studied abroad, technical students did not always have the connections or financing to organize their own projects, and many were simply not sufficiently trained to be so ambitious. Several studies suggest that students in engineering or the natural sciences who were successful upon return to China often made their way in finance, law, diplomacy, and politics, rather than in their specialized fields, and it is difficult to know whether this was due to insufficient training, a lack of opportunities, or a lingering sense that public service was the proper goal of higher education. See Wang Qisheng 王奇生, *Liuxue yu jiuguo* 留学与救国 (Study abroad and saving the nation) (Guilin, China: Guangxi shifan daxue chubanshe, 1995); and Zhang Yufa, "Returned Chinese Students from America and the Chinese Leadership (1846–1949)," *Chinese Studies in History* 35, no. 3 (April 2002), 52–86. In addition, many returned students faced suspicion in their homeland as other Chinese wondered about their loyalties. Learning from abroad had been their explicit goal, but learning too completely sometimes closed more doors than it opened. For more on the difficulties that returnees faced, see Stacey Bieler, *"Patriots" or "Traitors"? A History of American-Educated Chinese Students* (Armonk, NY: M. E. Sharpe, 2004); and John Wands Sacca, "Like Strangers in a Foreign Land: Chinese Officers Prepared at American Military Colleges, 1904–37," *Journal of Military History* 70, no. 3 (July 2006), 703–42.
30. See Edward J. M. Rhoads, *Stepping Forth into the World: The Chinese Educational Mission to the United States, 1872–81* (Hong Kong: Hong Kong University Press, 2011) for a study of the mission's impact in China and the US. A useful resource for tracking what is known of the mission's participants can be found at http://www.cemconnections.org/.
31. Zhang's thesis, "Geology of the Environs of Han-Cheou-Fu, China," included new information about the region's topography, structure, and stratigraphy, as well as several original maps. It was written in English and submitted under the guidance of Koto Bunjiro (小藤文次郎) in 1911. Before the 1920s, Zhang was only able to publish an incidental part of his thesis as an article in *Dixue zazhi* on "The Lure of Lime Fertilization in the Hangzhou Region" (Hangshu yi shihui daifei zhi yinyou) 杭屬以石灰代肥之引誘, *Dixue zazhi* 地學雜誌 (Earth studies journal) 3, no. 1 (1911): 1–8. As for earlier attempts by other returned students, Kuang Rongguang (鄺榮光), who took part in the Chinese Educational Mission, published rudimentary geological

maps in *Dixue zazhi* journal in 1910 (see chap. 3 for more), and later Gu Lang published information on the several Chinese-run mining ventures which included basic outlines of local structure and stratigraphy in addition to descriptions of the operations and output of the mines. Gu Lang 顧琅, *Zhongguo shi da kuangchang diaocha ji* 中國十大鑛廠調查記 (Record of investigation of China's ten largest mines) (Shanghai: Shangwu Shuju, 1916).

32. Zhang's thoughts on a Chinese geological survey can be found in Zhang Hongzhao 章鴻釗, "Zhonghua dizhi diaocha siyi" 中華地質調查私意 (My suggestions for China's geological investigations), pt. 1, *Dixue zazhi* 3, no. 1 (1912): 1–15; and Zhang Hongzhao, "Shijie geguo zhi dizhi diaocha shiye" 世界各國之地質調查事業 (The workings of the geological surveys of the countries of the world) pts. 1–4, *Dixue zazhi* 1, nos. 3–4 (1910): 1–4; 2, no. 11 (1911): 1–6; no. 12 (1911): 1–5; no. 13 (1911): 1–4. *Dixue zazhi* is discussed further in chap. 3. N.b.: pagination in this journal is not continuous but restarts with each article, and each number represents two facing pages. The inconsistency of printing conventions has caused some confusion about the citation for this article in the historical literature, and it seems as if a part 5 was originally anticipated but never published.
33. *Neige guanbao* 内閣官報 (Inner cabinet government news), 宣統三年九月初七 (October 10, 1918), no. 48, n.p. Aware of problems placing returned students and of returned students' lack of confidence in state recruitment, the Qing court issued orders in 1906 for local officials to round up students who had studied abroad and force (*qiangpo* 強迫) them to take subject-specific examinations conferring *jinshi* and *juren* degrees, equivalent to the highest two ranks of the old civil examination system. For more details and primary materials, see Wu Jing 武静, "Qingmo liuxuesheng guiguo kaoshi zhidu ji sikao" 清末留学生归国考试制度及思考 (Late Qing system and rationale for testing returned students), *Tianjinshi jiaokeyuan xuebao* 天津市教科院学报. *Journal of Tianjin Academy of Educational Science* 1 (2011): 25–28. Zhang Hongzhao and Ding Wenjiang were the only two geology students at the examination and their meeting in Beijing made later cooperation possible. Both earned highest honors in the exam. Coincidentally, Li Siguang also sat for this examination and earned highest honors as a *gongke* (工科 "engineering") *jinshi* for his shipbuilding studies in Japan. For details, see Ma Shengyun, *Li Siguang nianpu*, 22.
34. Song Guangbo 宋广波, "Dizhi yanjiusuo ruogan shishi bukao" 地质研究所若干史实补考 (Reexamination of certain historical facts about the Geological School), *Zhongguo kejishi zazhi* 27, no. 2 (2006), 163.
35. Zhang Hongzhao, *Liuliu zishu*, 31. After his departure from the Geological Section of the Ministry of Industry and Commerce, Zhang moved to the Ministry of Forestry and Agriculture as a technical specialist. That summer he conducted research on flooding in Hunan and Anhui provinces, and taught geology, mining, and topography at agricultural and normal schools in Beijing.
36. Hu Shi, *Ding Wenjiang zhuan*, 54.
37. Ding Wenjiang, "Xu" 序 (Introduction), *Dizhi huibao* 地質彙報 (Memoirs of the Geological Survey of China) 1 (1919), 2.
38. Ding Wenjiang, "Gongshangbu shiban dizhi diaocha shuomingshu" 工商部試辦地質調查說明書 (Synopsis of pilot geological investigations of the Ministry of Industry and Commerce) (Beijing: Gongshangbu, 1913), 5–6.
39. One Chinese *li* is approximately a third of a mile.
40. Ding Wenjiang, "Gongshangbu shiban dizhi diaocha shuomingshu," 9. This manuscript is dated as 1912 in the National Geological Library of China, but several

scholars have recently demonstrated that it was written in early 1913. See Li Xuetong 李学通, "Dizhi diaochasuo yange zhu wenti kao" 地质调查所沿革诸问题考" (Study of several problems concerning the development of the Geological Survey), *Zhongguo keji shiliao* 中国科技史料 (Chinese historical materials of science and technology), no. 4 (2003): 352, and Song Guangbo, "Dizhi yanjiusuo ruogan shishi bukao," 163. Both of these scholars offer very useful archival evidence to help evaluate the many contradictory accounts of the early years of the Geological Survey.

41. Dizhi diaochasuo 地質調查所, *Dizhi diaochasuo yange shilue* 地質調查所沿革事略 (Short account of the development of the Geological Survey), n.p., 1922: 2.

42. According to Hu Shi (p. 59), this reorganization occurred in October, but Cheng Yuqi 程裕淇 and Chen Mengxiong 陈梦熊 state that it was September 1913. Cheng Yuqi and Chen Mengxiong, eds., *Qian dizhi diaochasuo (1916–1950) de lishi huigu* 前地质调查所 (1916–1950) 的历史回顾 (Historical reminiscences of the former Geological Survey [1916–1950]) (Beijing: Dizhi chubanshe, 1996), 2. See also Li Xuetong, "Dizhi diaochasuo yange zhu wenti kao," 352.

43. I am translating 地質研究所 *dizhi yanjiusuo*, lit. "geology research institute" as the "Geological School" to clarify its function and distinguish it from the later Research Institute of Geology in the Academia Sinica, discussed in chap. 4. This liberty follows two official publications, The National Geological Survey of China, *The National Geological Survey of China 1916–1931: A Summary of Its Work during the First Fifteen Years of Its Establishment* (Peiping, March 1931), which uses "School of Geology" and J. G. Andersson, "The National Geological Survey of China." *Geografiska Annaler* 3 (January 1, 1921): 305–10.

44. Weng Wenhao 翁文灝, "Weng Xu" 翁序 (Preface by Weng) in *Dizhi yanjiusuo shidi xiuye ji* 地質研究所師弟修業記 (Student-teacher studies of the Geological School) (Beijing: Jinghua yinshu ju, 1916), 2.

45. The Peking University Geology Department originated in the prerevolutionary Imperial University of Peking, established as part of the 1898 reforms. A Geology Department (地質門, *dizhi men*) was proposed immediately but was deferred until 1909 when the Science College (格致科, *gezhi ke*) was founded with Chemistry and Geology as subdivisions. Five students (王烈 Wang Lie; 鄔友能 Wu Youneng; 裘傑 Qiu Jie; 陳祥翰 Chen Xianghan; and 路晉繼 Lu Jinji) entered the Geology Department in 1909 from the university's preparatory program in German language, and no further students applied. Only two students graduated in May 1913. Yu Guang 于洸, He Guoqi 何国琦, Liu Ruixin 刘瑞珣 (Liu Ruixun), Li Maosong, and Song Zhenqing 宋振清, "Hongyang chuantong, bawo jiyu, zaichuang huihuang: Qingzhu Beijing daxue jianli 100 zhounian, Beida dizhixue xi jianli 89 zhounian" 弘杨传统、把握机遇、再创辉煌—庆祝北京大学建立 100 周年、北大地质学系建立 89 周年 (Promoting tradition, grasping opportunities, recreating splendor: Celebrating 100 years of Peking University and 89 years since the founding of the Peking University Department of Geology), in *Beijing daxue guoji dizhi kexue xueshu yantaohui lunwenji* 北京大学国际地质科学学术研讨会论文集 (Collected papers of the Peking University International Academic Symposium of Geological Science), ed. Beijing daxue dizhixue xi 北京大学地质系 (Beijing: Dizhen chubanshe, 1998), 1.

46. Yu Guang et al., "Hongyang chuantong, bawo jiyu, zaichuang huihuang," 1. Weng seems to have a lot of mistakes in his chronological "Biographical Note of V. K. Ting," *BGSC* 16 (1936–37), i–xi.

47. This was buttressed by Ding Wenjiang's personal reputation, since several of the prospective Shanghai-area students (incl. 李學清 Li Xueqing and brothers 徐厚甫

Xu Houfu and 徐韋曼 Xu Weiman) knew of him as a dynamic young teacher at the Nanyang Middle School, where he taught biology for a year in 1911–12. "Dizhi diaocha suo yigu keji renyuan xiaozhuan" 地质调查所已故科技人员小传 (Deceased technical staff of the Geological Survey), appendix in *Qian dizhi diaochasuo (1916–1950) de lishi huigu*, ed. Cheng Yuqi and Chen Mengxiong, 242, 249. For other recruits from the Nanyang Middle School, see http://suny.nyschool.net.cn/blogarticle/article_pkId_1370_aid_56959.html.

48. Li Xuetong, *Shusheng congzheng*, 24.
49. Ding Wenjiang, "Gongshangbu shiban dizhi diaocha shuomingshu," 9; and "Bensuo zhangcheng" 本所章程," in *Nongshangbu dizhi yanjiusuo yilan* 農商部地質研究所一覽 (Overview of the Geological School of the Ministry of Agriculture and Commerce) (Beijing, Jinghua yinshuju, 1916), 1. Pagination restarts at the beginning of each preface, and then is continuous through the main body, with each set of facing pages given a new page number.
50. There are several conflicting reports of how many students were accepted into the program, though the standard view is likely based on Zhang Yiou's preface in *Nongshangbu dizhi yanjiusuo yilan*, 2. According to the official government announcement, however, there were twenty students accepted and nine named as alternates. "Gongshangbu luqu dizhi yanjiusuo shisheng chuangao" 工商部錄取地質研究所師生傳告 (Announcement of faculty and students selected for the Geological School of the Ministry of Industry and Commerce), *Zhengfu gongbao* 政府公報 (Government bulletin) (August 5, 1913). Interestingly the eighteen students who ultimately graduated from the Geological School were not all on this list, and the best accounts of which students declined acceptance, who dropped out or failed, and who later enrolled through make-up examinations can be found in Li Xuetong 李学通, "Nongshangbu dizhi yanjiusuo shimo kao" 农商部地质研究所始末考 (Investigation of the ins and outs of the Ministry of Agriculture and Commerce Geological School), *Zhongguo keji shiliao* 中国科技史料 (Chinese historical materials of science and technology) 22, no. 2 (2001), 22–23; and Song Guangbo, "Dizhi yanjiusuo ruogan shishi bukao," 167–68.
51. "Ben suo zhangcheng," 1.
52. "Xueke kecheng biao" 學科課程表 (Table of subjects and courses) in *Nongshangbu dizhi yanjiusuo yilan*, 2–3. Students used both English and German textbooks, and language study continued throughout all three years.
53. "Xueke kecheng biao," 3; and "Dizhi yanjiusuo biye ji" 地質研究所畢業記 (Geological school record of graduation), in *Nongshangbu dizhi yanjiusuo yilan*, 45.
54. "Xueke kecheng biao," 3.
55. Accounts of Solger's activities are quite diverse and often contradictory, though it is clear that he preferred fieldwork to teaching Chinese students (he had earlier taught at the Imperial University of Peking Department of Geology), and he was also on the payroll of the Geological Survey. Shellen Wu has uncovered archival material in the Bundesarchiv in Berlin-Lichterfelde, which gives the German side of the story. According to this material, German authorities believed that Solger had been selected to direct a Chinese geological school, and they later mistook references to the school (*dizhi yanjiusuo*) for a Chinese plan for a Prussian-influenced geological institute. Though the German archives reveal some general confusion about Chinese intentions, they are clear about their own interests. As one German official in China wrote: "Participation in such a comprehensive planned geological survey . . . could be used in numerous ways. Through the survey, we would be in possession

of valuable information about the mineral treasures of China earlier than everyone else, and with this work we would receive leadership of the mining section of the ministry of agriculture and commerce" (quoted from Shellen Wu, "Underground Empires," 172). Germany's confusion about its role in China's proposed geological institutions might have been compounded by the success of German geologist Edmund Naumann in outlining the structure for the Geological Survey of Japan in 1877. His suggestions, paired with those of native geologist Wada Tsunashirō (和田維四郎), were adopted in 1897 and resulted in the founding of the Geological Survey of Japan in 1882, with Wada as director but all scientific staff under the guidance of Naumann himself. Chishitsu Chōsajo (Japan), *Imperial Geological Survey of Japan: With a Catalogue of Articles Exhibited at the Panama-Pacific International Exposition Held at San Francisco, United States of America in 1915* (Tokyo: Imperial Geological Survey of Japan, 1915), 1.

56. See Song Guangbo, "Dizhi yanjiusuo ruogan shishi bukao," 163–65. Many of the teaching staff that Zhang employed were also returned students who had various degrees of training abroad, and this highlights how underused technical returnees were at this time. According to J. G. Andersson, these men were eager enough to put their training to use that they were often willing to teach at the school without pay (Andersson, "National Geological Survey of China," 305–6).
57. Zhang Hongzhao, "Zhang xu" 章序 (Preface by Zhang), in *Dizhi yanjiusuo shidi xiuye ji*, 1.
58. Zhang Hongzhao, "Zhang xu," in *Nongshangbu dizhi yanjiusuo yilan*, 2.
59. Zhang Hongzhao, *Liuliu zishu*, 33. Ding was actually in favor of moving geological education elsewhere, and the closing of the school in 1916 was not overly contentious.
60. Weng Wenhao, "Weng xu," in *Dizhi yanjiusuo shidi xiuye ji*, 2. Though the Chineseness of the school was a point of pride, as Solger's original appointment shows, Ding and Zhang were not opposed to non-Chinese instructors. They simply insisted on training local talent in China, and ultimately considered the overwhelmingly Chinese bent of the school to have marked an important milestone.
61. For a listing of courses and faculty, see Second Historical Archives, Nanjing 1038 (2581 and 2582). Additional faculty covered courses such as chemistry, physics, mathematics, geography, and German. One of the few geology courses that Zhang Hongzhao did not lead himself was structural geology, which was taught by Wang Lie, who was one of three students in the original Department of Geology at the Imperial University of Peking. Wang completed coursework but left the department in 1909, taught briefly in a Chinese middle school, and then won a government scholarship to study at the Freiberg Mining Academy in 1911. He returned in 1913. Wang later became a professor at Peking University when the department reinvented itself in 1918, under the chancellorship of Cai Yuanpei, and he also participated informally in the Geological Survey in China. Though he received first rate training, he did not have a vision of how to shape Chinese geology through fieldwork (see sec. 3) and did not engage in much original research.
62. Zhang Hongzhao, *Liuliu zishu*, 33.
63. Weng Wenhao, "Weng xu," in *Dizhi yanjiusuo shidi xiuye ji*, 2.
64. Ibid., 1.
65. Ibid., 2. In the winter of 1915, for instance, the most advanced students at the Geological School were divided into groups and led on a month-long field excursion. One of these groups was led by Swedish geologist Johan Gunnar Andersson, who

was working as an advisor to the Geological Survey, and this group worked in and around Jiangsu Province. Andersson was not an official member of the Geological School. Li Xuetong 李学通 "Zhongguo dizhi shiye chuqi ruogan shishi kao" 中国地质事业初期若干史实考 (Investigation of certain historical facts concerning the early period of China's geological endeavors), *Zhongguo keji shi zazhi* 中国科技史杂志 (The Chinese journal for the history of science and technology) 27, no. 1 (2006): 63.

66. Zhang Hongzhao, "Zhang xu," in *Dizhi yanjiusuo shidi xiuye ji*, 2.
67. Zhang Hongzhao, *Zhongguo dizhixue fazhan xiaoshi*, 16.
68. *Nongshangbu dizhi yanjiusuo yilan*, 46–47.
69. *Nongshangbu dizhi yanjiusuo yilan*, 46.
70. *Nongshangbu dizhi yanjiusuo yilan*, 4–7; and Second Historical Archives, Nanjing, 1038 (2582). Metallurgy was taught by Zhang Yiou himself and Chinese mining engineers Zhu Kun (朱焜) and Li Bin (李彬) taught mining (*Nongshangbu dizhi yanjiusuo yilan*, 13).
71. *Nongshangbu dizhi yanjiusuo yilan*, 7.
72. Drafting was taught by Wang Shaoying (王紹瀛), while Zhang Hongzhao, Ding, and Weng all taught report writing. Each field group or individual was also assigned an advisor who either led students in the field or oversaw their final projects. *Nongshangbu dizhi yanjiusuo yilan*, 7, 13.
73. Weng Wenhao, "Weng xu," in *Dizhi yanjiusuo shidi xiuye ji*, 2.
74. Because of the influence of the 1915 vernacular movement spearheaded by Hu Shi and Chen Duxiu, this book was also one of a dying breed of modern scientific texts written in the classical language.
75. *Nongshangbu dizhi yanjiusuo yilan*, 45. Page 8 describes the introduction of the advanced degree and its qualifications, based on independent fieldwork and test results. Five of the eighteen advanced students completed more than one independent project (pp. 16–18). Secondary sources almost uniformly claim that twenty-one students graduated from the Geological School, and most references in the primary materials (e.g., graduation speeches corroborate this number). However, there are twenty-two students listed in the *Nongshangbu dizhi yanjiusuo yilan*, 14–15). These seem to be in rank order, but there is no explanation of the status of the extra student, and in 1940, Zhang Hongzhao reports that twenty-two students graduated from the school (Zhang Hongzhao, *Zhongguo dizhixue fazhan xiaoshi*, 39). The eighteen advanced degree graduates are not contested.
76. Most secondary sources list all eighteen advanced graduates as investigators in the Geological Survey, but the original constitution of the survey (October 24, 1916, Institute of Modern History, Academia Sinica Archives 08-24 5 [1]) lists six technical experts (技師, *jishi*) and twelve investigators (調査員, *diaochayuan*). The *Dizhi diaochasuo yange shilue* (p. 1) states that these eighteen graduates were hired as investigators, interns (學習員, *xuexiyuan*), "and other" positions, implying even that the basic graduates of the school might have taken up administrative and support positions. (The entire survey had twenty-five members in 1916, which might comprise Zhang, Ding, Weng, and all twenty-one graduates). The survey published no personnel roster in 1916, but a ministry memo suggests that less qualified students were "retained for further study." "Nongshangbu ling" 農商部令 (Ministry of Agriculture and Commerce decrees), *Nongshang gongbao* 農商公報 (Ministry of Agriculture and Commerce bulletin) 3, no. 5 (1916). For the brief period from January to October 1916, at the instigation of Yuan Shikai as Emperor Hongxian, the Geological

Survey was called the *dizhi diaochaju* (地質調查局, Geological Survey Office or Geological Research Bureau) and was directly under the executive branch. It reverted to its original name (*dizhi diaocha suo*, 地質調查所, Geological Survey) and administrative position in October, when it was also granted a formal budget and constitution (see *Dizhi diaochasuo yange shilue*, 1; and *Nongshangbu gongbao* 農商部公報 (Ministry of Agriculture and Commerce bulletin) 2, no. 7 (February 15, 1916), 49. I have rendered both Chinese offices as the "Geological Survey" to avoid confusion.

77. Weng Wenhao, "Duiyu Ding Zaijun xiansheng de zhuiyi" 對於丁在君先生的追憶 (Recollections of Mr. Ding Zaijun), in *Ding Wenjiang zhuanji ziliao* 丁文江傳記資料 (Biographical materials of Ding Wenjiang), ed. Zhu Chuanyu 朱傳譽 (Taibei: Tianyi chubanshe, 1979), 111.
78. Y. C. Wang, *Chinese Intellectuals and the West, 1872–1949* (Chapel Hill, NC: University of North Carolina Press, 1966), 371.
79. *Nongshangbu dizhi yanjiusuo yilan*, 44. For an interesting testimonial to the success of the Geological School and Geological Survey in instilling the value of fieldwork, see the 1921 letter from Andersson inviting the Crown Prince of Sweden to China: "The fact is that decadent mandarinism has created a bias here against the participation of members of the upper class in any work that taxes the strength of the body. We have encountered this difficulty, but we have been able to train a team of young government geologists, who see it as a natural thing to make use of their legs and arms in field work." Magnus Fiskesjö and Chen Xingcan, *China Before China: The Discovery of China's Prehistory* (Stockholm: Museum of Far Eastern Antiquities, 2004), 38.
80. For more on Cai Yuanpei's educational values, in particular his insistence on intellectual diversity, see Liang Zhu 梁柱, "Cai Yuanpei de xueshu guan ji qi daxue kecheng jianshe sixiang tanxi" 蔡元培的学术观及其大学课程建设思想探析 (Analysis of Cai Yuanpei's academic views and his thoughts on building a university curriculum), *北京大学学报 (哲学社会科学版)* (Journal of Peking University [Humanities and Social Sciences]), no. 3 (2004): 132–40. For Cai's impact on Peking University and the ways that he challenged narrow views of utilitarian education, see Xiaoqing Diana Lin, *Peking University: Chinese Scholarship and Intellectuals, 1898–1937* (Albany, NY: State University of New York Press, 2006).
81. Some secondary sources claim that Beida's Geology Department began in 1918, but those sources most directly related to Beida itself uniformly place the department's founding in 1917, which conforms much more reasonably to the timing of Cai Yuanpei's chancellorship and the fact that the department's first eight students graduated in 1920 (Hu Bosu 胡伯素, "Dizhixi biyesheng ji zaixiao xuesheng renshu zhi tongji" 地質系畢業生及在校學生人數之統計 (Statistics of Department of Geology graduates and enrolled students), *Guoli Beijing daxue dizhi xuehui huikan* 4 (April 1930), 275. According to Amadeus Grabau, courses in geology were offered at Beida in 1916 ("A Decade of Research in Chinese Geology," *Guoli Beijing daxue dizhi xuehui huikan* 4 [April 1930]: 1). If so, these were not in any formal geology department, but it is not unlikely, since several of the 1920 graduating class entered the department in 1917 with credits from Beida's preparatory program, and He Jie was already teaching at Beida in the College of Engineering. Preparatory programs were common in the much of the Republican period, since national standards of middle school education were very uneven. According to Yu Guang, Beida students generally took three years of preparatory courses (heavily emphasizing fundamentals such as mathematics and foreign languages) and three years of university classwork until 1917 when this was

changed to two years of preparatory work and four years of departmental courses. The preparatory program was closed in 1929 (Yu Guang et al., "Hongyang chuantong, bawo jiyu, zaichuang huihuang," 2).

82. He Jie won a Boxer Indemnity scholarship in 1909 and received an undergraduate degree in mining engineering from the Colorado School of Mines. He then received a scholarship to Lehigh University and returned to China in 1914 with a masters degree in geology. He taught in the College of Engineering at Peking University until the founding of the Department of Geology, which he headed from 1917 to 1924. Wang Lie taught both German and structural geology at the Geological School before joining the faculty at Peking University. He chaired the Department of Geology from 1924 to 1927 and again from 1928 to 1931. See note 61 for more on Wang.

83. When the Geology Department was established in 1917, several students entered from Beida's preparatory program. Others, such as the star of the 1920 graduating class, Sun Yunzhu (孫雲鑄), were enrolled in mining courses at other institutions and transferred to Beida to focus on geological theory (Jiangsusheng zhengxie wenshi ziliao weiyuanhui 江苏省政协文史资料委员会, ed., *Yidai zongshi—Sun Yunzhu jiaoshou jinian zhuanji* 一代宗师—孙云铸教授纪念专辑 [Master of a generation—Prof. Sun Yunzhu commemorative collection] [Nanjing: Huadong shiyou dizhiju yinshuachang, 1995]). Within a decade, several other geology departments were founded throughout China, often by graduates or affiliates of the Beida program. The first of these was the Earth Sciences Department of Nanjing Southeastern University, which encompassed geology, geography, and meteorology, and later gave rise to a stand-alone Department of Geology in 1930 (as part of National Central University). Guangdong University followed with a Department of Geology in 1925 (reorganized in 1928 as part of Guangdong Zhongshan University). Other universities, such as Qinghua, did not have a dedicated geology department, but provided quite rigorous geological training. Though other geological programs were quite successful during the Republican period, Peking University remained the preeminent department with the closest association to the Geological Survey. See Wang Genyuan 王根元, "Zhongguo dizhi jiaoyu shilue" 中国地质教育史略 (Brief history of geological education in China), *Diqiu kexue—Zhongguo dizhi daxue xuebao* 地球科学-中国地质大学学报 (Earth science-journal of China University of Geosciences) 11, no. 2 (March 1986): 207–15.

84. Yu Guang et al., "Hongyang chuantong, bawo jiyu, zaichuang huihuang," 2.

85. Wang Gongmu 王恭睦, "Beida dizhi yanjiuhui xiaji lüxingtuan de tiyi" 北大地質研究會夏季旅行團的提議 (Proposal for a Beida Geology Society summer field trip group), *Guoli Beijing daxue dizhi yanjiuhui niankan* 國立北京大學地質研究會年刊 (The annual of the geological society of the university, Peking) 1 (October 10, 1921), 1. (Journal repaginated after each article.)

86. Wang Gongmu, "Beida dizhi yanjiuhui xiaji lüxingtuan de tiyi," 1; Zhao Guobin 趙國賓 "Benhui yinianlai de huigu he nianlai niban de shixiang" 本會一年來的回顧和年來擬辦的事項" (Reflections on the past year of our aociety and planned items for the coming year), *Guoli Beijing daxue dizhi yanjiuhui niankan* 1 (October 10, 1921), 1; and Yu Guang et al., "Hongyang chuantong, bawo jiyu, zaichuang huihuang," 6.

87. *Nongshangbu dizhi yanjiusuo yilan*, 46–47.

88. Ferdinand von Richthofen et al., *China*, 1:38, as quoted in V. K. Ting (Ding Wenjiang), "Foreword," *Bulletin of the Geological Survey of China* 1 (July 1919), i. This quotation was repeated many times among republican geological circles, and suggests the ways that Chinese geologists self-consciously tried to distance themselves from

earlier generations of Chinese intellectuals. This was at once a critical part of their identification with the plight of China and their claim to disciplinary authority. See Grace Yen Shen, "Taking to the Field: Geological Fieldwork and National Identity in Republican China," *Osiris* 24, no. 1 (January 1, 2009): 231–52, for more.

89. Richthofen's notebooks show clear admiration for hardworking Chinese peasants and suggest the complexity of his experiences in Asia. However, it is very clear that the samurai attitude he observed in Japan struck him as more masculine and effective than Chinese literati culture. It is interesting to note that while Richthofen showed himself an adventurous field geologist in other contexts, in China he traveled by horse with servants and, according to Jürgen Osterhammel "did not lift a finger himself" as part of his expression of racial superiority. (Osterhammel, "Forschungreise und Kolonialprogramm," 179). Bailey Willis, whose Carnegie expedition in 1903–4 was largely on foot, recalls being advised that walking would be a loss of face, but he chose to do so anyway.
90. The original quotation is "勞心者治人, 勞力者治于人" ("Laoxinzhe zhi ren, laolizhe zhi yu ren"). *Mengzi* 孟子, Teng Wengong 滕文公 上 (pt. 1). (Classic texts are cited by chap. heading.)
91. The social meaning of "shi" (士) has a long and involved history, which for simplicity's sake I have collapsed into "minor nobility." At the time of Confucius, *shi* were a martial class and it is significant that over time, the influence of Confucian *wen* (literary or civil) culture realigned *shi* toward the scholar-official.
92. The original quotation is "孔子能舉北門之関, 而不以力自章, 知夫筋骨之力, 不如仁義之力榮也" ("Kongzi neng ju beimen zhi guan, er bu yi li zizhang, zhi fu jingu zhi li, buru renyi zhi li rong ye"). Wang Chong 王充, *Lunheng* 論衡, *xiaoli* 效力 chap. Benevolence and righteousness are two of the five virtues of the *junzi* (君子), or superior man.
93. A Legalist-influenced brand of Confucianism was officially recognized as the governing philosophy of the Han dynasty, and later in the Sung dynasty a Buddhist- and Taoist-inflected form was canonized in the Four Books of the civil examination system.
94. Several historical figures demonstrated both physical and intellectual achievements, but these men, like Wang Yangming, were exceptions that proved the rule. Interestingly, as Joanna Waley-Cohen notes, the Qing were particularly interested in giving more weight to the martial (Joanna Waley-Cohen, *The Culture of War in China: Empire and the Military under the Qing Dynasty* [London: I. B.Tauris, 2006], 4). However, among Han Chinese the imbalance embedded in the *wen* and *wu* distinction still carried great weight.
95. "Shuigong er tianxia zhi" 垂拱而天下治, in *Shangshu* 尚書, Wu cheng 武成 chap.
96. Weng Wenhao, "Duiyu Ding Zaijun xiansheng de zhuiyi," 112
97. See Kam Louie, *Theorising Chinese Masculinity: Society and Gender in China* (Cambridge: Cambridge University Press, 2002) for an excellent analysis of the *wen-wu* dyad and its many social and cultural manifestations. For more on the ways that this image of gender roles in marriage and the nuclear family was challenged by New Culture Movement reformers, see Susan Glosser, *Chinese Visions of Family and State, 1915–1953* (Berkeley: University of California Press, 2003).
98. For additional ways in which concepts of femininity and masculinity shaped each other and defied the expectations of corresponding Western notions, see Martin Huang, *Negotiating Masculinities in Late Imperial China* (Honolulu: University of Hawaii Press, 2006).

99. Chen Duxiu 陳獨秀, "Jinri zhi jiaoyu fangzhen" 今日之教育方針 (Today's educational policy), *Qingnian zazhi* 青年雜誌 (*La jeunesse*) 1, no. 2 (October 15, 1915), 118. *Qingnian zazhi* became *Xin Qingnian* (新青年 [New youth]) in 1916, and this is how it is more widely remembered.

100. The Opium War commonly refers to the Anglo-Chinese War which ended in the Treaty of Nanjing and set the precedent for future "unequal treaties." The "second" Opium War is also known as the Arrow War or Anglo-French War, and produced the treaties of Tianjin (1858) that established the contentious principle of extraterritoriality. Hostilities quieted in 1860 after the sacking of the Summer Palace and the Qing dynasty's acceptance of the Peking Convention.

101. The Self-Strengthening movement (*ziqiang yundong,* 自強運動) and Foreign Affairs movement (*yangwu yundong,* 洋務運動) are largely synonymous and refer to the efforts of several prominent scholar-officials such as Zhang Zidong, Li Hongzhang, Zeng Guofan, and Zou Zongtang to promote Western technology in answer to Chinese military weakness. Their efforts are associated with several military modernization as well as educational and commercial enterprises such as the Jiangnan Arsenal, the Fuzhou Shipyards, and the China Merchants Shipping Company. See also chap. 1, note 100.

102. Yan Fu followed his classical training with several years at the Fuzhou Shipyard School studying navigation and two years of study in England, where he became interested in British social, political, and economic thought. Yan's essays and interpretive translations were "probably read by every eager student in China at the beginning of the [twentieth] century." (Wang, *Chinese Intellectuals and the West,* 206). For more information on Yan Fu, see Benjamin Schwartz, *In Search of Wealth and Power: Yen Fu and the West* (Cambridge, MA: Harvard University Press, 1964).

103. Translation follows Wang, *Chinese Intellectuals and the West,* 196.

104. Bodily vigor, intelligence, and virtue were Yan's rendering of Herbert Spencer's three energies: the physical, intellectual, and moral. According to Yan, the same measures which had made the West strong (including the Self-Strengtheners' arsenals, railroads, and battle fleets) were ineffective in China because the people themselves were unfit. As Yan Fu is discussed in most treatments of modern Chinese history, but for a comprehensive analysis of Darwinism's impact on China, see James Reeve Pusey, *China and Charles Darwin* (Cambridge, MA: Harvard University Press, 1983). An even longer perspective is offered by Susan Brownell's work on "body culture" and sport in contemporary China, and she highlights the legacy of Yan Fu realization that traditional "body as a state" could be powerful inverted by seeing the "state as a body." *Training the Body for China: Sports in the Moral Order of the People's Republic* (Chicago: University of Chicago Press, 1995), 45.

105. In Zhou Zhenfu 周振甫, *Yan Fu shiwen xuan* 嚴復詩文選 (Selections from Yan Fu's poems and prose) (Taibei: Taida chubanshe, 1964), 14–32.

106. Schwartz, *In Search of Wealth and Power,* 86.

107. For more on the reconceptualization of and struggle for control of the body in China's struggle for modernity, see Huang Jinlin 黃金麟, *Lishi, shenti, guojia: Jindai Zhongguo de shenti xingcheng, 1895–1937* 歷史, 身體, 國家: 近代中國的身體形成 1895–1937 (History, the body, and nation: The formation of the body in modern China) (Taibei: Lianjing chuban shiye gongsi, 2001).

108. Liang Qichao 梁啟超, *Yinbingshi quanji* 飲冰室全集 (Complete works from the ice drinker's studio) vol. 3, pt. 4 (Kunming, China: Yunnan daxue chubanshe 1941), 109–18.

109. The YMCA, which ran many government and Christian school athletic programs in China, noted that participation in competitions and practices was seriously hampered by traditional bias against physical activity. See chap. 1 of Jonathan Kolatch, *Sports, Politics, and Ideology in China* (New York: Jonathan David Publishers, 1972).

110. Cai Yuanpei 蔡元培, "Zai Nankai xuexiao quanxiao huanyinghui yanshuo ci" 在南開學校全校歡迎會演説辭" (Speech at the welcome meeting and Nankai School), May 23, 1917, *Cai Yuanpei jiaoyu wenxuan* 蔡元培教育文選 (Selected writings on education by Cai Yuanpei) (Taibei: Fuxing Shuju, 1956). These views were shared by the young Mao Zedong, who published "Study of Physical Education" in *New Youth* 3, no. 2 (April 1917). In this article he claimed that "it is the body that contains knowledge and houses virtue," and "when the body is strong, then one can advance speedily in knowledge and morality" (trans. follows Stuart Schram, in *The Political Thought of Mao Tse-tung* (Harmondsworth, UK: Penguin Books, 1969), 153–54. However, Mao's article was not influential at the time of its publication.

111. This did not mean that geologists were not concerned with the same nation-strengthening issues as their contemporaries. Much of their geological research was aimed at providing the resources and materials for national development. My point here is simply that the physicality of fieldwork was seen as part of good geological methodology, rather than as a form of abstract self-cultivation. Ding's more mainstream interests in strengthening the bodies of the people manifested themselves in side projects, such as his promotion of eugenics, rather than in his geological work.

112. Ding and other geologists' views occupy a middle ground between what Andrew Morris calls the transformative aspect of physical culture (*tiyu* 體育), in which physicality and body cultivation sharpened one's senses and values and what Kapil Raj has analyzed as the training of the colonial body as a precision instrument. Andrew Morris, *Marrow of the Nation: A History of Sport and Physical Culture in Republican China* (Los Angeles: University of California Press, 2004); and Kapil Raj, "When Human Travellers Become Instruments: The Indo-British Exploration of Central Asia in the Nineteenth Century," in *Relocating Modern Science: Circulation and the Construction of Knowledge in South Asia and Europe, 1650–1900* (New York: Palgrave Macmillan, 2007), 181–222. For Ding, the physical exertion of fieldwork inculcated the values of good science at the same time that it honed the body and mind into sensitive tools and reshaped the objects of physical engagement. Thus fieldwork transformed—and affirmed—the individual, the land, and knowledge making in a single act. To see the difference implied here between geology and mathematics or physics, consider the significance of physicality in Andrew Warwick, "Exercising the Student Body: Mathematics and Athleticism in Victorian Cambridge," in *Science Incarnate: Historical Embodiments of Natural Knowledge*, ed. Christopher Lawrence and Steven Shapin (Chicago: University of Chicago Press, 1998), 288–326

113. Weng Wenhao, "Duiyu Ding Zaijun xiansheng de zhuiyi," 111; Hu Shi 胡适, *Ding Wenjiang zhuan*, 85; Ji Rongsen 計榮森, transcriber, *Jiangxuehui jishi* 講學會記事 (Minutes of lecture meetings), vol. 1 (handwritten) (8/26/1930) meeting notes.

114. Weng Wenhao, "Duiyu Ding Zaijun xiansheng de zhuiyi," 112.

115. Gao Zhenxi 高振西, "Zuo jiaoshi de Ding Wenjiang xiansheng" 做教師的丁文江先生" (Mr. Ding Wenjiang as a teacher), in *Ding Wenjiang zhuanji ziliao* 丁文江傳記資料 (Biographical materials of Ding Wenjiang), vol. 1, ed. 朱傳譽 Zhu Chuanyu (Taibei: Tianyi chubanshe, 1979), 173.

116. Weng Wenhao, "Duiyu Ding Zaijun xiansheng de zhuiyi," 111.

117. Ma Shengyun, *Li Siguang nianpu*, 36.

118. See chap. 3 for more information about Amadeus Grabau.
119. Hu Shi, *Ding Wenjiang zhuan*, 56.
120. Hu Shi is considered the foremost liberal intellectual of the Republican period, one of the main promoters of the vernacular movement, and the leader of the so-called Chinese Renaissance. He also served as ambassador to the US Hu and Ding were lifelong friends and collaborators, and Hu is Ding's most famous biographer. A good overview of Hu's contributions to modern Chinese society and culture can be found in. For more of a prosopographical take on Hu Shi's intellectuals and political friends, see Tao Fangxuan 陶方宣, *Hu Shi de quanzi* 胡适的圈子 (Hu Shi's circle) (Jinan, China: Shandong huabao chubanshe, 2010).
121. Because Grabau came to China as both professor at Peking University and chief paleontologist of the Geological Survey, his generous salary of $1,600 each month was guaranteed by funds paid out of the American Boxer Indemnity Fund rather than through the University itself. Allan Mazur, *A Romance in Natural History: The Lives and Works of Amadeus Grabau and Mary Antin* (Syracuse, NY: Garret, 2004), 224.
122. Ma Shengyun, *Li Siguang nianpu*, 45; J. S. Lee, "The Stratigraphy of the Anthracolithic Formation in the Liuhokou Coalfield, N. China," *Guoli Beijing daxue dizhi yanjiuhui niankan* (October 10, 1921), 1:1.
123. Ma Shengyun, *Li Siguang nianpu*, 47; Yang Zhongjian 杨钟健, "Li Siguang laoshi huiyilu" 李四光老师回忆录 (Recollections of teacher Li Siguang), in *Li Siguang jinian wenji* 李四光纪念文集 (Collection of papers commemorating Li Siguang), ed. Li Siguang yanjiuhui choubeizu, Dizhi xuehui dizhi lixue zhuanye weiyuanhui 李四光研究会筹备组, 地质学会地质力学专业委员会 (Beijing: Dizhi chubanshe, 1981), 8.
124. Before returning to China, Li gave a speech before the Diligent Work-Thrift Study Society in Paris on February 28, 1920 entitled "Xiandai fanhua yu tan" ("現代繁華與炭") that stressed the important of energy sources for industry and modernization and describing the role of geology in identifying new stores of coal (Li Siguang, *Chuanguo dipingxian*, 95–117).
125. Yu Guang 于洸, "Li Siguang jiaoshou zai Beijing daxue 李四光教授在北京大学 (Professor Li Siguang and Peking University), *Hebei dizhi xueyuan xuebao* 河北地質學院學報 (Academic journal of the Hebei Geological Institute) 15, no. 5 (November 2, 1992), 535. The department received further support for its laboratory facilities in May 1922, and by January 1924; when participants of the third annual meeting of the Geological Society of China visited the Beida facilities, one attendee remarked that specimens and equipment surpassed that of the Geology Department of the University of Paris. Yu Guang et al., "Hongyang chuantong, bawo jiyu, zaichuang huihuang," 6).
126. Ma Shengyun, *Li Siguang nianpu*, 50; Yu Guang et al., "Hongyang chuantong, bawo jiyu zaichuang huihuang," 6.
127. A prolific worker, Grabau produced over 150 books and articles before leaving the United States, most of which were based on detailed fieldwork in northern and western New York, Michigan, Ohio, and southeastern Canada. Grabau is credited with bringing the facies concept to America, despite staunch opposition for E. O. Ulrich of the US Geological Survey.
128. These guidebooks were so cherished by Chinese students that several students studying in the United States in the 1940s made special trips to the Eighteen Mile Creek area to see the famous region in person.
129. Mazur, *Romance in Natural History*, 236–37.

130. From letter to Ray Bassler in Mazur, *Romance in Natural History*, 250, 299. The Kaiping Basin lies east of Beijing, in Hebei Province and is a coal-bearing region known at the time for the Kaiping Coal Mines.
131. Grabau's research at Tang Shan produced his paper "Ordovician Fossils from North China," *Paleontologica Sinica*, series B, vol. 1 (1922): 3–127.
132. Tsui-hua Yang Lee, "Geological Sciences in Republican China, 1912–1937," 85.
133. V. K. Ting, "Presidential Address," *BGSC* 3, no. 1 (1924): 9–11.
134. "Dizhixue kecheng" 地質系課程 (1927), Peking University Archives, BD 1927018.
135. Ding Wenjiang's January 6, 1924, speech "The Training of a Geologist for Working in China (中國地質工作者指培養)" (English), cited in Xia Xiangrong and Wang Genyuan, *Zhongguo dizhi xuehui shi*, 61. See also Tao Menghe 陶孟和, "Zhuiyi Zaijun" 追憶在君 (Remembering Zaijun), *Duli Pinglun* 獨立評論 (Independent review) 188 (February 16, 1936), 33–34. By the late 1920s foreign students interested in the geology of China actually began applying to Beida to do postgraduate research under local professors. Because of the uncertain political and military conditions in Beiping, only two students (one of whom became the editor of the Geological Survey of the Manchurian Railway and the other of whom earned a doctorate in Prague based on his research) actually studied in the Beida department, though several applied. The Beida department did not itself confer advanced degrees in the Republican period. Amadeus Grabau, "Contributions to Geologic Science by Graduates of the National University," *Guoli Beijing daxue ziran kexue jikan* 國立北京大學自然科學季刊 (Science quarterly of the National University of Peking) 1, no. 3 (April 1, 1930), 240.
136. Parkin Wong, "The Mineral Resources of China" (MA thesis, Cornell University, 1914); and Wah Seyle Lee, "Bibliography of the Geology of China" (MA thesis, Stanford University, 1916). Wah remained in the US and worked for the Chinese-American Mercantile Company, while Wong eventually moved to Hong Kong. Both attended high school in the US.
137. Ding Wenjiang, "Xuanxue yu renshengguan" 玄學與人生觀 (Metaphysics and the philosophy of life), in *Kexue yu renshengguan* 科學與人生觀 (Science and the philosophy of life), ed. Hu Shi 胡适 (Shanghai: Yadong tushuguan, 1925): 15–44; and Weng Wenhao, "Yige dapo fanmen de fangfa" 一個打破煩悶的方法 (A method for cracking depression), *Weng Wenhao xiansheng yanlun ji*, 27–33.
138. "Government by good men" was cited in many forms in several articles in the *Nuli zhoubao* (努力周報, *Endeavor Weekly*), founded by Ding and others (including Hu Shi and Cai Yuanpei) in 1922. The notion of vigorous action in geology resonated with broader ideas of strenuous activity among public intellectuals. See for instance Hu Shi's "Song of Hard Work," cited in Timothy Weston, *The Power of Position: Beijing University, Intellectuals, and Chinese Political Culture, 1898–1929* (Berkeley: University of California Press, 2004), 221. For more on the "government by good men" ideal and geology, see chap. 4.
139. See chap. 4 for more details of this development.

CHAPTER THREE

1. William Kirby, "The Internationalization of China." *China Quarterly* 150 (June 1997): 443–58.
2. To get a contrasting idea of the impact of the Great War and its wake in an explicitly colonial context, see Teresa Salomé Alves Da Mota, "A Mere Shadow of an

Institution: The Unhappy Story of the Portuguese Geological Survey (PGS) in the Period Between the Two World Wars," *Annals of Science* 64, no. 1 (2007): 19–40.

3. Andersson was recommended by fellow Swede, Erik Nyström, who had been collecting and teaching in Shanxi Province. Andersson was the secretary general of the 11th International Geological Congress in 1910 and had expressed interest in the iron ores of China.

4. Hu Shi, who was very aware of the effort put into these endeavors, was one of the most effusive, writing in the *Nuli zhoubao*, "This week the big event in China isn't the beating of Finance Minister Dong Kang. It also isn't the resignation of the cabinet, and it isn't even the warfare in Sichuan. Rather it is the opening on the 17th of the Beijing Geological Survey's museum and library." Hu Shi, "Zhe yi zhou" 這一周 (This week) *Nuli zhoubao*, July 23, 1922, cited in Li Xuetong 李学通, "Zhongguo dizhi shiye chuqi ruogan shishi Kao," 68. In addition nominal president of the Republic, Li Yuanhong (黎元洪), personally donated 1,000 yuan to the library and attended the opening ceremony. Wu Ningsheng 武宁生 and Li Shuxian 李淑贤, "Qian dizhi diaochasuo dizhi tushuguan shi" 前地质调查所地质图书馆史 (History of the Library of the Former Geological Survey), in *Qian dizhi diaochasuo (1916–1950) de lishi huigu*.

5. This suggestion that it would be "natural" to assign responsibility for the geological mapping of East Asia, in particular China, to Britain, Russia, the Netherlands, or Japan was first made by Emmanuel de Margerie at the International Geological Congress of 1913 in Toronto. After the war, at the First Pan-Pacific Science Conference held at Honolulu in 1920, members noted that a Geological Survey was active in China but still recommended that Naomasa Yamasaki of Japan report on the geological mapping of Japan, China, and Korea, and Chinese representatives were not invited to the meeting. Yamasaki's report included no data from the Chinese survey. E. de Margerie, "The Geological Map of the World," in *Compte-rendu de la XIIe session du Congrès géologique international, Canada, 1913*, 185; T. Wayland Vaughan, "Status of Areal Geological Mapping in the Pacific Region," in *Proceedings of the First Pan-Pacific Scientific Conference, Honolulu, Hawaii, August 2 to 20, 1920* (Honolulu: Honolulu Star-Bulletin, 1921), 639–40; N. Yamasaki, "Geological Mapping in Japan, Korea, and China," *Proceedings of the First Pan-Pacific Scientific Conference* (Honolulu: Honolulu Star-Bulletin, Ltd., 1921), 677–79.

6. J. S. Lee, "Reflections on Twenty Years' Experience," *BGSC*, 22, no. 1–2 (June 1942): 22.

7. There is surprisingly little available about the development of geology in Japan, even in Japanese sources. However, good starting points for the Geological Survey of Japan and the Geological Society of Japan (originally the Tokyo Geological Society) are the Survey's own *Imperial Geological Survey of Japan: with a Catalogue of Articles Exhibited at the Panama-Pacific International Exposition Held at San Francisco, United States of America in 1915* and Working Group of INHIGEO Japan Members, "Special Issue: An Introduction to the History of Geological Sciences in Japan," *JAHIGEO Newsletter*, no. 13 (May 2011): 2–26. http://www.geocities.co.jp/bhermes001/jahige01.html. See also chap. 2, n. 55.

8. *Dixue* is often translated as "geography," but it generally refers to a very varied set of practices and traditions pertaining to the territory (including but not limited to topography, mapping, landscape studies broadly construed, geomancy, and even certain agricultural studies). I am rendering *Dixue zazhi* as the *Earth Studies Journal*

to reflect the ways in which the translation of *dixue* remained open-ended in the late Qing and even the early Republican period. In 1920, after debate on the role of the journal and the Earth Studies Society in Chinese public life, the journal added the English title "Geographical Magazine" and the Society chose as its English equivalent "Chinese Geographical Society"—not to be confused with the Geographical Society of China (中國地理學會, *Zhongguo dili xuehui*), founded by Zhu Kezhen in 1934. These moves were part of a wider shift in focus and direction that took place after the period in which the society and journal were critical to the development of geology under consideration in this book. In contemporary usage, the terms *dilixue* (地理學, lit. "study of earth/land principles") and *dizhixue* (地質學, lit. "study of earth/land substance") correspond to "geography" and "geology," respectively.

9. The promise of constitutional assemblies and a parliamentary monarchy after nine years was made as part of the late Qing New Policy reforms proposed in 1901. These New policies included a revamping of both central and provincial administration as well as changes in public education.
10. Bai Yukun gave his life in the unsuccessful January 1912 Luanzhou uprising, one of several incidents in which local gentry supported the new Nanjing government against entrenched military powers. See Lu Kaiyu 卢开宇, "Zhongguo dixue hui zai Xinhai geming qianhou de huodong" 中国地学会在辛亥革命前后的活动 (Activities of the Chinese Earth Studies Society before and after the Xinhai Revolution), *Shi lin* 史林 (Historical circles) 1 (2003): 65–72; and Li Chuanyong 李传永, "Gexin Zhongguo dilixue de xianqu—Zhang Xiangwen" 革新中国地理学的先驱—张相文 (Pioneer who broke new ground in Chinese geography—Zhang Xiangwen), *Keji renwu* 科技人物 (Sci-tech people) 5 (2003): 93–95.
11. "Bianji buzhang Luanzhou xunnan Bai lieshi xiaozhuan" 編輯部長灤州殉難白烈士小傳 (Profile of editorial chief Bai who gave his life at Luanzhou), *Dixue zazhi* 3, no. 1 (1912): 85.
12. Zhang Xiangwen 張相文, "Zhongguo dixue hui qi" 中國地學會啓 (Opening of the Chinese Earth Studies Society), *Dixue zazhi* 1, no. 1 (1910): 1–2. See chap. 1 for more on Wei Yuan and Xu Jiyu.
13. Kuang Rongguang was one of the first cohort of the Chinese Educational Mission organized by Yale graduate Rong Hong (容閎, better known as Yung Wing in English) in 1872. Kuang's father was a mine worker from Guangdong, and Kuang studied mining at Lafayette College. In 1881 he was recalled to China when the mission was discontinued by conservative officials who felt that mission students were being deracinated and were losing their Chinese culture. Kuang did not graduate with a degree but, unlike other students who could not find an outlet for their specialized learning, he managed to find several posts related to mining and prospecting, and he later collaborated with other members of the mission, like Zhan Tianyou (詹天佑, widely known as Jeme Tien Yow), China's most famous railway engineer. For more information on Kuang, see Mao Shizhen 茆诗珍 and Xu Fei 徐飞, "Liu Mei youtong yu Zhongguo zaoqi dizhixue" 留美幼童与中国早期地质学 (American-educated youth and China's early geology), *Kexue jishu yu bianzhengfa* 科学技术与辩证法 (Science, technology and dialectics) 22, no. 6 (December 2005): 88–91.
14. Kuang's maps included little explanatory material and were not followed by detailed information on his research methods or findings. Kuang Rongguang, "Zhili dizhi tu" 直隸地質圖 (Geological map of Zhili), *Dixue zazhi* 1, no. 1 (1910); Kuang Rongguang, "Zhili kuangchan tu" 直隸礦產圖 (Mining map of Zhili), *Dixue zazhi* 1, no. 2

(1910); Kuang Rongguang, "Zhili shiceng guji" 直隸石層古跡 (Paleontological relics of Zhili strata), *Dixue zazhi* 1, nos. 3–4 (1910).

15. Zhang Hongzhao, "Shijie geguo zhi dizhi diaocha shiye," pt. 1, *Dixue zazhi* 3–4 (1910): 1. According to his autobiography, Zhang drafted "Shijie geguo zhi dizhi diaocha shiye" during his second year at Imperial University (Zhang Hongzhao, *Liuliu zishu*, 30) but the full manuscript was not published until late 1911. Parts 2, 3, and 4 are continued in numbers 11, 12, and 13 of *Dixue Zazhi*.
16. Zhang Hongzhao, "Shijie geguo zhi dizhi diaocha shiye," pt. 1, p. 3. It is unclear what sources Zhang used since no citations are given and the information is quite general, but his focus is predominantly on what he calls the "Geological Survey of the United Kingdom" rather than on the Society.
17. Zhang Hongzhao, "Zhonghua dizhi diaocha siyi." Pagination in this issue is still restarted with each article, but numbers represent a single side, rather than a set of facing pages. Part 2 of the article is in *Dixue zazhi* 3, nos. 3–4 (1912): 14–20. In this issue, pagination is continuous, but once again numbering represents a set of facing pages.
18. Zhang Hongzhao, *Liuliu zishu*, 30. A short introduction to "My Suggestions for China's Geological Investigations" was provided by a friend in the Ministry of Industry of the provisional republican government, agricultural specialist Tao Changshan (陶昌善).
19. Zhang Hongzhao, "Zhonghua dizhi diaocha siyi," pt. 1, p. 7.
20. Ibid., 6–7.
21. Ibid., 7. Zhang also argued that China could not maintain its identity as a nation founded upon agriculture if it did not even understand its own soil.
22. Ibid., 6, 9.
23. Zhang Hongzhao, "Dixue hui ying xing shiwu zhi shangque" 地學會應行事務之商榷 (Discussions on the proper duties of a geosciences society), *Dixue zazhi* 3, nos. 3–4 (1912): 2–3. I render *dixue* as "geosciences" here to set it slightly apart, since the article seems to imagine *dixue hui* as a broad category, rather than as the Earth Studies Society in specific.
24. The first government of the Republic was set up in Nanjing before moving to Beijing in 1912. This Geological Section was the first reference to "geology" (*dizhixue*, 地質學) in any official government organization in Chinese history.
25. Zhang Hongzhao, "Dixue hui ying xing shiwu zhi shangque," 2.
26. Zhang was careful to limit this help to groups that were seeking knowledge, and not just conducting geological research for private reasons, such as for a mining company, etc. Ibid., 3.
27. Ibid.
28. For more on study and mutual aid societies in the Qing and Republican eras and their role in modern state-society relations, see Wang Ermin 王爾敏, "Zhongguo jindai xuehui yuelun" 中國近代學會約論 (On the study societies in modern China), *Shihuo yuekan* 食貨月刊 (Social and economic conditions monthly) 1, no. 5 (1971): 287–95; Joanna F. Handlin Smith, "Benevolent Societies: The Reshaping of Charity During the Late Ming and Early Ch'ing," *Journal of Asian Studies* 46, no. 2 (May 1, 1987): 309–37.
29. "Dixue hui ying xing shiwu zhi shangque," 2. Zhang refers to the Society's members as "scholars" but later refers to "specialists" by using a different term.
30. Zhang used "Geological Society" (*dizhi xuehui*) when discussing institutions in England, France, and Prussia in "Shijie geguo zhi dizhi diaocha shiye," pt. 1, p. 2.

31. "Dixue hui ying xing shiwu zhi shangque," 2.
32. The Republican government moved its capital from Nanjing to Beijing in April 1912, and the Geological Section was folded into the Bureau of Mines under the newly reorganized Ministry of Industry and Commerce, where it was largely ignored.
33. The Earth Studies Society continued its previous activities without much change, though they moved their headquarters from Tianjin to Beijing. "Benhui jishi" 本會紀事 (Records of our society), *Dixue zazhi* 1, nos. 3–4 (1910): 4.
34. As the government reorganized, the only task Zhang was able to push through the uninterested bureaucracy was the mailing of a questionnaire to provincial authorities. Zhang Hongzhao, *Liuliu zishu*, 31.
35. "Xu yi" 序一 (Preface one), *Guoli Beijing daxue dizhi yanjiuhui niankan* 國立北京大學地質研究會年刊 (The annual of the geological society of the university, Peking) 1 (Oct. 10, 1921): 1. The Beida Geological Society's publication was an annual until 1928, when it became *Guoli Beijing daxue dizhi yanjiu xuehui huikan* (國立北京大學地質研究學會會刊, Bulletin of the Geological Society of the National University, Peking). In 1930, it changed its Chinese title to *Guoli Beijing daxue dizhi xuehui huikan* (國立北京大學地質學會會刊) but retained its English name. Pagination in the *Annual* begins anew with each article.
36. For more, see chap. 2.
37. The Geological School's equipment and facilities originally belonged to Beida's precursor, the Imperial College.
38. When this new class graduated, there were twenty-seven students, many of whom transferred in due to positive word of mouth, recommendations from coprovincials, and interest in Grabau and Lee. (Personal interview with Cui Kexin.)
39. Many of the leaders of the May Fourth Incident were Peking University students, and Beijing was the epicenter of the uprisings, which spread to Shanghai and other cities.
40. There is a vast body of literature on the May Fourth, or New Culture, Movement, which though universally accepted as the major cultural movement in early twentieth-century China, is nevertheless quite controversial. There is little agreement on what parameters define the movement and different scholars give different dates for the movement based on particular aspects relevant to their own work. One leading authority, Chow Tse-tsung, offers 1917–21 as the span of the movement, but I think it can be extended backward to the publication of *New Youth* magazine. The precise details are less important, however, than the understanding that this was the dominant intellectual tide for several years around the 1919 incident. The content of the movement is equally difficult to pinpoint, since it comprised liberal thought, anarchism, communism, and even certain forms of fascism. The common elements of these trends were an openness to foreign ideologies (often as a radical break with tradition) and an emphasis on the rejuvenation of Chinese society, culture, and politics through youth activism. As a starting point, see Chow Tse-tsung, *The May 4th Movement: Intellectual Revolution in Modern China* (Cambridge, MA: Harvard University Press, 1960). More on recent mainland Chinese scholarship can be found in Qian Zhao, "A Review of Studies of the May Fourth Movement in China over the Past Decade," *Chinese Studies in History* 43, no. 4 (Summer 2010): 73–89.
41. Yang Xinxiao杨新孝, "Ji Beida dixhi yanjiuhui" 记北大地质研究会 (Remembering Beida's Geological Society), 27 (author's photocopy).
42. Yang Zhongjian, spearhead of Beida's Geological Society and subsequent leader of China's Peking Man project, was personally very active in May Fourth activism and

publishing widely. For more on his activities, see his autobiography Yang Zhongjian, *Yang Zhongjian huiyilu* 杨钟健回忆录 (Memoir of Yang Zhongjian) (Beijing: Dizhi chubanshe, 1983).

43. Zhao Guobin, "Benhui yinianlai de huigu he nianlai niban de shixiang," 1.
44. Tian Qirui 田奇瑞 et al., "Ni zuzhi 'Beijing daxue dizhi yanjiuhui' gongqi" 擬組織『北京大學地質研究會』公啓 (Public announcement of the plan to organize a Beijing University Geological Research Society), *Guoli Beijing daxue dizhi yanjiuhui niankan* 1 (October 10, 1921), 4.
45. The first issue of the Society's *Annual* was also published on National (or Double-Tenth) Day, 1921.
46. Zhao Guobin, "Benhui choubei shidai jiyao" 本會籌備時代紀要 (Essential notes from the preparatory phase of our society), *Guoli Beijing daxue dizhi yanjiuhui niankan* 1 (October 10, 1921), 4.
47. Yang Chuan 楊銓, "Xuehui yu kexue" 學會與科學 (Learned societies and science), *Kexue* 科學 (Science) 1, no. 7 (1915): 707–11. Founded at Cornell University by a group of enterprising students studying abroad, the Science Society quickly became Republican China's premier scientific association when it relocated from overseas. The Science Society was not only an encouraging model of success for Beida students, but its journal *Science* (科學, *Ke Xue*) was also the template for the Beida Geological Society's anticipated quarterly. For more on the Science Society, see Wang, Zuoyue, "Saving China Through Science: The Science Society of China, Scientific Nationalism, and Civil Society in Republican China," *Osiris* 17 (2002): 291–322; and Zhang Jian 张剑, *Kexue shetuan zai jindai Zhongguo de mingyun: Yi Zhongguo kexueshe wei zhongxin* 科学社团在近代中国的命运: 以中国科学社为中心 (The Science Association and the change of society in modern China: A study on the Science Society of China [sic]) (Jinan: Shandong jiaoyu chubanshe, 2005).
48. For more on this general trend in education, see Wen-Hsin Yeh, *The Alienated Academy: Culture and Politics in Republican China, 1919–1937* (Cambridge, MA.: Harvard University Press, 1990), 7–42.
49. Zhao Guobin, "Benhui yinianlai de huigu he nianlai niban de shixiang," 1.
50. After it was officially founded, Geology Department students far outnumbered those from the Prospecting Division, but in its origins, the Beida Geological Society was dominated by students from the Prospecting Division. Of the two prime movers, Zhao Guobin and Yang Zhongjian, Zhao Guobin was in Prospecting, and of the other five leaders, four (Luo Yunlin, Li Fangzhou, Zeng Qinying, and Wu Guozian) were his classmates. The Geological Society (this is its official English translation, but a literal rendering would be the "Geological Research Society") had twenty-nine student members in its first year, including one from the Chemistry Department, an undeclared student that later transferred into Geology, and two students in the preparatory program of the university.
51. Zhao, "Benhui yinianlai de huigu he nianlai niban de shixiang," 1.
52. This view likely influenced the department's initial emphasis on textbook learning over fieldwork, since fieldwork was often associated with applications such as mining. The shift in understanding of fieldwork represented by Ding, Weng, Zhang, and Li, among others, was a major turning point in the development of Chinese geology (see chap. 2).
53. "Benhui shisheng chahuahui jishi" 本會師生茶話會記事 (Notes from our society's teacher-student tea party), *Guoli Beijing daxue dizhi yanjiuhui niankan* 1 (October 10, 1921): 7–9.

54. "Benhui shisheng chahuahui jishi," 8; "Benhui chengli hui yanshuo ci: He Jie jiaoshou yanshuo ci" 本會成立會演説詞: 何杰教授演説詞 (Opening lecture of our Society's inaugural meeting: Professor He Jie's text). *Guoli Beijing daxue dizhi yanjiuhui niankan* 1 (October 10, 1921), 5.
55. These activities were very similar to those suggested by Zhang Hongzhao for a geosciences society, but with a greater emphasis on original research and less stress on an advisory role. Tian Qirui et al., "Ni zuzhi 'Beijing daxue dizhi yanjiuhui' gongqi," 2–3. The Society also hoped to coordinate with the University Popular Education Lecture Corps to have members give talks on general geological topics every Saturday evening.
56. Solicitation for donations, Peking University Archives, BD 1922011/1.
57. "Kuangshi shi tushu shi yuelan guize" 礦石室圖書室閱覽規則 (Regulation for use of the specimen room and library), *Guoli Beijing daxue dizhi yanjiuhui niankan* 1 (October 10, 1921): 12.
58. "Dizhi yanjiuhui tushu zhanji biao" 地質研究會圖書暫記表 (Provisional inventory of Geological Society book and map collection), *Guoli Beijing daxue dizhi yanjiuhui niankan* 1 (October 10, 1921): 15.
59. Members estimated that at least 400 yuan would be needed to start a regular journal. The annual budget of the Society in its first year was 154 yuan, most of which was collected by Professor He Jie from various faculty members. The *Annual* alone cost one hundred yuan to publish. "Benhui chuang kan dizhi zazhi jihua shu" 本會創刊地質雜誌計劃書 (Plan for the founding of our society's geological journal), *Guoli Beijing daxue dizhi yanjiuhui niankan* 1 (October 10, 1921): 12–14. Many of the miscellaneous reports were digests of foreign material on advanced geological topics. In 1924 the department began publishing *Contributions from the Geological Institute, National University of Peking*, which was devoted to student field and lab work. Similar journals were published by several universities with active geology departments by the 1930s.
60. Wang Gongmu (Class of 1923) angrily complained about the lack of resources at the university and the lack of opportunities for fieldwork. He blamed the university in part but also pointed out that students themselves were not able to coordinate a proper summer expedition. This report contrasts with the experiences of students from the old Geological School where practical fieldwork was emphasized, and students who entered Beida just a year later reported plentiful field research in the basic course requirements (e.g., see graduation requirements of 1927). "Beida dizhi yanjiuhui xiaji lüxing tuan de tiyi," 1–3. Students also published details of their field trips in other journals, such as *Dixue zazhi* and *Magazine of Natural History* (*Bowu zazhi*, 博物雜誌), and general university publications.
61. Besides broadly fitting the goals of standardization and popularization, students hoped that beginning textbooks on geology would support Chancellor Cai Yuanpei's attempt to have geology added to the standard middle school curriculum. "Benhui shisheng chahuahui jishi," 8.
62. Quoted in Yang Zunyi 杨遵仪, *Tao li man tianxia: Jinian Yuan Fuli jiaoshou bainian danchen* 桃李满天：纪念袁复礼教授百年诞辰 (Students all over the world: Commemorating the hundredth anniversary of Professor Yuan Fuli's birth) (Wuhan: Zhongguo dizhi daxue chubanshe, 1993), 288, and cited as part of an interview with Yuan on November 9, 1980, in footnote 2 of Xia Xiangrong and Wang Genyuan, *Zhongguo dizhi xuehui shi*, 7.
63. Xie Jiarong came back from Stanford and the University of Wisconsin–Madison in

1920. He was especially active at Madison and took part in the Geology Club of the Geology Department. Yuan Fuli came back from Columbia in the summer of 1921, opting not to continue with a PhD after completing both his bachelor's and master's degrees in the Geology Department of Columbia University and participating in its Geological Journal Club.

64. More museum space for paleontological specimens was added through private donations in 1927 as well. Pan Jiang 潘江, "Qian dizhi diaochasuo kuangchan chenlieguan gaikuang" 前地质调查所矿产陈列馆概况 (Overview of the mineral exhibition rooms of the former Geological Survey), in *Qian dizhi diaochasuo (1916–1950) de lishi huigu*, 84.
65. H. T. Chang (Zhang Hongzhao), "On the History of the Geological Science in China," *BGSC*, 1, nos. 1–4 (1922): 6.
66. After 1921, when he took up the general manager's post at the Beipiao Coal Mining Company, Ding was technically honorary director, with Weng specially appointed as acting director, but etiquette among Chinese demanded that Ding still be acknowledged as director.
67. Twenty-two of the twenty-six charters members worked in the Geological Survey. Besides Amadeus Grabau, two other foreigners, J. G. Andersson of the Geological Survey and Luella Miner of Yanjing University, also participated in the founding of the Geological Society. Andersson was mining advisor to the Chinese government since 1914 and Luella Miner wrote the first Chinese-language textbook of geology *Dizhixue* 地質學 (Geology) (Beijing: Xiehe nü shuyuan, 1911) under her Chinese name, Mai Meide 麥美德.
68. The first officers of the Geological Society also included Weng Wenhao and Li Siguang as vice presidents; Xie Jiarong as secretary; Li Xueqing as treasurer; Ding Wenjiang as editor of *BGSC*; and Andersson, Dong Chang, Ding Wenjiang, Wang Chongyou, Wang Lie, and Grabau as councilors.
69. Davidson Black, "The Geological Society and Science in China," *BGSC* 1, nos. 1–4 (1922): 10.
70. The Science Society was founded by Chinese students studying abroad in 1915, at Cornell University, and its headquarters moved to Shanghai in 1918. It comprised twelve specialized fields and also had many social scientists as members. In applied fields, Chinese organized much earlier than in the sciences. The Chinese Medical Association was also founded in 1915, but dealt more with licensing and professional issues than research. Similarly, the Chinese Institute of Engineers was established in 1912 by famed engineer Zhan Tianyou. For more on the rise of modern professional through specialist organizations in Republican China, see Xu Xiaoqun. *Chinese Professionals and the Republican State: The Rise of Professional Associations in Shanghai, 1912–1937* (Cambridge: Cambridge University Press, 2001).
71. The Anatomical and Anthropological Association was founded February 26, 1920, and Davidson Black was its first president.
72. The Meteorological Society was founded in 1924 by Zhu Kezhen (竺可楨), one of the original members of the Geological Society. The Paleontological Society of China was founded in 1929 by Yang Zhongjian (one of the founders of the Beida Geological Society) and Sun Yunzhu, who began planning two years earlier while studying together in Germany. The Chinese Society of Mathematical Sciences (precursor to the Chinese Mathematical Society, established 1935) was also founded in 1929. China's Chemical Society formed in 1932 and its Geographical, Botanical, and Zoological Societies all began in 1934. The meteorologist Zhu Kezhen was also instrumental in

establishment of the Geographical Society of China, and several prominent geologists were charter members. Yin Zanxun 尹贊勛. "Zhongguo dizhi xuehui" 中国地質學會 (Geological Society of China), *Kexue dazhong* 科學大衆 (Popular science) 4, no. 6 (1948): 271.

73. Y. C. Sun, "Proceedings of the Second Annual Meeting," *BGSC* 3, no. 1 (1924): 3–4. In 1929 the constitution was amended once more to create nonvoting memberships for "persons interested in geology and allied sciences." "Constitution of the Geological Society of China," *BGSC* 8, no. 1 (1929): 11–13 (Chinese version unpaginated facing pages 10 and 11). Subsequently, the title of "fellows" was abandoned in favor of simply "members," with actively interested nongeologists grouped with students under the category of "associates."
74. Of these non-Chinese members, twenty-two were fellows and one was an associate. Twenty-four of the members were also in the Geological Survey of China. "Membership, 1922," *BGSC* 1, no. 1–4 (1922): 97–99.
75. In 1924, the constitution was amended to make any person holding the office of president a permanent member of the council, and this was later changed to a three-year term. In 1931, the vice president's position was abolished.
76. Article XI of the Society's constitution states that the Society "shall hold one general meeting each year" and Article XII allows for "special meetings." However in the *BGSC* the former are referred to as "annual meetings" and occasional meetings are either called "general meetings" or "special meetings," depending on whether or not they are associated with an event, such as the visit of the Crown Prince of Sweden. After the Geological Society moved its official headquarters to Nanjing in 1936 and the Chinese geological community was no longer as heavily concentrated in Beijing, occasional meetings were no longer called "general meetings" and happened much less frequently.
77. Government ministries did offer donations at various times, but these were not always received and such donations were never regular sources of income. For more on the significance of private funding institutions in the development of Republican-era science, see Laurence Schneider, "The Rockefeller Foundation, the China Foundation, and the Development of Modern Science in China," *Social Science & Medicine* 16 (1982): 1217–21; and Yang Cuihua 楊翠華, *Zhongjihui dui kexue de zannzhu* 中基會對科學的贊助 (The China Foundation's sponsorship of science) (Taibei: Zhongyang yangjiuyuan jindaishi yanjiusuo, 1991). Specifics of foundation contributions to the Survey, which in turn made contributions to the Society, can be found in Yang Tsuihua Lee, "Geological Sciences in Republican China, 1912–1937" (PhD diss., State University of New York at Buffalo, 1985), 60–63, 133–34.
78. Corporate membership dues were reduced to $200, but with fewer benefits than the $400 level.
79. V. K. Ting, "The Aims of the Geological Society of China," *BGSC* 1, nos. 1–4 (1922): 8.
80. Roy Chapman Andrews, "China as a Field for Scientific Research," *BGSC* 1, nos. 1–4 (1922): 8.
81. Most of these articles were in the vein of travel descriptions, native place reports, or event narratives. Student field trips were sometimes reported as news items with scientific interest. Local gazetteers still largely concentrated on physical, political, and economic geography rather than incorporating modern geology.
82. Because of repeated reorganizations in the new Republican government, the Geological Survey of China published reports first in the Ministry of Industry and Commerce *gongbao* and then in the Ministry of Agriculture and Commerce *gongbao*.

83. *Kexue* was established in 1915, but because it was published overseas before 1918, authors in China found it difficult to publish in the Science Society's journal until after it relocated to Shanghai and later Nanjing.
84. *Science* dealt with the promotion of science and scientific spirit rather than technical material, though substantive articles increased after the late 1920s.
85. Typical of these contributions were introductory pieces such as "Zhongguo dizhi gangyao" "中國地質綱要" (Essentials of Chinese geology) or informal discussions, such as "Mantan Sichaun de dizhi" "漫談四川的地質" (Musings on the geology of Sichuan). Often these articles did highlight new insights into China's geology, but they summed up relatively solid advances rather than the cutting edge of active research.
86. There were several "natural history" magazines in China around this time, but the most active in Beijing from 1920 to 1925 was published by the Beijing Upper Normal Society for Natural History (北京高等師範博物學會). In 1926 the *Peking Natural History Bulletin*, published by the Peking Natural History Society, which shared significant membership with the Geological Society of China, became the leading publication in the field. For more on this new Society, see William Joseph Haas, *China Voyager: Gist Gee's Life in Science* (Armonk, NY: M. E. Sharpe, 1996).
87. Can be found as part of the Hangzhou geological reports at the Library of the University of Tokyo and is a rebuttal of Ishii Yamajiro's 1909 theory that West Lake was a volcanic formation. Ishii Yamajiro (石井八万次郎), "清國浙江杭州附近地質調查概報" (General report of the geological investigation of the Hangzhou, Zhejiang area of the Qing nation), *Tōkyō chishitsugaku zasshi* 京地質學雜誌 (Journal of the Geological Society of Japan) 16, no. 185 (1909): 53–65. Zhu Kezhen's 1921 theory, based on fieldwork in 1920 is a more widely known response to Ishii.
88. Chung Yu Wang (Wang Chongyou), *Bibliography of the Mineral Wealth and Geology of China* (London: C. Griffin & Co., 1912).
89. Wang Chongyou published a continuation of his earlier work in Shanghai in 1917, but this also had limited circulation. More influential were T. I. Young (楊遵儀), *Bibliography of Chinese Geology up to 1934* (中國地質文獻目錄) (Peiping: National Academy of Peiping, 1935); Y. S. Chi, (計榮森), *Science Bibliography of China, V. Section of Geology* (中國科學著作目錄, 第五組: 地質學) (Nanking: National Research Council, Academia Sinica, 1936); and Y. S. Chi, *Bibliography of Chinese Geology for the Years 1936–1940* (中國地質文獻目錄 民国二十五年至二十九年) (Pehpe: National Geological Survey of China, 1942). For more on Wong and Lee, see chap. 2, n. 136.
90. Ding Wenjiang (as V. K. Ting) had a few articles from the *North China Herald* reprinted in the *Far Eastern Review*. Some of these were in turn reprinted in specialist journals, such as V. K. Ting, "Mineral Resources of China," *Mining Magazine* 17 (1917): 188–90, which did reach a significant audience and was noted for its challenge to the coal estimates of Noah Drake.
91. Wang Chongyou also published or sent news to the *Transactions of American Institute of Mining, Metallurgical, and Petroleum Engineers*, and the *Engineering and Mining Journal*.
92. Weng Wenhao's encyclopedia entry (W. H. Wang, "Geology of China, The" in *The Encyclopaedia Sinica*, ed. Samuel Couling [London: Oxford University Press, 1917], 201–4) opens with the statement, "Unlike astronomy, geology as a science was quite unknown to the ancient Chinese," and this suggests the ways that, despite its open boundaries as a new discipline in China, modern geology was understood to be fundamentally different from the forms of natural historical observation and geo-

graphical knowledge that were so common in earlier Chinese intellectual activity. Based on the article, much of this has to do with the idea of the materiality of the land, which could be used to deduce the historical developments that linked China to global geological phenomena. In contrast to the descriptive nature of this entry and its modest assessment of Chinese geological understanding, Weng Wenhao's acclaimed doctoral dissertation from University of Louvain, 1914, was published in Belgium and widely cited. However, this was a local study of Belgium rather than a contribution to the geology of China, and it was published in French as part of ongoing local debates within an active geological community.

93. Li Siguang's (J. S. Lee's) master's thesis was published as "An Outline of Chinese Geology" in *Geological Magazine* 58, no. 6 (May 1921): 259–65, 324–29, 70–77, 409–20, with the assistance of William Savage Boulton. This also formed the basis of his structural geology class but could not reach beyond that small audience.

94. As Shellen Xiao Wu points out, even though Richthofen's work was the most widely cited study of Chinese geology and geography from the late nineteenth to early twentieth centuries, it was really only his English-language work that had any broad impact in China. This was partially due to the local availability of his 1870–72 letters to the Shanghai Chamber of Commerce (published by the *North China Herald*) and partially due to the wider legibility of English in China, compared to Richthofen's native German. China's leading geologists were all able to work with German sources, but even for them, English was often preferred. Language affected the ability of foreign work to travel in as well as Chinese work to travel out. See Shellen Wu, "Underground Empires," 68.

95. See Davidson Black, *Asia and the Dispersal of Primates: A Study in Ancient Geography of Asia and Its Bearing on the Ancestry of Man* (Peking: Geological Survey of China, 1925); William Diller Matthew, *Climate and Evolution* (New York: New York Academy of Sciences, 1915), which strongly influenced Black.

96. See Émile Argand, "La Tectonique de l'Asie," *International Geological Congress XIII*, vol. 1, no. 5 (1922): 171–372.

97. A few other examples include the *Journal of the West China Border Research Society* (1922) and *The China Journal of Science and Arts* (1923) which served foreign communities different regions of China. There is also a considerable transnational missionary literature on the geology of China.

98. See Nakano Takamasa, "Some Prevailing Trends of the Historical Development of Geosciences in the Far East," *Geoforum* 1, no. 3 (1970): 63–80; and Liang Bo (梁波) and Feng Hui (冯炜), "Mantie dizhi diaochasuo" 满铁地质调查所 (Geological Survey Institute of South Manchuria Railway Company), *Kexue yanjiu* 科学研究 (Scientific research) 20, no. 3 (2002): 251–55.

99. See T. I. Young, *Bibliography of Chinese Geology up to 1934* for a sampling of this scattering.

100. Luella Miner, for instance, was quite capable in Chinese, as was Erik Nyström. Several other Western geologists recorded in their letters, surveys, and diaries that they studied Chinese, but almost none of them were able to work in Chinese or read Chinese geological reports, and they worked with Chinese colleagues either directly in Western languages or indirectly through translators. This includes the most famous foreigners working most closely with Chinese geologists, such as Grabau and de Chardin.

101. Eliot Blackwelder, "Review of *Geology of the Yang-tze Valley (China)*" *Science* 40, no. 1026 (August 28, 1914): 312.

102. In the first few issues, a few speeches were published in Chinese, but Western languages, particularly English, became dominant for all but a few administrative reports, and, pointedly, articles were never abstracted or translated into any other language. The *BGSC*'s policy was to publish exclusively in the author's chosen scientific language.
103. J. S. Lee, "Reflections on Twenty Years' Experience," 23–24. Interestingly, even foreign geologists who lived in China for years, like Grabau, did not learn Chinese.
104. Li Siguang actually considered China's short history of modern science as an asset because it freed Chinese scientists from entrenched schools of thought and conservative tradition.
105. Though its annual compilation of publications ended in 1916, a quick scan of the *International Catalogue of Scientific Literature* from the late nineteenth to early twentieth century will demonstrate the paucity of included material in languages other than English, French, German, and Italian. The researchers from other nations who are listed in the *Catalogue* were those who published in one of the major scientific languages. After World War II, even German declined, for many international scientific conferences renewed their meetings, but German scientists and presentations were also excluded for quite some time.
106. In contrast to the Geological Society of China, the Geological Society of Japan encouraged Japanese scientists to study Asian territories outside of Japan but did not publish the original research of foreign geologists until after World War II, with the exception of a few Chinese students working with Japanese mentors. Japanese Society members did, however, translate and digest important foreign findings of interest to the group.
107. These estimates are based on libraries presently holding the *BGSC*, archival material on library exchanges, and references to the *BGSC* in international conference proceedings. Many more libraries hold individual volumes, especially regarding Peking Man. For detailed exchange stats, see *Zhongguo dizhixue gaikuang* 中國地質學會概況 (*Overview of the Geological Society of China*), ed. Ji Rongsen (Chongqing: Zhongguo dizhi xuehui, 1942).
108. Lee, "Reflections on Twenty Years' Experience," 23.
109. Chang, "On the History of the Geological Science," 7.
110. Black, "Geological Society and Science in China," 10.
111. Andrews, "China as a Field," 8.
112. E. E. Ahnert, "The Geological Society and Science in Asia," *BGSC* 1, nos. 1–4 (1922): 10. Ahnert's language reflects his use of the Geological Society meeting to air his position regarding Soviet, Chinese, and British contacts in Central Asia, and suggests that one of the Society's most important functions was as a meeting ground for foreign researchers with competing imperial ambitions.
113. Ahnert, "Geological Society and Science in Asia," 10.
114. A. W. Grabau, "The Annual Meeting of the Geological Society of China," *Peking Leader*, Sunday, January 14, 1923, p. 4.
115. W. H. Wong, "Proceedings of the Fifth Annual Meeting of the Geological Society of China," *BGSC* 6, no. 1 (1927): 4.
116. Most of these foreign members lived in Beijing, but even those who did not still traveled to meetings and were active in China. C. Y. Hsieh, "Proceedings of the First Annual Meeting of the Geological Society of China," *BGSC* 2, nos. 1–2 (1923): 2.
117. These members were no longer active in Asia, and resident in North America, Eu-

rope, or Japan. Y. C. Sun, "Proceedings of the Second Annual Meeting of the Geological Society of China," *BGSC* 3, no. 1 (1924): 3.

118. With corrections from "Annual Meeting of Geological Society Holds First Day's Session," *Peking Leader*, Sunday, January 6, 1924, p. 7; Sun, "Proceedings of the Second Annual Meeting," 4. Friedrich Solger was an exception because he was not a notable German geologist, but he rendered years of service to Chinese geological education before returning to Germany during World War I. See chap. 2, n. 55 for more.
119. Letter dated June 10, 1922, Second Historical Archives, Nanjing 1038 (2221).
120. Grabau, "Annual Meeting," 5.
121. Ibid.
122. "Distinguished Scientists Speak at Final Session Monday of Geological Society," *Peking Leader*, Wednesday, January 9, 1924, p. 5.
123. "A Feast of Wit at Annual Dinner of Geological Society Saturday," *Peking Leader* Wednesday, January 7, 1925, p. 3.
124. "Feast of Wit," 3.
125. Grabau, "Annual Meeting," 5.
126. No one—not even his many political opponents—would call a man like Ding an *er maozi*, or foreign lackey. Ding, who spent 1925–26 successfully fighting with treaty powers for native rights in Shanghai while feting international geologists in Beijing, derived a great deal of his political influence with both foreigners and Chinese from his reputation as a scientist. Weng and Li were similarly active in patriotic, anti-imperial efforts, and younger men like Yang Zhongjian, Yuan Fuli, Xie Jiarong, and Huang Jiqing all managed to balance their ease with foreigners with reputations for national feeling.
127. C. Y. Hsieh, "Proceedings of the Society" *BGSC* 1, nos. 1–4 (1922): 32. Hayasaka later taught at Taihoku University, on Taiwan.
128. Japanese geologists had, for example, represented Chinese geology at the preceding two International Geological Congresses, and they were called upon to provide information for a Geological Map of the World proposed in 1913.
129. Hsieh, "Proceedings of the Society," 37.
130. A. W. Grabau, "The First Year of the Third Asiatic Expedition's Activities: Report of Last Friday's Meeting," *Peking Leader*, Tuesday, October 3, 1922, p. 2. According to Charles Gallenkamp's biography, however, Andrews seems to have actually done a significant amount of military reconnaissance. *Dragon Hunter: Roy Chapman Andrews and the Central Asiatic Expeditions* (New York: Viking, 2001).
131. This was done with a Bausch and Lomb Opaque/Slide Balopticon. See http://www.balopticon.com for more information.
132. Grabau, "First Year of the Third Asiatic Expedition's Activities," 2.
133. Again, this exchange highlighted the tense geopolitical relationship between the USSR and China concerning both Central Asia and Mongolia. Roy Chapman Andrews and his crew very pragmatically worked with both Chinese and Mongolian authorities. Though Mongolia was officially recognized as part of China, the USSR supported an "independent" Mongolian administration whose day to day activities Chinese authorities had little power to contest.
134. Four papers were submitted by the Geological Survey including Ding's work on the tectonic geology of Yunnan, Weng's paper on the geological conditions of earthquake centers in China, Weng and Grabau's joint paper on Carboniferous formations in China, and J. G. Andersson's report on the Cenozoic deposits of China.

Weng read the first three papers at the Congress, in French and English, and Andersson presented his own work.

135. China participated nominally in three previous Congresses. In 1906 the Imperial government appointed Clarence Day as its representative but he did not appear at the actual meeting. In 1910, the government sent Djin Da Min, a minor diplomatic official from the Qing Empire's European legation. He was elected a vice president and councillor of the Congress as a representative of China, but did not participate in any scientific sessions. In 1913, the new Republic of China was formally represented by Parkin Wong, a graduate of Cornell University with a master's degree in mining. Wong was also elected a vice president and councillor without giving any scientific papers. Too much has been made of Weng Wenhao's election as vice president and councillor at the 13th International Congress in 1922, since these were ceremonial titles granted to all official delegates, but the achievement of presenting four papers from China's newly founded Geological Survey was an important milestone and its effect on Chinese confidence is worthy of note. Weng's activities at the Congress aroused great curiosity and were very well received, perhaps because of his familiarity and affection for Belgium, where he completed his undergraduate and doctoral degrees almost a decade earlier. See Li Xuetong, "Zhongguo dizhi shiye chuqi ruogan shishi kao," 68–69.
136. "Mongolian Explorers Meet and Tell of Work Accomplished and Expedition Plans" *Peking Leader*, Wednesday, April 15, 1925, p. 3.
137. Raphael Pumpelly did some work for the imperial government, and Ferdinand von Richthofen completed much more detailed reconnaissance for the foreign-run Shanghai Chamber of Commerce (see chap. 1). Sven Hedin was supported by powerful provincial governor Li Hongzhang, and Richthofen later donated copies of his multivolume *China* to the Qing government, but none of these materials reached either the public or native specialists. Materials were either very selective (Richthofen) or just reports of conclusions without raw data. "Geological Society Hedin Presents Complete Published Collection of His Tibet Exploration to Survey," *Peking Leader*, November 10, 1923.
138. "Much Enthusiasm When Third Asiatic Expedition Reports to Geological Society" *Peking Leader*, Friday, September 25, 1925, p. 1.
139. The Central Asiatic Expedition team also showed field books and explained that the expedition's use of a motor car to gather data for geological sections could not have been detailed enough for military purposes.
140. "Much Enthusiasm," 5.
141. Ibid. Further donations were also made in 1927, (Walter Granger letter to George Sherwood, dated September 21, 1927, in AMNH DVP Spec. Coll. RC Andrews Papers).
142. H. Osborn letter to Childs Frick, dated June 13, 1929, in AMNH Spec. Coll. HF Osborn Papers: "I am still giving an immense amount of time to the Chinese problem. It is one of the most complicated I have ever known, starting with the absolute ignorance of the Chinese, combined with cupidity, in August last when it was first taken up with our government, now involving the Committee on the Preservation of Ancient Objects in Peking, the Chinese Government in Nanjing, the personalities of Granger, Andrews and Minister MacMurray, of the Chinese Minister here and our Secretary of State who is doing splendidly." For more on Henry Fairfield Osborn, see Brian Regal's very informative study *Henry Fairfield Osborn : Race, and the Search for the Origins of Man* (Burlington, VT: Ashgate, 2002).

143. Roy Chapman Andrews, "Andrews Tells About Deadlock" *Peking Leader*, April 20, 1929, pp. 1, 6.
144. Jane Camerini, "Wallace in the Field," *Osiris* 11 (1996): 44–65.
145. See chap. 4 for more on Pleistocene glaciation.
146. C. Y. Hsieh, "Proceedings of the Society," 33. The full paper was published as "L'influence seismogenique de certaines structures geologiques en Chine" in the Congres geol. Int. 13me Session, Belgique, *Comptes Rendus*, fas. 2, 1161–97.
147. "Discussion" (appended to J. S. Lee, "The Fundamental Cause of Evolution of the Earth's Surface Features"), *BGSC* 5, nos. 3–4 (December 1926): 259.
148. See "Hedin Thrills Audiences Here," *Peking Leader* April 26, 1929, pp. 1, 7–8, for an example of the romanticization of exploration.
149. The term "obligatory passage point" refers to the Actor-Network Theory (ANT) of Michel Callon and Bruno Latour. Put simply, in ANT, a focal actor (here the Chinese geologists) performs an act of "translation" (e.g., making Chinese geology a cosmopolitan project) that defines the interests of a heterogeneous group in such a way that those interests align with its own. The focal actor makes itself an obligatory passage point (in this case through the Geological Society) and renders itself "indispensible." For a concise explanation of ANT, see Michel Callon, "Some Elements of a Sociology of Translation: Domestication of the Scallops and the Fishermen of St Brieuc Bay," in *Power, Action and Belief: A New Sociology of Knowledge?*, ed. J. Law (London: Routledge & Kegan Paul, 1986): 196–233. What is interesting here is that it was the Chinese land itself, as a "mutable immobile," so to speak, that made the hosting role of Chinese geologists possible, for anyone interested had to come to China, and invariably, that brought people to Beijing.

CHAPTER FOUR

1. For a thought-provoking look at Republican research on Peking Man in the context of paleoanthropology and humanism, see Sigrid Schmalzer, *The People's Peking Man: Popular Science and Human Identity in Twentieth-Century China* (Chicago: University of Chicago Press, 2008).
2. "Letter to the Research Institute of Geology, Academia Sinica, the Liangguang Geological Survey, and the Geological Survey of the Hunan Construction Bureau," Academia Sinica, Institute of Modern History Archives 17–24 1(2).
3. Additional plans were proposed over the next three years, through the various departments in charge of geology and mining, but ultimately coordination of Chinese geological agencies was managed independently by the institutions themselves, often through the mediation of the survey. See details on file Second Historical Archives, Nanjing 393 (178).
4. *Guomin zhengfu gongbao* 國民政府公報 (Nationalist government bulletin), Fagui di shiwu 法规第 15 (Regulation no. 15) (November 10, 1928): 2–3, cited in Shiwei Chen, "Government and Academy in Republican China: History of Academia Sinica, 1927–1949" (PhD diss., Harvard University, 1998), 76.
5. Yang Tsui-hua, "Geological Sciences in Republican China, 1912–1937," 63.
6. Letter from Weng Wenhao to ministry, November 15, 1928, Academia Sinica, Institute of Modern History Archives 17(4) 1(1).
7. Letter received on April 30, 1929, and undated letter to Ministry of Foreign Affairs, Academia Sinica, Institute of Modern History Archives 17(4) 1(1).
8. Letter from Weng Wenhao to ministry, November 15, 1928.
9. Ibid.

10. Report from Weng Wenhao to ministry, November 19, 1928, Academia Sinica, Institute of Modern History Archives 17–24 1(1).
11. See *Annual Report* of the China Foundation, 1926–29.
12. See "Dizhi kuangchan chenlieguan kuochong jianzhu juankuan diyinian baogao" 地質礦產陳列館擴充建築捐款第一年報告 (First year's report on fundraising for the building expansion of the geological museum) and "Dizhi kuangzhan chenlieguan kuochong jianzhu juankuan diernian baogao" 地質礦產陳列館擴充建築捐款第二年報告 (Second year's report on fundraising for the building expansion of the geological museum), Academia Sinica, Institute of Modern History Archives 17–24 1(1).
13. "Qinyuan ranliao yanjiushi jianzhu luocheng baogao" 沁園燃料研究室建築落成報告 (Report on the completion of the Sinyuan Fuel Laboratory), Academia Sinica, Institute of Modern History Archives 17–24 2(3); and Yang Tsui-hua, "Geological Sciences in Republican China, 1912–1937," 64.
14. Most of the funding for the National Academy of Beiping's Institute of Geology was also from the China Foundation. "Baogao" 報告 (Report), *Guoli Beiping yanjiuyuan yuanwu huibao* 國立北平研究院院務彙報 (National Academy of Beiping Institutional Affairs Report) 1, no. 2 (July 1930): 1–10.
15. Letter from Weng Wenhao to Davidson Black, dated October 18, 1926, in Jia Lanpo and Huang Weiwen, *The Story of Peking Man: From Archaeology to Mystery* (New York: Oxford University Press, 1990), 38.
16. "Cooperation between the National Geological Survey of China and the Peking Union Medical College for Research on Tertiary and Quaternary Deposits in North China," February 14, 1927, IVPP Archives, Davidson Black.
17. Dora Hood, *Davidson Black: A Biography* (Toronto: University of Toronto, 1964), 99.
18. "Peiping Union Medical College: Black, 1925–1944" in Rockefeller Foundation Archives, 601D, Box 39; and Hood, *Davidson Black*, 100.
19. Pei Wenzhong was born January 19, 1904, and died September 18, 1982. He graduated from Peking University Department of Geology in 1927 and entered the Geological Survey. Though trained as a geologist, he is best remembered for his paleoanthropological work on the Peking Man project, especially his discovery of the first skullcap at the end of the 1929 excavation season. Jia Lanpo was born on November 25, 1908 (died July 8, 2001), and graduated from the Huiwen Academy in 1929. He was not formally trained as a geologist but joined the Zhoukoudian excavations as an assistant and trainee in 1931 and eventually became superintendent of the project. Yang, Pei, and later Jia were considered the founders of Chinese archaeology and paleoanthropology.
20. For a good discussion of some of the problems encountered by Hedin, see Li Xuetong, "Zhongrui xibei kexue kaochatuan zujian de zhengyi" 中瑞西北科学考察团组建的争议 (The controversy over the establishment of the Sino-Swedish scientific expedition to northwestern China), *Zhongguo keji shiliao* 25, no. 2 (2004): 95–105. A more complete treatment of the Expedition, its scientific achievements and the dynamics of international collaboration can be found in Zhang Jiuchen and Xu Fengxian 徐凤先, *Zhongguo xibei kexue kaochatuan zhuanlun* 中国西北科学考查团专论 (A focused study of the Chinese expedition to the Northwest) (Beijing: Zhongguo ke xue ji shu chu ban she, 2009). Details on Andrews's difficulties can be found in *Dragon Hunter* and scattered news reports, such as "Andrews Is Hit in Statement by Group in Peking," *Peking Leader*, April 30, 1929, p. 10.

21. Davidson Black was allowed to publish in foreign journals with the consent of the Director of the Survey, but all such papers would also be published in the survey's *Paleontologia Sinica* for domestic readers. "Constitution of the Cenozoic Research Laboratory of the Geological Survey of China," in Lanpo and Weiwen's *Story of Peking Man*, 53–58.
22. See Mary Bullock Brown, *An American Transplant: The Rockefeller Foundation and Peking Union Medical College* (Berkeley: University of California Press, 1980); Peter Buck, *American Science and Modern China, 1876–1936* (Cambridge: Cambridge University Press, 1980); Laurence A. Schneider, "The Rockefeller Foundation, the China Foundation, and the Development of Modern Science in China," *Social Science and Medicine* 16, no. 12 (1982): 1217–21; Lu Yiyi 卢宜宜, "Luokefeile jijinhui de Zhongguo xiangmu (1913–1941)" 洛克菲勒基金会的中国项目(1913～1941) (The Rockefeller Foundation's China projects, 1913–1941), *Zhongguo keji shiliao* 19, no. 2 (1998): 24–28; Zi Zhongyun, "The Rockefeller Foundation and China" *American Studies in China* 2 (1995): 84–121.
23. George Barbour, "Peiping Scientists Tell at Meeting Here of New Approach to 'First Man,'" *Peking Leader*, December 29, 1929, p. 1.
24. Julia Strauss, *Strong Institutions in Weak Polities: State Building in Republican China, 1927–1940* (Oxford: Clarendon Press, 1998), 8.
25. V. K. Ting, "Modern Science in China," *Asia* 36, no. 2 (February 1936): 131–34.
26. Weng spent four months in the field observing the aftermath of the 1920 Haiyuan earthquake in Gansu. Weng Wenhao, "L'influence sèismogenique de certaines structures géologiques en Chine."
27. These included several stations in Japan, which was a world leader in seismology, on Taiwan, in the Phillipines, and in the Dutch East Indies. "Ni ju jianshe dizhen yanjiusuo banfa chengqing" 擬具建設地震研究所辦法呈請 (Submission of proposed plan to build an earthquake research institute), Academia Sinica, Institute of Modern History Archives 17-24 2(1).
28. For more, see Joseph Needham, *Science and Civilisation in China*, vol. 3, *Mathematics and the Sciences of the Heavens and the Earth* (Cambridge: Cambridge University Press, 1959), 624–35.
29. "Ni ju jianshe dizhen yanjiusuo banfa chengqing," Academia Sinica, Institute of Modern History Archives 17-24 2(1).
30. Weng Wenhao, *Dizhen* 地震 (Earthquakes) (Shanghai: Shangwu yinshuguan kexue xiao congshu, 1929).
31. Interestingly, the survey's contract with Lin Xinggui also stipulated that at least one assistant at the seismic station should also receive some training in forestry and help with cultivation of Lin's forest land, in which the Jiufeng station was situated. "Qiyue" 契約 (Contract) dated September 1, 1930, Academia Sinica, Institute of Modern History Archives 17-24 2(1).
32. National Geological Survey of China, *National Geological Survey of China, 1916–1931*, 11.
33. From 1934 to 1936, Li Shanbang also trained at the California Institute of Technology and the Geophysical Institute at Potsdam, where he studied both seismology and geophysical prospecting. During the War of Resistance, the seismic division of the survey became the geophysical prospecting division and Li led the wartime search for lead and zinc. In 1943, Li and fellow geophysicist Qin Xinling designed and constructed a "Yong-Ni" model seismograph (named after Weng Wenhao's honorific name) from scrap materials at Chongqing. This basic design (which impressed

Joseph Needham with its ingenuity) was used to study tremors in Sichuan and Yunnan until the end of the war. "Li Shanbang," "Qian dizhi diaochashuo de dizhen, wutan he diqiu wuli gongzuo" 前地质调查所的地震、物探和地球物理公所 (The seismology, geophysical exploration and geophysics work of the former Geological Survey of China), and "Qian dizhi diaochasuo (1916–1950) de dizhen gongzuo" 前地质调查所 (1916–1950) 的地震工作 (The seismological work of the former Geological Survey of China), in *Qian dizhi diaochasuo (1916–1950) de lishi huigu*, 267–68, 130–40, 144–46; Wang Yangzhi, *Zhongguo dizhi diaochasuo shi*, 31–33).

34. The *Seismological Bulletin of the Chiufeng (Jiufeng) Seismic Station* was established by the Institute of Geology at the National Academy of Beiping in 1930 and published in English for ease of international use. This is not to be confused with the quarterly *Seismological Bulletin* (1932–1936) put out by the Institute of Meteorology of the Academia Sinica. The publication included results from the Institute of Meteorology's twenty ton Weichert seismograph, which became active at the Nanjing Observatory in July 1932. "Qian dizhi diaochashuo de dizhen, wutan he diqiu wuli gongzuo," 131–32.

35. "Ni ju juankuan jianshe ranliao yanjiushi banfa chengqing he shi" 擬具捐款建設燃料研究室辦法呈請核示 (Request for review of proposed plan to fundraise for the construction of a fuel laboratory) dated September 30, 1929 in Academia Sinica, Institute of Modern History Archives 17–24 2(1). See also Weng Wenhao, "Zhongguo shitan zhi fenlei 中國石炭之分類 (Classification of Chinese coal), *Bulletin of the Geological Survey of China* 8 (1926): 59–82; and "Benyuan yu nongkuangbu jianxia Beiping dizhi diaochasuo zuijin gongzuo baogao" 本院與農礦部兼轄北平地質調查所最近工作報告 (Report on recent activities of the Joint Academia Sinica-Ministry of Agriculture and Mines Geological Survey at Beiping), *Guoli zhongyang yanjiuyuan yuanwu yuebao* 國立中央研究院院務月報 (Monthly bulletin of the Academia Sinica) 1, no. 3 (September 1929), 11–12.

36. Li Erkang 李爾康, "Woguo huaxue gongye gaikuang ji fazhan tujing" 我國化學工業概況及發展途徑 (Overview and development of our country's chemical industry), *Gongye zhongxin* 工業中心 (Industrial center) 6, nos. 7–8 (1937): 268–80; and 6, nos. 9–12 (1937): 339–57.

37. "A Feast of Wit at Annual Dinner of Geological Society Saturday" *Peking Leader* Wednesday, January 7, 1925, p. 3.

38. Jin Kaiying studying at both University of Wisconsin at Madison and Columbia University, from which he received a master's degree in chemical engineering. "Qinyuan ranliao yanjiushi jianzhu luocheng baogao" 沁園燃料研究室建築落成報告 (Report of the completion of the Qinyuan fuel laboratory) in Academia Sinica, Institute of Modern History Archives 17–24 2(3).

39. The central government hoped that the survey would gradually move to the national capital, but Weng explained that the National Academy of Beiping helped pay for laboratory expenditures and thus, the facility should remain in Beiping in the same complex as the rest of the survey. "Ni ju juankuan jianshe ranliao yanjiushi banfa chengqing he shi," in Academia Sinica, Institute of Modern History Archives 17–24 2(1); and "Qinyuan ranliao yanjiushi jianzhu luocheng baogao," in Academia Sinica, Institute of Modern History Archives 17–24 2(3).

40. In addition to almost 10,000 yuan of new instruments, equipment and supplies from the survey's existing chemical analysis laboratory were also moved to the new fuel laboratory, where electrical and water supply, temperature and particle regulation, and storage were specially adapted to experimental research. "Plan of the

Fuel Laboratory," in Academia Sinica, Institute of Modern History Archives 17–24 2(3).

41. Most speculations on northern Chinese loess, including the hypotheses of Pumpelly, Richthofen, Barbour, Wissmann, and others, focused on parent material and mode of transport rather than on loess as a soil per se.

42. Letter from Neustreuv to Wong, dated Janurary 30, 1928, in Academia Sinica, Institute of Modern History Archives 08–24 5(4).

43. Xiong Yi 熊毅, "Turang gongzuo shiwu nian" 土壤工作十五年 (Fifteen years of soil research), *Turang jikan* 土壤季刊 (Soils quarterly) 5, no. 3 (1946): 144.

44. Yang Tsui-hua, "Geological Sciences in Republican China," 62; and Wang Yangzhi, *Zhongguo dizhi diaochasuo shi*, 36.

45. Charles F. Shaw, *The Soils of China* (Shanghai: Geological Survey of China, 1931). Interestingly, John Lossing Buck, whose *Land Utilization in China* is the most comprehensive study of Chinese agricultural practice in the Republican period, was married to Pearl S. Buck, author of *The Good Earth*.

46. James Thorp, *Geography of the Soils of China* (Nanjing: National Geological Survey of China, 1936); *Instructions to Chinese Soil Survey Field Parties* (Peiping: National Geological Survey of China, 1935).

47. At various points, the soil laboratory was also supported by limited funds from the Rockefeller Foundation and regional agricultural bureaus.

48. Weng Wenhao was also eager to offer a general map that would add specific advantages over the existing maps consulted by international researchers. In his report from the 1929 Pan Pacific Science Congress, Weng discussed what he considered the important features of a broadly usable geological map, and noted that contrary to the international preference for Japan's maps of south China, none of the maps shown at the Congress were as good as the ones China had for the north and middle of the country. See "Disi ci taipingyang kexue huiyi jilue" 第四次太平洋科學會議紀略 (General notes on the Fourth Pan Pacific Science Congress), 14–16 (Academia Sinica Modern History Institute 中央研究院近史所檔案 17–24 1[2]) on new maps around the China Sea area. This draft was later adapted for publication in *Kexue*.

49. Previously geologists and surveyors were responsible for producing their own maps with the help of draftsmen, but no separate department existed. "Qian dizhi diaochasuo dizhi tushuguan shi" 前地质调查所地质图书馆史 (History of the library of the former Geological Survey of China) and "Zeng Shiying" 曾世英 in *Qian dizhi diaochasuo (1916–1950) de lishi huigu*, 81–83, 266.

50. Zeng Shiying, "Bianzuan liyan" 编纂例言 (Introductory remarks on compilation), *Zhonghua minguo xin ditu* 中華民國新地圖 (*New Atlas of the Republic of China*) (Shanghai: Shanghai shenbao guan, 1934), 5.

51. 1:200,000 in China proper and 1:500,000 in northern and western frontier regions. Ding, Weng, and Zeng also produced a compact *Provincial Atlas of China* (*Zhongguo fensheng xintu 中國分省新圖*) in August 1933, which sold 50,000 copies within half a year and was reprinted three times between 1933 and 1939, with updates as needed. Sun Guanlong 孙关龙, "*Zhonghua mingguo xin ditu* ji qi bianzhizhe zhiyi Zeng Shiying xiansheng"中华民国新地图及其编制者之一曾世英先生 (Zeng Shiying, compiler of the new atlas of the Republic of China), *Zhongguo keji shiliao* 11, no. 2 (1990): 49.

52. Weng Wenhao, "Zhonghua mingguo xin ditu xu" 中華民國新地圖序 (Preface to the New Atlas of the Republic of China), in *Zhonghua Minguo xin di tu* 中華民國新地圖 (New Atlas of the Republic of China), compiled by Ding Wenjiang, Weng

Wenhao, Zeng Shiying, and Shenbao guan 申報館 (Shanghai: Shenbao guan, 1934), 1–10.

53. Japanese forces invaded Shanghai on January 28, 1932, forcing the Nanjing government to retreat to Luoyang in Henan Province until a truce was mediated in May and Japan agreed to pull its forces out of southern China.
54. Japanese advances briefly slowed after May 31, 1933, when the Tanggu Truce made eastern Hebei a demilitarized zone, and Japanese militarists engaged in surveys and planning for later advances.
55. Kirby, *Germany and Republican China*, 86.
56. Ibid., 82–84.
57. Li Siguang did not join Ding and Weng on the *Duli pinglun*. See Ma Shengyun, *Li Siguang nianpu*, 104.
58. Weng's decision was likely influenced by his native place connection to Jiang and his friendship with Ding Wenjiang. Kirby (*Germany and Republican China*, 95) suggests that Weng stayed on in Nanjing but Weng's diaries and administrative records from the survey indicate that he was fully in charge of the day to day activities of the survey and was based in Beijing. See Li Xuetong, *Weng Wenhao nianpu*.
59. William Kirby, "Engineering China: Birth of the Developmental State, 1928–1937," in *Becoming Chinese, passages to Modernity and Beyond*, ed. Wen-hsin Yeh (Berkeley: University of California Press, 2000), 150.
60. Ibid., 153. See Deng Lilan邓丽兰, "Nanjing zhengfu shiqi de zhuanjia zhengzhi lun: sichao yu shijian" 南京政府时期的专家政治论: 思潮与实践 (The rhetoric of expert government during the Nanjing Decade: Theoretical currents and realities), *Tianjin shehui kexue* 天津社会科学 (Tianjin social sciences) 2 (2002), for another view of technocracy in Republican China.
61. In the early 1940s, Weng Wenhao actually became president of the Society of Chinese Engineers, and the fact that a geologist held this position gives some indication of geology's stature in technical circles.
62. Weng Wenhao, "Jianshe yu jihua" 建設與計劃 (Reconstruction and planning), in *Weng Wenhao xiansheng yanlun ji*, 25.
63. Ibid., 23.
64. Ibid., 23.
65. Ibid., 26.
66. Ibid., 23.
67. "Wei lichen jingfei kunnan qingxing qing su ding weichi banfa" 為歷陳經費困難清醒請速定維持辦法 (A sober itemization of funding problems and a request for a speedy determination of a maintenance plan) Academia Sinica, Institute of Modern History Archives 17(4) 1(1).
68. Letter to Research Institute of Geology, Academia Sinica, Liangguang Geological Survey, Geological Survey, Hunan Construction Bureau, Academia Sinica, Institute of Modern History Archives 17–24 1(2).
69. "Chenshu dizhi diaochasuo guanyu nongkuang shiye zhi yide jieguo ji jinxing jihua" 陳述地質調查關於農礦事業之已得結果及進行計劃 (Statement of the Geological Survey's past achievements and future plans concerning agriculture and mining activities), November 5, 1930, Academia Sinica, Institute of Modern History Archives 17–24 2(3).
70. Weng Wenhao, "Jianshe yu jihua."
71. Interview with Cui Kexin; *Qian dizhi diaochasuo (1916–1950) de lishi huigu*, 17.

72. Weng Wenhao, "Wode yijian buguo ruci" 我的意見不過如此 (My views are just these), in *Weng Wenhao xiansheng yanlun ji*, 34–40.
73. Zhu Jiahua completed his doctoral degree in geology at Berlin University in 1920 and, in addition to his many positions at the provincial and national ministerial levels, he played a prominent role in several Chinese geological institutions, including the Geological Society of China, the Liangguang Geological Survey (which he founded and directed), and several university geology departments, most notably those of Zhongshan University and National Central University. However, Zhu's involvement in politics began much earlier, during the Xinhai Revolution of 1911, and unlike Weng Wenhao, Zhu is widely regarded as a Guomindang official rather than a scientist. In fact, so much of his impact on geology was connected to his influence as an administrator and politician that when he was elected as an academician of the Academia Sinica in 1948, he was cited as a "research specialist" rather than a geologist, and he could hardly be called anything else, for his curriculum vitae listed only two scientific papers from his graduate student days. Despite much criticism that Zhu's academic status was largely the result of political clout and patronage, it seems fair to follow historians Zhang Jian and Huang Ting in crediting Zhu as a sincere promoter of geology rather than just a party stalwart. For more on Zhu's complex relationship to the sciences and his understanding of "scientific values" as central to Sun Zhongshan's ongoing revolution, see Zhang Jian 张剑 and Huang Ting 黄婷, "Zhu Jiahua de kexue guannian yu guomin zhengfu shiqi kexue jishu de fazhan" 朱家骅的科学观念与国民政府时期科学技术的发展 (Zhu Jiahua's scientific views and Republican-era development of science and technology), *Jindai Zhongguo* 近代中国 (Contemporary China) no. 14 (2004): 291–321.
74. Weng Wenhao, "Guofang sheji weiyuanhui zhi mudi ji shuoming" 國防設計委員會之目的及說明 (Purpose and clarification of the National Defense Planning Commission), December 1932, Shanghai Academy of Social Sciences, NRC 47 (4) 7.
75. Letter to Research Institute of Geology, Academia Sinica, Liangguang Geological Survey, Geological Survey, Hunan Construction Bureau, Academia Sinica, Institute of Modern History Archives 17–24 1(2).
76. Tungsten was a critical wartime metal used in the production of heat-resistant steel and antimony was vital for hardening lead alloys. Both were critical to munitions manufacture in the first and second world wars. "Guofang gongye chubu jihua caoan" 國防工業初步計劃草案 (Preliminary proposal for the early planning of national defense industries) 1933–34, Shanghai Academy of Social Sciences, NRC 47 (2) 18.
77. The NRC was so secret it was referred to only by address in official documents.
78. Yang Zhongjian, *Yang Zhongjian huiyilu*, 90; Gao Zhenxi and Liu Yi 刘毅, "Zhongguo dizhi bowuguan shiye de fazhan gaikuang" 中国地质博物馆事业的发展概况 (Summary of the development of the Geological Survey of China's museum project) *Dizhixue shi luncong* 3 (1995): 71; de Chardin to Granger, February 18, 1936, Granger-de Chardin papers, box 1, folder 16.
79. The Geological Society moved to Nanjing with the survey, although it raised funds for separate Society offices and no longer shared the survey's space. Beiping geologists not only formed the first regional branch of the Geological Society in Beiping, but they also spearheaded the Society's publication of a Chinese-language journal in 1936. See chap. 5 for more details.
80. Academia Sinica, *Academia Sinica, with Its Research Institutes* (Shanghai: Science

Press, 1929), 2. The complicated history of the Academia Sinica is well-documented by Chen Shiwei's doctoral dissertation "Government and Academy in Republican China: History of the Academia Sinica, 1927–1949," which explains the many shifts in structure, leadership, political support, and research philosophy that inflected the academy and its relationship with the state. For a review of research on the development of the Academia Sinica, especially as a scientific institution, see Zhang Peifu 张培富 and Xia Wenhua 夏文华. "Guojia keyan jigou yu Zhongguo xiandai kexue wenhua—yi 80 nianlai zhongyang yanjiuyuan yanjiu zhe zhuangkuang yu jinlu wei kaocha neirong" 国家科研机构与中国现代科学文化——以80年来对中央研究院研究之状况与进路为考察内容 (National research institutions and modern scientific culture in China: Taking a review of research on the conditions and development of the Academia Sinica as a study object), *Shanxi daxue xuebao (Zhexue shehui kexue ban)* 山西大学学报(哲学社会科学版) (Journal of Shanxi University [Philosophy and social science edition]) 33, no. 6 (2010): 11–17.

81. The official anniversary of the Academia Sinica is generally marked on June 9, 1928, when it held its first annual meeting, but it was approved in late 1927. The institutes of Physical Science, Social Science, Geology, and Meteorology opened in January 1928, and the academy's Organic Law passed in November. Five new institutes were established in 1929 and by 1947, the Academia Sinica had thirteen research divisions and several hundred members.
82. "Guoli zhongyang yanjiuyuan dizhi yanjiusuo zhangcheng" 國立中央研究院地質研究所章程 (Constitution of the Research Institute of Geology, National Academia Sinica), *Guoli zhongyang yanjiuyuan zong baogao* 國立中央研究院総報告 (Annual report) (1929), 14.
83. The fossil record of Class Foraminifera extends back at least to the early Cambrian and both benthic and planktic forms are still plentiful in modern seas. However, Order Fusulinida developed in the late Devonian and thrived until the mass extinction at the end of the Permian, after which they are no longer found.
84. J. S. Lee, "A Graphical Method to Aid Specific Determination of Fusulinoids and Some Results of Its Application to the Fusulinae from N. China," *BGSC* 2, nos. 3–4 (1923): 51–86.
85. For example, "New Terms in the Description of the Fusulinidae," *BGSC* 3, no. 1 (1924): 13–14; "Fusulina from the Pingting Basin, Northeastern Shansi," *BGSC* 3, no. 1 (1924): 15–16; "Grabauina, A Transitional Form between Fusulinella and Fusulina," *BGSC* 3, no. 1 (1924): 51–54; "Classification and Correlation of the Paleozoic Coal-bearing Formation in North China," *BGSC* 5, no. 2 (1925): 113–34, etc.
86. J. S. Lee, "Fusulinidae of North China," *Paleontologia Sinica*, ser. B., vol. 4, fasc. 1 (1927) reprinted in *Li Siguang quanji* 李四光全集 (Complete works of Li Siguang), vol. 3 (Wuhan: Hubei renmin chubanshe, 1996), 86.
87. Of particular note were "Note on Traces of Recent Ice-action in N. China," *Geological Magazine* 59 (Jan 1922): 14–21; "Canon of Marine Transgression in Post-Paleozoic Times," *BGSC* 7, no. 1 (1928): 81–128, "Some Characteristic Structural Types in Eastern Asia and Their Bearing upon the Problem of Continental Movement," *Geological Magazine* 66, no. 782 (August 1929): 358–75; 66, no. 783 (September 1929): 413–31; 66, no. 784 (October 1929): 457–522.
88. "Dizhi diaochasuo xian renyuan yilan" 地質調查所現人員一覽 (Present staff of the Geological Survey at a glance), Second Historical Archives, 393 (135) and "Guoli zhongyang yanjiuyuan ben niandu yanjiu renyuan tongji biao" 國立中央研究院本

年度研究人員統計表 (Statistical chart of this year's staff at the National Academia Sinica), *Annual Report 1929*, 47.

89. Some notable cases were Zhu Sen and He Zuolin. See Chen Qun 陳群 et al., *Li Siguang zhuan* 李四光傳 (Biography of Li Siguang) (Beijing: Renmin chubanshe, 1984), 82–83; and Yang Tsui-hua, "Geological Sciences in Republican China," 87–88.
90. Li Siguang, letter to the Ministry of Industry in April 1931 (Second Historical Archives, 393 [178]), was an explanation of the Research Institute's position that it supported but was not restricted by the 1929 Regulations for Management and Evaluation of National Geological Investigation. See "Guoli zhongyang yanjiuyuan dizhi yanjiusuo shijiu niandu baogao" 國立中央研究院地質研究所十九年度報告 (Year 18 [of the republic] report of the Research Institute of Geology, National Academia Sinica), *Guoli zhongyang yanjiuyuan shijiu niandu zong baogao* 國立中央研究院十九年度総報告 (Year 19 annual report of the Academia Sinica), 194.
91. Letter from Research Institute of Geology to Secretariat of the Central Political Committee, September 11, 1929, Second Historical Archives, 393 (178).
92. Chen Qun, *Li Siguang zhuan*, 82.
93. See, for example, Li Siguang, "Xianzai fanhua yu tan" 現在繁華與炭 (Modern prosperity and coal) (Speech to the Sino-French Work Study School, February 1920), in *Chuanguo dipingxian*, 95–117; "Ranliao de wenti" 燃料的問題 (The fuel problem), *Xiandai pinglun* 現代評論 (Contemporary commentary) 7 (1926): 157–81; "Dongya konghuang zhong Zhongguo mei tie gongji wenti" 東亞恐慌中中國煤鉄供給問題 (The problem of China's coal and iron supplies during the East Asian crisis), *Wuhan daxue like jikan* 武漢大學理科季刊 (Wuhan University science and engineering quarterly) 5, no. 2 (1934): 173–78.
94. Letter to Research Institute of Geology, Academia Sinica, Liangguang Geological Survey, Geological Survey, and Hunan Construction Bureau, Academia Sinica, Institute of Modern History Archives 17–24 1(2).
95. "Zhongyang yanjiuyuan minguo shiqi niandu linshifei yusuan biao" 中央研究院民國十七年度臨時費預算表 (Year 17 [of the republic] interim expenditure calculations chart for the Academia Sinica), *Guoli zhongyang yanjiuyuan shiqi niandu zong baogao* 國立中央研究院十七年度総報告 (Year 17 annual report of the Academia Sinica) (1928): 56–60; and Chen Shiwei, "Government and Academy in Modern China," 98.
96. This siting not only placed these institutes near China's modern industrial center, it also served notice to the Japanese "Shanghai Natural Science Institute" (1923–45) that Chinese science and technology were on the rise. For more information on the Shanghai Natural Science Institute and its research and intelligence activities during the Republican period and the War of Resistance, see text of Juliette Chung, "Transnational Science: The Japanese Establishment of Shanghai Natural Science Institute and the Knowledge of Taxonomy in China, 1923–1945" (paper presented at the annual meeting for the History of Science Society, Vancouver, British Columbia, November 2–5, 2000).
97. "Guoli zhongyang yanjiuyuan dizhi yanjiusuo shijiu niandu baogao," 134.
98. Chen Shiwei, "Government and Academy in Modern China," 177–79.
99. "Dizhi yanjiusuo yaowen" 地質研究所要聞 (Important news from the Research Institute of Geology), *Guoli zhongyang yanjiuyuan yuanwu yuebao*, 1, nos. 5–6 (1929): 47; and Chen Qun, *Li Siguang zhuan*, 81–82.
100. "Guoli zhongyang yanjiuyuan dizhi yanjiusuo shijiu niandu baogao," 140.
101. "Benyuan ge yanjiusuo tushu yiqi shebei zhi jinkuang ji yaowen" 本院個研究所圖

書儀器設備之近況及要聞 (Recent status and important news regarding the books, instruments, and facilities of our research institutes), *Guoli zhongyang yanjiuyuan yuanwu yuebao* 1, no. 9 (March 1930): 22.

102. Chen Qun, *Li Siguang zhuan*, 83.
103. In the early 1930s the Research Institute of Geology also arranged to share laboratory facilities with the institutes of Physics and of Chemistry.
104. In the Communist era, Li's work would be most associated with "geomechanics" but this field did not develop until late in the War of Resistance, and quaternary glaciation elicited significantly more international and domestic debate during the Republican period.
105. Compare, for example, Peng Guangqin 彭光欽, "Kexue de yingyong" 科學的應用 (The application of science), *Duli pinglun* 199 (May 3, 1996): 11–13; Ding Wenjiang, "Zhongyang yanjiuyuan de shiming" 中央研究院的使命 (Mission of the Academia Sinica), *Dongfang zazhi* 東方雜誌 (Eastern miscellany) 32, no. 2 (January 1935): 5–8; and Weng Wenhao, "Yige dapo fanmen de fangfa," 27–33.
106. Li Siguang, *Li Siguang quanji*, vol. 1, *Geology of China*, 65.
107. Ibid., 68.
108. These concerns were a result of Britain and Russia's ongoing Great Game in Central Asia and Britain's negotiations with Tibetan authorities in the Simla Conference in 1913.
109. Letter from Li Siguang to He Shi, November 2, 1953, quoted in Chen Qun, *Li Siguang zhuan*, 113.
110. *Guoli zhongyang yanjiuyuan shiqi niandu zong baogao*, 163.
111. While this is a very loose usage of the term "double bind," it does draw upon interesting resemblances to the communicative paradox proposed by Gregory Bateson in his work on schizophrenia. His theory is quite complex, but in essence a double bind occurs in intense relationships when one party is desperate for the approval of another, and the second offers two (or more) messages at different orders of abstraction that either contradict or negate one another. A classic example is, "You must love me" spoken by a mother to her child. The explicit injunction collides with the implicit demand "of your own free will" and creates a dissonance which, if not resolved, may produce psychological trauma. See Gregory Bateson, *Steps to an Ecology of Mind: Collected Essays in Anthropology, Psychiatry, Evolution, and Epistemology* (Chicago: University of Chicago Press, 1972).
112. "The Third General Meeting," *BGSC* 1, nos. 1–4 (1922): 37; Chen Qun, *Li Siguang zhuan*, 90–91.
113. "Quaternary Glaciation in the Yangtze Valley," in Li Siguang, *Li Siguang quanji*, vol. 2, *Glacial Geology*, 16.
114. After taking a break from Beida to set up the Academia Sinica's Research Institute of Geology in Shanghai, Li resumed teaching at the end of 1929 and from 1931 to 1936 Li balanced duties as institute director and chairman of the Peking University Department of Geology. See Ma Shengyun, *Li Siguang nianpu*, 96, 103.
115. "Proceedings of the Tenth Annual Meeting of the Geological Society," *BGSC* 13, no. 1 (December 1933): 6.
116. J. S. Lee, "Data Relating to the Study of the Problem of Glaciation in the Lower Yangtze Valley," *BGSC* 13, no. 3 (1934): 395; Zhou Yaohua 周耀华, "Lushan disiji bingchuan de san ci da lunzheng" 庐山第四纪冰川的三次大论争 (Three great debates on quaternary glaciation on Lushan), in *Tiandi zongheng* 天地縱横 (Heaven and earth, warp and weft), ed. Yin Peiji 尹培基 (Beijing: Dizhi chubanshe, 1992), 60.

117. Li suggests that the last period of glaciation in Lushan may be correlated to the Würm glaciation in the Alps. "Quaternary Glaciation in the Yangtze Valley," 54; Hermann von Wissmann, "The Pleistocene Glaciation in China," *BGSC* 17, no. 1 (March 1937): 160.

118. In many ways, Agassiz's original concept was indeed incorrect, but it brought the possibility of coordinated global glaciation and climate change into the spotlight. Subsequent theories of multiple glaciation that seem quite prescient from our present point of view were very reasonably disputed both for lack of conclusive evidence and absence of convincing causation. By the 1920s and 1930s alpine glaciation was part of the geological canon, and several theories of widespread continent glaciation were very well respected.

119. Weng Wenhao, "Zhongguo de kexue gongzuo"中國的科學工作 (China's scientific Work), *Duli pinglun* 34 (January 8, 1933): 8.

120. "Quaternary Glaciation in the Yangtze Valley," 18.

121. Cited in Zhou Yaohua, "Lushan disiji bingchuan de san ci da lunzheng," 64.

122. Li was accompanied by former students Li Chunyu and Wang Hengsheng in the Alps (photograph), and he briefly visited the Rockies on his return from England (Ma Shengyun, *Li Siguang nianpu*, 118–20). When debates about Pleistocene glaciation became purely domestic after 1937, Chinese researchers began to turn the issue of expertise and experience on themselves, and often pit those who had studied abroad against those who had not. See, for example, the 1944 papers on Guangxi glaciation in *Dizhi lunping* by Sun Dianqing, Xu Yujian, and Dong Su.

123. "Quaternary Glaciation in the Lushan Area," in *Li Siguang quanji*, 145; and "Quaternary Glaciation in the Yangtze Valley,"18.

124. In a sense, Chinese geologists outside of Li's research institute were caught in a different sort of double bind. If they sided with the dominant foreign view, it could smack of conformism or an acknowledgement of Chinese inexperience, but if they sided with Li, it could be viewed as equally partisan. Given the entanglement of national pride, scientific authority, and geopolitics, an "independent opinion" might not seem very independent. As a result, the Geological Survey's leaders did not publish directly on the problem of Quaternary glaciation, though it was well known that they were skeptical of Li's ideas. Anecdotal evidence from younger members of the geological community indicates that the controversy was a source of persistent tension between the Survey and Research Institute, and Chinese-language publications from junior geologists in the late 1930s and 1940s (as well as disciplinary politics during the Communist era) bear this out. China's geological leaders, however, avoided direct confrontation and worked hard to maintain strong ties across native geological organizations in order to strengthen Republican China's reputation abroad. For a fascinating reinterpretation of these tensions in light of post-1949 developments, see Li Siguang, "Dizhi gongzuozhe zai kexue zhanxian shang zuo le yixie shenme?" 地質工作者在科學戰綫上做了一些什麼? (What have geological workers done on the frontlines of science?) *Xinhua yuebao* 新華月報 (New China monthly) 29 (March 1952): 167–72.

125. Chen Qun, *Li Siguang zhuan*, 100; Thomas M. King, John Sullivan, and Mary Wood Gilbert, eds., *The Letters of Teilhard de Chardin and Lucile Swan* (Washington, DC: Georgetown University Press, 1993), 13–14.

126. Weng Wenhao was recovering from a near-fatal car accident and could not participate. Davidson Black died March 15th.

127. Li Siguang, "Dizhi gongzuozhe zai kexue zhanxian shang zuo le yixie shenme?"

Around the same time, W. Brian Harland, a young Oxford graduate doing missionary work in China published a paper on evidence of recent glaciation in Sichuan, but published in an obscure journal, this work did not garner much attention. Later, Harland became an influential geologist known for his early support of continental drift, his arctic research, and radical ideas that led to the "snowball earth" theory. He maintained contact with the Chinese geological community and visited Li Siguang in the 1970s when Chinese scientists were still very isolated from non-Communist colleagues. W. B. Harland, "On the Physiographical History of Western Szechwan, with Special Reference to the Ice Age in the Red Basin," *Journal of the West China Border Research Society*, ser. B, vol. 15 (1945): 1–19.

128. J. S. Lee, "Confirmatory Evidence of Pleistocene Glaciation from the Huangshan, Southern Anhui," *BGSC* 15, no. 3 (1936): 279. Decades later, Li called this attitude "scientific suspicion." Li Siguang, *Zhongguo disiji bingchuan* 中國第四紀冰川 (Quaternary glaciation in China) (Beijing: Kexue chubanshe, 1975), 50.
129. These claims were expressed in individual papers by Li and other Research Institute members and also integrated in *The Geology of China*.
130. Hermann von Wissmann, "Die quartäre Vergletscherung in China," *Zeitschrift der Gesellschaft für Erdkunde zu Berlin* 7/8: 241–62; Hermann von Wissmann, "The Pleistocene Glaciation in China."
131. The full story of debates on Pleistocene glaciation extends to the present day and is beyond the scope of this book, but the problem was one of the first areas in which Chinese geologists were willing to split from accepted international theories, and it remains a dividing line between many Chinese geologists and their foreign counterparts today.
132. Sun's Revolutionary Alliance formed in 1905 as the fusion of three revolutionary groups calling for the overthrow of the Manchu dynasty and the Wuchang uprising of October 10, 1911, was attributed to branch members of this organization. Sun is reputed to have told Li, then the youngest party member, "Work hard in your studies and you will be a great asset to the nation." Ma Shengyun, *Li Siguang nianpu*, 14.
133. Ma Shengyun, *Li Siguang nianpu*, 25; Wang Zancheng 王贊承, "Xinhai shouyi yangxia guangfu jishi" 辛亥首義陽夏光復紀實 (Eyewitness account of the recovery of Xinhai uprising at Hanyang and Hankou), *Xinhai geming huiyilu* 辛亥革命回憶錄 (Recollections of the Xinhai Revolution), vol. 2 (Beijing: Zhonghua shuju, 1961), 45–46.
134. Li switched to geology from mining in 1915. Ma, *Li Siguang nianpu*, 26–29.
135. Li Xuetong, *Shusheng congzheng*, 108.
136. Ding Wenjiang, "Wo suo zhidao de Weng Yongni" 我所知道的翁詠霓 (The Weng Yongni that I know) *Duli pinglun* 97 (April 22, 1934): 2–21.
137. Quotation from American journalists Theodore White and Annalee Jacoby, *Thunder Out of China* (New York, 1946), 55, cited in William Kirby, *Germany and Republican China*, 95. For more on scientism in the May Fourth period, see Daniel Kwok, *Scientism in Chinese Thought, 1900–1950* (New York: Biblo, 1971).
138. Weng and Jiang were linked by native place ties, and Jiang took such particular care of Weng during his recovery from his 1934 car accident that Weng felt he owed Jiang his life.
139. See Li Siguang, "Guofang yu beifang" 國防與北防 (National defense and the North) *Xiandai pinglun* 8, no. 191 (1928): 183–86.
140. For an example of Li's attitude, see Li Siguang, "Zhanguo hou Zhongguo neizhan de tongji he zhiluan de zhouqi" 戰國後中國内戰的統計和治亂的周期 (Statistics of

China's post-warring states' internal wars and the period of pacification), *Qingzu Cai Yuanpei xiansheng liushiwu sui lunwenji* 慶祝蔡元培先生六十五歲論文集 (Essay collection in honor of Cai Yuanpei's sixty-fifth birthday), vol. 1 (Beiping: Zhongyang yanjiuyuan lishi yuyan suo, 1933): 157–66.

141. Ma Shengyun, *Li Siguang nianpu*, 104.
142. Widow of Sun Zhongshan, Song Qingling, was active in Chinese political affairs and later a staunch supporter of the Chinese Communist movement, which she believed to be the true heirs of her husband's republican revolution. Yang Quan was a founding member of the Science Society of China and a trusted secretary of Sun who, though committed to the Guomindang cause, became sympathetic to Communists during the first United Front, and later joined the Third Party opposition to Jiang Jieshi's political domination. The Chinese League for the Guarantee of Civil Rights was divided between those who simply wanted the organization to serve as a watchdog against flagrant abuses of civil liberties and those who hoped to use it to protect the Communist movement, and its history remains contentious. For varying perspectives, see Chen Shiwei's treatment in "Government and Academy in Modern China," chap. 6; Chen Shuyu 陳淑渝 and Tao Xin 陶忻, eds., *Zhongguo minquan baozhang tongmeng* 中國民權保障同盟 (Chinese League for the Guarantee of Civil Liberties) (Beijing: Shehui kexue chubanshe, 1979); Zhou Tiandu 周天度, *Cai Yuanpei zhuan* 蔡元培傳 (Biography of Cai Yuanpei) (Taibei: Xinchao she, 1994), 380–418; Guan Guoxuan 関国宣, "Hu Shi yu Zhongguo minquan baozhang tongmeng" 胡适與中國民權保障同盟 (Hu Shi and the Chinese League for the Guarantee of Civil Liberties), *Zhuanji wenxuan* 傳記文選 (Selected biographies) 52, no. 6: 36–42.
143. Li Siguang, "Tingqiao fenlei biaozhun ji erdieji zhi qi xinshu" 蜓殼分類標準及二叠紀之七新屬 (Classification standards for Fusulinida shells and seven new Permian genuses), *Guoli zhongyang yanjiuyuan dizhi yanjiusuo xiwen jikan* 國立中央研究院地質研究所西文集刊 (Western language journal of the Research Institute of Geology, National Academia Sinica) no. 14 (November 1933): 1–21.
144. Yu Deyuan, who was expelled from Peking University for organizing rail workers after the April 12, 1927, suppression of Communist operatives, was one case, and Li not only hired him but helped him get readmitted to the university in 1929. Ma, *Li Siguang nianpu*, 89–90, 95.
145. By early 1937, after the Xi'an Incident, Jiang had agreed to work with the Communists to fight Japan and establish a second United Front, but Li was already too disillusioned with the Guomindang to meet with Jiang, even though he was actually conducting research on Lushan at the time of Jiang's conference. Ma Shengyun, *Li Siguang nianpu*, 121–22.
146. Several other institutes relocated to Yunnan and funding was restored during the war. Ma Shengyun, *Li Siguang nianpu*, 126; Fan Bozhang范柏樟, "Li Siguang chuangli de Guilin kexue shiyanguan" 李四光创立的桂林科学实验馆 (The Guilin Scientific Laboratory established by Li Siguang), *Zhongguo keji shiliao* 11, no. 1 (1990): 75–78.
147. "Guoli zhongyang yanjiuyuan diyici gongzuo baogao" 國立中央研究院第一次工作報告 (First report of the work of the National Academia Sinica), *Guoli zhongyang yanjiuyuan zong baogao* 國立中央研究院総報告 (Annual report of the National Academia Sinica), vol. 1 (1928): 273. Even among scientists committed to "science for science's sake," very few accepted the view that science should be "aloof from the world," and such opinions were overwhelmingly criticized. For an example, see Gu Yuxiu 顧毓琇, "Xueshu yu jiuguo" 學術與救國 (Scholarship and national salvation), *Duli pinglun* 134 (January 6, 1935): 6–8.

CHAPTER FIVE

1. Sun Zhongshan 孫中山, *Sanmin zhuyi* 三民主義 (Three Principles of the People), (Changsha, China: Yuelu shushe, 2000), 99.
2. This facilitated the Allies' 1941–42 Europe First policy. Chinese were not able to launch a counter offensive until 1944.
3. Prewar trading space for time referred to the GMD policy of making concessions to the Japanese to buy time for Chinese economic and military development. From July 7, 1937, to October 1938, it indicated the military strategy of giving way to the strongest points of Japanese attack and avoiding decisive battles in the eastern flatlands where Japanese mechanical superiority provided the greatest advantage, and closing in on support/communication flanks instead, to slow the lead column's advance. This strategy made it difficult for Japanese forces to actually advance, even when they were moving forward rapidly and winning at every major engagement. For more, see chap. 3 of Lloyd Eastman, *Nationalist Era in China, 1927–1949* (Cambridge: Cambridge University Press, 1986) and chap. 5 of Sun Youli, *China and the Origins of the Pacific War, 1931–41* (London: St. Martin's Press, 1993). As Sun points out, Jiang only committed to the trading space for time strategy after his defeat at Shanghai, and later the idea of a protracted war actually helped the Communists under Mao Zedong gain traction in north China, where they perservered as guerrilla troops.
4. Even after 1941 and foreign-aid programs, fuel and material was often simply unable to reach China's interior except through difficult northern routes from the USSR and by air over the famed "hump."
5. English-language reports by Chinese scholars can be found in *Science*. See, for example, Chung Yu Wang, "Science in China" *Science* 98, no. 2548, new ser. (October 29, 1943): 376–78.
6. James Reardon-Anderson notes that while Chinese engineers were gratified for the opportunity to put their training to use, they often found that their studies were too theoretical or too dependent on state of the art facilities and equipment to be of immediate use under the emergency conditions of China's wartime research program. Many required significant reorientation to adapt, and often they complained that national demands were too short-sightedly utilitarian. James Reardon-Anderson , "Science in Wartime China" in *China's Bitter Victory: The War With Japan, 1937–1945*, ed. Steven Levine and James Chieh Hsiung (Armonk, NY: M. E. Sharpe, 1992), 213–306.
7. Interview with Dr. Chen Ning Yang, by John Israel and Larry Schneider, in Stony Brook, New York, November 1, 1980. Niels Bohr Library & Archives, American Institute of Physics, College Park, MD, USA, http://www.aip.org/history/ohilist/4984.html.
8. Joseph Needham and Dorothy Needham, eds., *Science Outpost: Papers, 1942–1946* (London: Pilot Press, 1948), 13. Romanization follows the original.
9. For entry points into the vast literature on this phenomenon, see Derek de Solla Price, *Little Science, Big Science . . . and Beyond* (New York: Columbia University Press, 1986); and Peter Galison and Bruce Hevly, eds., *Big Science: The Growth of Large Scale Research* (Stanford, CA: Stanford University Press, 1992).
10. See, for example, Liu Shaomin, *Zhonghua dizhixue shi*, 528–35; Wang Yangzhi, *Zhongguo dizhi diaochasuo shi*, 175–92; Huang Jiqing, "Zhongguo dizhi kexue de zhuyao chengjiu (a)" 中國地質科學的主要成就 (a) (Main achievements of Chinese geology [a]), *Zhongguo keji shiliao* 4, no. 3 (1983): 1–11.
11. Weng Wenhao, "Wode yijian buguo ruci," 34–40. See chap. 4, n. 72.

12. Zhu Hanguo 朱汉国, *Nanjing guomin zhengfu jishi* 南京国民政府纪实 (Records of the nationalist government at Nanjing) (Hefei: Anhui renmin chubanshe, 1993), 563.
13. For more on international reaction and the Shanghai offensive, see Sun Youli, *China and the Origins*, 92–97.
14. In fact, the area has harbored several dynasties under threat from invaders or rebels. Perhaps the most famous case is that of Shu-Han during the Three Kingdoms period (220–80 CE).
15. For more on Lianda, see John Israel, *Lianda: A Chinese University in War and Revolution* (Stanford, CA.: Stanford University Press, 1998).
16. Wang Yangzhi, "Yang Zhongjian nianpu" 杨钟健年谱 (Chronicle of Yang Zhongjian), *Xibei daxue xuebao (ziran kexue ban)* 西北大学学报(自然科学版) (Journal of Northwest University [Natural science edition]) 39, no. 2 (1983): 119.
17. Zhang Hongzhao, *Liuliu zishu*, 70; Mazur, *Romance in Natural History*, 399.
18. "Dizhijie xiaoxi" 地質界消息 (Geological community news), *Dizhi lunping* 2, no. 6 (1937): 586; This marked the beginning of wartime cooperation between the Geological Survey and the Science Institute of West China, which aided in procuring boats for the move. See Hou Jiang 侯江, "Kangzhan neiqian Beibei de zhongyang dizhi diaochasuo yu Zhongguo xibu kexueyuan" 抗战内迁北碚的中央地质调查所与中国西部科学院 (Geological Survey of China and Science Institute of West China as relocated to Beibei during the War of Resistance), *Dizhixue kan* 地质学刊 (Geology magazine) 32, no. 4 (2008): 312.
19. "Dizhijie xiaoxi" 地質界消息 (Geological community news), *Dizhi lunping* 3, no. 2 (1938), 205–6. Internal reporting was also stepped up in the form of *jianbao* to speed up transmission of information from the field and keep sensitive information private.
20. C. Y. Hsieh, "The Petroleum Resources of China," *Bulletin of the Geological Survey of China*, no. 30 (1937): English 53–67, Chinese 17–24.
21. "Dizhijie xiaoxi" 地質界消息 (Geological community news), *Dizhi lunping* 3, no. 1 (1938): 95–96.
22. "Dizhijie xiaoxi" 地質界消息 (Geological community news), *Dizhi lunping* 3, no. 4 (1938): 464–65.
23. William Kirby, "The Chinese War Economy," in *China's Bitter Victory: The War with Japan, 1937–1945*, 190.
24. Chinese Ministry of Information, *China Handbook, 1937–1943* (New York: Macmillan Co., 1943), 240.
25. Ibid., 237.
26. Ibid., 243.
27. For more on the logistics and politics surrounding this ambitious airlift program, see John D. Plating, *The Hump: America's Strategy for Keeping China in World War II* (College Station, TX: Texas A&M University Press, 2011).
28. Chinese Ministry of Information, *China Handbook*, 370.
29. Wu Yuexing 武月星, *Zhongguo xiandaishi dituji, 1911–1949* 中國現代史地圖集, 1911–1949 (Atlas of contemporary Chinese history) (Beijing: Zhongguo ditu chubanshe, 1999), 100, 103.
30. *Zhongyang yanjiuyuan chengli wushi zhounian jinian lunwenji* 中央研究院成立五十周年紀念論文集 (Collected papers commemorating the fifteeth anniversary of the founding of the Academia Sinica), vol. 1 (Taibei: Zhongyang yanjiuyuan, 1978), 56–59, 71, 77, 84. See *Acta Brevia Sinensia* for more examples.
31. Kirby, "Chinese War Economy," 198.

32. James Reardon-Anderson, *The Study of Change: Chemistry in China, 1840–1949* (Cambridge: Cambridge University Press, 1991), 310–11.
33. William S. Cooper and Tang Pei-sung, "Science in China," *Science*, 91, no. 2358 (March 8, 1940): 239. For more on Tang's experiences, see Tang Pei-sung, "Aspirations, Reality, and Circumstances: The Devious Trail of a Roaming Plant Physiologist," *Annual Review of Plant Physiology* 34, no. 1 (1983): 1–20.
34. James Reardon-Anderson, "Science in Wartime China," 215.
35. Second Historical Archives, 393 (02872) Weng Wenhao yu Zhu Kezhen de wenjian 翁文灝與竺可楨的文件 (Files of Weng Wenhao and Zhu Kezhen); see also excerpt from letter from Tang Pei-sung to Wm. H. Adolph in "Science Conditions in China," *Science* 100, no. 2590 (August 18, 1944), 142.
36. Joseph Needham, *Chinese Science* (London: Pilot Press, 1945), 13.
37. Qin Xinling 秦馨菱, "Qian dizhi diaochashuo de dizhen, wutan he diqiu wuli gongzuo" 前地质调查所的地震、物探和地球物理公所 (The seismology, geophysical exploration and geophysics work of the former Geological Survey of China), *Qian dizhi diaocha suo (1916–1950) de lishi huigu*, 134.
38. Sun Benwen 孫本文, "Kangzhan qinianlai zhi kexuejie" 抗戰七年來之科學界 (The scientific community of the past seven years of the War of Resistance), in Zhongguo zhanshi xueshu 中國戰時學術 (China's wartime academics) (Beijing: Tiandi chubanshe, 1945), 171; Yang Zhongjian, Letter to wife, December 14, 1941 (personal collection).
39. Needham and Needham, eds., *Science Outpost*, 241–42.
40. Cooper and Tang Pei-sung, "Science in China," 239.
41. The Guangxi clique retained power in Guangxi Province and parts of Guizhou throughout the war and though they were very supportive of the resistance effort against the Japanese, they were wary of Jiang Jieshi's power and created an environment where several of his enemies, both CCP and GMD, could operate with relative impunity. See T'ien-wei Wu, "Contending Political Forces," in *China's Bitter Victory*, 51–78.
42. See Second Historical Archives, 374 (39) Li Siguang de xinjian 李四光的信件 (Letters of Li Siguang), and 375 (427) Cheng Houzhi, Cheng Yuqi, Zhang Boziang, Li Siguang deng yu Huang Jiqing laiwang xinjian 1938–1940 程厚之程裕琪張伯香李四光等與黃汲清來往信件(Correspondence between Huang Jiqing and Cheng Houzhi, Cheng Yuqi, Zhang Boziang, Li Siguang, et al., 1938–1940), for numerous examples and details of collaboration with other geological organizations.
43. "Kangzhan shiqi gongchang neiqian shiliao xuanji (1) 抗战时期工厂内迁史料选辑(一) (Selected historical documents on War of Resistance relocation of factories to the Interior [1]), *Minguo dang'an* 民國檔案 (Republican archives) 2 (1987): 36–37.
44. Li Xuetong, *Shusheng congzheng*, 168. Weng was officially conferred as Minister of Economic Affairs on May 23, 1938, but the ministry was fully operational as of January 1, 1938.
45. Huang Jiqing and He Shaoxun 何绍勋, eds., *Zhongguo xiandai dizhixuejia zhuan* 中国现代地质学家传 (Biographies of contemporary Chinese geologists) (Changsha: Hunan kexue jishu chubanshe, 1990), 278; Wang Yangzhi, *Zhongguo dizhi diaochasuo shi*, 182.
46. Yang Tsui-hua, "The Development of Geology in Republican China, 1912–1937," in John Z. Bowers, J. William Hess, and Nathan Sivin, *Science and Medicine in Twentieth-Century China: Research and Education* (Ann Arbor: University of Michigan Press, 1989), 87.

47. Second Historical Archives, Nanjing 375 Suozhang fachu hangao 所長發出函稿 (Outgoing letter drafts of the survey director) 1–11; Second Historical Archives, Nanjing 375 (460) Weng Wenhao, Huang Jiqing laiwang xinjian 翁文灝、黃汲清來往信件 (Correspondence between Weng Wenhao and Huang Jiqing).
48. The Academia Sinica's Research Institute of Geology functioned wholly independently, though it also cooperated with the survey and regional geological programs from time to time.
49. Wang Yangzhi, "Xie Jiarong—Kancha Zhongguo shiyou dizhi de xianqu" 謝家榮—勘查中國石油地質的先驅 (Xie Jiarong: Pioneer in the exploration of China's petroleum geology), *Zhongguo keji shiliao* 12, no. 3 (1991): 54–55, 59. The NRC established the Xukun Railway Prospecting Bureau in June 1940. This was soon renamed the Southwest Mineral Prospecting Office and, under Xie Jiarong's direction, became the national Mineral Prospecting Office in 1942.
50. Yang Zhongjian, *Yang Zhongjian huiyilu*, 89–129.
51. Wang Yangzhi, *Zhongguo dizhi diaochasuo*, 43, 181.
52. "Dizhijie xiaoxi," *Dizhi lunping* 5, no. 5 (1940), 460.
53. This was part of the "develop the northwest" (開發西北 *kaifa xibei*) movement initiated in the early 1930s and intensified after Japanese attacks on the southwestern provinces in the early 1940s. This program was in many ways the precursor of the contemporary "Go West" (西部大开发 xibu da kaifa, lit. "great opening and development of the west) program that was started in 2000 to spur economic growth in the western provinces and autonomous regions. For more, see, for example, Arvinder Singh, "Understanding China's 'Go-West' Campaign," *China Report* 38, no. 1 (February 1, 2002): 121–27.
54. "Dizhijie xiaoxi," *Dizhi lunping* 8, nos. 1–6 (1943): 242.
55. Zhongyang dizhi diaochasuo xibei fensuo 中央地質調查所西北分所 (Northwest Branch of the National Geological Survey), *Zhongyang dizhi diaochasuo xibei fensuo gaikuang* 中央地質調查所西北分所概況 (Overview of the Northwest Branch of the National Geological Survey) (Nanjing: Zhongyang dizhi diaochasuo, 1948), 6–7.
56. Chen Mengxiong 陈梦熊, "Kangri zhanzheng shiqi xibei diqu de dizhi kuangchan diaocha gongzuo" 抗日战争时期西北地区的地质矿产调查工作 (Geological work in northwestern China during the anti-Japanese war period), *Hebei dizhi xueyuan xuebao* 河北地质学院学报 (Journal of Hebei College of Geology) 17, no. 1 (1994): 105–6; Qiao Zuoshi 喬作拭, "Qian dizhi diachasuo xibei fensuo de chuangye jingshen" 前地質調查所西北分所的創業精神 (Pioneering spirit of the Northwest Branch of the former Geological Survey), in *Qian dizhi diaocha suo (1916–1950) de lishi huigu*, 64.
57. *Zhongyang dizhi diaochasuo xibei fensuo gaikuang*, 2–4.
58. Xie Jiarong, "Dizhi diaocha hezuo banfa" 地質調查合作辦法 (Geological survey collaboration methods), *Dizhi lunping* 1, no. 5 (1936): 521–22.
59. *Hunan dizhi diaochasuo gaikuang* 湖南地質調查所概況 (Overview of the Hunan Geological Survey), MS n.d.; Liu Shaomin, *Zhonghua dizhixue shi*, 532.
60. *Jiangxisheng dizhi diaochasuo gaikuang* 江西省地質調查所概況 (Overview of the Jiangxi provincial geological survey) (Nanchang: Jiangxi zhengfu jianshe ting, 1948); Liu Shaomin, *Zhonghua dizhixue shi*, 533.
61. Second Historical Archives 375 (121).
62. Second Historical Archives 375 (124), "Jingjibu ling gedi diaocha 1943–1948 經濟部令各地調查 1943–1948" (Ministry of Economic Affairs Directives, regional surveys 1943–1948), Letter from Xinjiang Construction Bureau, March 29, 1943. Under

Wang Hengsheng, the Xinjiang Survey sent field reports and other geological information to the Ministry of Economic Affairs as secret documents and did not openly publish research until after 1945.

63. For more on Lianda geological education, see Gong Zhaomin 宫兆敏, Sun Jing 孙晶, and Chen Baoguo 陈宝国, "Cong 'juejiao yizai' kan xinan lianda de rencai peiyang" 从'绝徼移栽'看西南联大的人才培养 (Considering the cultivation of talent from the perspective of "transplants to the far periphery") *Zhongguo dizhi jiaoyu* 中国地质教育 (Chinese geological education), no. 1 (2007): 105–9.
64. There were many advertisements and calls for new geological workers. See, for example, "Jingjibu gongkuang tiaozhengchu zhengji jishu renyuan zhanxing banfa" 經濟部工礦調整処徵集技術人員暫行辦法 (Ministry of Economic Affairs industrial and mining adjustment administration provisional personnel recruitment measures), *Dizhi lunping* 4, no. 2 (1939): 151–52.
65. "Proceedings of the 10th Annual Meeting of the Geological Society," *BGSC* 13, no. 1 (December 1933): 9.
66. "Proceedings of the 11th Annual Meeting of the Geological Society," *BGSC* 14, no. 1 (March 1935): 5–6.
67. "Proceedings of the 9th Annual Meeting of the Geological Society," *BGSC* 12, no. 1 (December 1932): 9.
68. "Proceedings of the 12th Annual Meeting of the Geological Society," *BGSC* 15 (1936): 13. The board of editors was divided into departments for dynamic geology, stratigraphy, structural geology, petrology and mineralogy, vertebrate paleontology, invertebrate paleontology, paleobotany, economic geology, and physiography.
69. "Zhongguo dizhi xuehui chubanpin" 中國地質學會出版品 (Publications of the Geological Society of China), *Dizhi lunping* 2, no. 3 (1937): 304. It also welcomed contributions from interested nonmembers. "*Dizhi lunping* shegao jianzhang" 地質論評投稿簡章 (General regulations for draft submissions to the *Dizhi lunping*), *Dizhi lunping* 2, no. 2 (1937): 222.
70. "Zhongguo dizhi xuehui chubanpin," 304.
71. Many foreign papers read by title only at annual meetings were included in the *Bulletin*, and over time foreign participation in the Society became dominated by publication rather than attendance at meetings. The *Bulletin* thus kept China in touch with the outside world, even through times of physical inaccessibility such as the War of Resistance and the later civil war.
72. Weng Wenhao, "Xu" 序 (Introduction), in *Zhongguo dizhi xuehui gaikuang*, 1.
73. Ibid.
74. Xie Jiarong, "Fakan ci" 發刊辭 (Statement on the founding of the journal), *Dizhi lunping* 1, no. 1 (1936): 1.
75. Ibid., 1–2.
76. The Geological Society and *Bulletin* followed the government to Changsha and then Chongqing, but had difficulty finding local printers who could produce high-quality plates.
77. The *Bulletin* had news and notes section at one time, but stopped because it was too "local."
78. Because the Geological Society framed itself as cosmopolitan, the *Bulletin* was in many ways quite sanitized. Unlike established foreign scientific journals, which often encouraged heated debate on matters of scientific importance, the *Bulletin* was designed to present Chinese geology in its best, most welcoming light. Though Chi-

nese geologists had made great advances, they were still concerned about projecting the image of decorum and erudition to the international scientific community.

79. A good example of one of the more productive wartime exchanges can be found in the discussion of conventions for a 1:200,000 geological map of China, which began with Li Chunyu, "Guanyu cezhi ershiwan fenzhiyi dizhitu zhi shangtao (fu taolun san)" 關於測制二十萬分之一地質圖之商討 (附討論三) (Discussion concerning the surveying and production of a 1:200,000 geological map), *Dizhi luping* 5, no. 6 (1940): 549–63. However, the trend continued into the postwar period, when topics such as uranium in Guangxi took on new significance. This discussion further demonstrated the international/outsider–domestic/insider division of labor between the Geological Society's two journals. When Chinese geologists wanted to challenge "Note on Some Uranium Minerals from Eastern Kuangsi" (published in English by Nan Tingzong 南延宗 [Y. T. Nan] and Wu Leibo吴磊伯 [L. P. Wu] in *BGSC* 23, nos. 3–4 (1943): 169–72), they did so within the Chinese readership of the *Lunping* rather than in the *Bulletin*, and the dialogue lasted through vol. 11, nos. 1–2, in 1946.
80. "Proceedings of the 14th Annual Meeting of the Geological Society," *BGSC* 18, no. 1 (March 1938): 1.
81. "Proceedings of the 17th Annual Meeting," *BGSC* 21, nos. 2–4 (August 1941): 119.
82. Weng Wenhao, "Weng Wenhao xiansheng zai zhi dizhi diaochasuo tongren shu" 翁文灝先生再致地質調查所同人書 (Second letter from Weng Wenhao to his colleagues at the Geological Survey), *Dizhi lunping* 3, no. 1 (1938): 98.
83. See, for example, Yang Zhongjian, "Feichang shiqi zhi dizhijie" 非常時期之地質界 (The geological community in extraordinary times), *Dizhi lunping* 2, no. 6 (1937): 509–14.
84. Yang Zhongjian, "Dui rexin caiji guhuashi tongzhi jin yiyan" 對熱心採集骨化石同志進一言 (A word to those comrades who are eager to collect fossil bones), *Dizhi lunping* 5, no. 6 (1940): 543–48.
85. For example, Robert Swinhoe, "Notes on the Island of Formosa," *Journal of the Royal Geographical Society of London* 34 (1864): 6–18; and Francis H. Nichols, "Notes from the Diary of the Late Francis H. Nichols in China," *Bulletin of the American Geographical Society* 37, no. 6 (1905): 339–56.
86. Wang Yangzhi, *Zhongguo shiyou biannianshi* 中国石油编年史 (Annals of Chinese petroleum) (Beijing: Shiyou gongye chubanshe, 1996), 20.
87. Zhang Shuyan 张叔岩, "Woguo zaoqi shiyou gongye de fazhan yu yinjin xishou guowai xianjin jishu" 我国早期石油工业的发展与引进吸收国外先进技术 (Early development of the petroleum industry in China and the introduction and assimilation of advanced foreign technologies), in *Zhongguo shiyou shi yanjiu* 中国石油史研究 (Researches in Chinese petroleum history), ed. Shi Baoheng 石宝珩, Wang Yangzhi, and Liu Bingyi 刘炳义 (Beijing: Shiyou gongye chubanshe, 1992), 189.
88. Noel Pugach, "Standard Oil and Petroleum Development in Early Republican China," *Business History Review* 45, no. 4 (Winter 1971): 452–73.
89. M. L. Fuller, "Explorations in China," *Bulletin of the American Association of Petroleum Geologists* 3 (1919): 116; M. L. Fuller, "Oil Prospects in Northeastern China," *Bulletin of the American Association of Petroleum Geologists* 10, no. 11 (1926): 1103.
90. Eliot Blackwelder, "Moving Underground Water in the Accumulation of Oil and Gas," *Economic Geology* 17, no. 3 (1922): 217.
91. Personal communication, Xie Xuejin.
92. Weng Wenhao and Xie Jiarong, "Gansu yumenkuang baogao" 甘肅玉門油礦報告

(Gansu Province Yumen Min report), *Hunan shiye zazhi* 湖南實業雜誌 (Hunan industrial journal) 54 (1922).

93. Du Rulin 杜汝霖 and Liu Yamin 刘亚民, "Shenqie huainian laoyibei dizhi xuejia Zhang Renjian xiansheng" 深切怀念老一辈地质学家张人鉴先生 (Heartfelt reminiscenses of elder geologist Zhang Renjian), *Shijiazhuang jingji xueyuan xuebao* 石家庄经济学院学报 (Journal of Shijiazhuang University of Economics) 24, no. 2 (2001): 194–208.
94. J. M. Weller, F. A. Sutton, and C. C. Sun, "Geological Reconnaissance in Kansu and Ching-Hai Provinces, Northwest China," MSS 1938 (National Geological Library); K. F. Dallmus, "Weller trip to China, 1937," University of Illinois at Urbana-Champaign Archives, J. Marvin Weller Papers, 1898–1976, record ser. no.: 44/3/21, box 2, J. Marvin Weller Correspondence, Folder D; Zhang Jiangyi 張江一 et al., *Sun Jianchu zhuan* 孫健初傳 (Biography of Sun Jianchu) (Beijing: Shiyou gongye chubanshe, 1989), 62–65.
95. V. A. Obruchev, "Expedition to the Barlyk and Tarbagatai in 1905," MSS 1907.
96. Shen Lisheng 申力生, *Zhongguo shiyou gongye fazhan shi* 中國石油工業發展史 (History of the development of China's petroleum industry) (Beijing: Shiyou gongye chubanshe, 1988), 27.
97. Wong Wen-hao, "Mineral Resources of China," *Memoirs of the Geological Survey of China*, ser. B, vol. 1(1919), was followed by seven volumes of *Zhongguo kuangye jiyao* 中國鑛業紀要 (Summary of mineral resources in China), published from 1921 to 1945.
98. Wang Yangzhi, *Zhongguo shiyou biannianshi*, 29.
99. C. Y. Hsieh, "Petroleum Resources of China," *Report of the XVIII International Geological Congress, USSR* 4 (1940): 283–93.
100. Sun Jianchu recalled one such example in Zhang Jiangyi, *Sun Jianchu zhuan*, 62–63.
101. See, for example, Sun Ronggui 孙荣圭, "Guanyu Zhongguo dizhijie changqi zhenglun de wenti" 关于中国地质界长期争论的问题 (Regarding issues long-debated by Chinese geological circles), *Zhongguo keji bao* 中国科技报 (Chinese science and technology newspaper) July 23, 1986, p. 3; Tao Shilong, "Guanyu 'Zhongguo pinyu lun' " 关于 '中国贫油论' (On the "China is petroleum-poor" theory), *Renmin ribao* 人民日报 (People's daily) September 25, 1986, p. 1. In contrast research in the early 1990s demonstrated that "oil poor" was not a term used in China's petroleum discourse until post-1949. See Shi Baoheng 石宝珩, Wang Yangzhi, and Liu Bingyi 刘炳义, *Zhongguo shiyou shi yanjiu* 中国石油史研究 (Researches on China's petroleum history) (Beijing: Shiyou gongye chubanshe, 1992).
102. C. C. Wang, "On the Oil Geology of N. Shensi," *Bulletin of the Geological Survey of China* 12, no. 20 (1933): English 65–82, Chinese 45–58.
103. Louderback's study of Sichuan was incorporated into Weng's 1919 "Mineral Resources of China."
104. H. C. T'an and C. Y. Lee, "Oil Fields in Szechuan Province," *Bulletin of the Geological Survey of China*, no. 22 (1933): English 1–31, Chinese 1–38.
105. Huang Jiqing to Weng Wenhao letter, January 30, 1941.
106. Liu Lifan 刘立范, Zhong Yuanqiang 张元强, and Zhang Shuyan 张叔岩, *Zhongguo shiyou tongshi* 中国石油通史 (General history of Chinese petroleum), vol. 2 (Beijing: Zhongguo shihua chubanshe, 2003), 163–69. Researchers did locate several natural gas deposits, however.
107. This cooperation was particularly unexpected given how jealously local warlord Ma Fuxiang guarded his province's mineral resources when Sun Jianchu visited in the

late 1930s, and it reflected the urgency of wartime solidarity. For information on Huang's Gansu fieldwork, see "Gansu yongchang jiaoshui yicai dizhi jianbao" 甘肃永昌窖水一带地质简报 (Brief geological report of the Gansu Yongchang-Jiaoshui region) (1942) MSS, National Geological Library Special Collections.

108. For a personal account of this expedition describing the cultural encounter between the scientists of the survey and local leaders, technicians, and natives, see Huang Jiqing, *Tianshan zhi lu* 天山之麓 (Foothills of the Tianshan range) (Chongqing: Duli chubanshe, 1945).

109. T. K. Huang et al., "Report on Geological Investigation of Some Oil-Fields in Sinkiang," *Memoirs of the Geological Survey of China*, ser. A, no. 21 (February 1947); Huang Jiqing et al., "Wusu dushanzi youtian" 乌苏独山子油田 (Wusu Duzhanzi oilfield) (1943) MSS, National Geological Library Special Collections; Liu Lifan et al., *Zhongguo shiyou tongshi*, vol. 2, pp. 180–82.

110. Sun Jianchu, "Gansu yumen youtian dizhi baogao" 甘肃玉门油田地质报告 (Gansu Yumen oil field geological report) Second Historical Archives 375 (122); Shen Lisheng, *Zhongguo shiyou gongye fazhan shi*, 87. For more on this cooperation, see Li Yuanqing 李元卿, "Yumen youtian de kaifa yu guogong hezuo" 玉门油矿的开发与国共合作 (The opening of Yumen Oilfield and Nationalist-Communist cooperation), *Shiyou daxue xuebao (shehui kexue ban)* 石油大学学报 (社会科学版) (Journal of China University of Petroleum [Social Sciences edition]) 16, no. 2. 文化与石油文化 (February 1, 2000): 37–39.

111. Zhang Jiangyi, *Sun Jianchu zhuan*, 97–101.

112. C. H. Pan (Pan Zhongxiang), "Non-Marine Origin of Petroleum in North Shensi, and the Cretaceous of Szechuan, China," *Bulletin of the American Association of Petroleum Geologists* 25, no. 11 (November 1941): 2049–68.

113. "Fazhan Zhongguo youkuang jihua gangyao" 發展中國油礦計劃綱要 (Outline of a plan to develop China's petroleum resources) was written in 1943, but not published until the first volume of *Dizhi zaxun* 地質雜訊 (Miscellaneous news of geology) in 1947. See reprint in Sun Jianchu, *Sun Jianchu di zhi lun wen xuan ji* 孙健初地质论文选集 (Selected geological papers of Sun Jianchu) (Beijing: Shiyou gongye chubanshe, 1998), 311–19; "Cong Zhongguo dili dizhi shuodao shiyou zhi fenbu" 從中国地理地質說到石油之分佈 (Discussion of petroleum distribution based on China's geography and geology) *Ziyuan weiyuanhui jikan* 资源委员会季刊 (National Resources Commission quarterly) 5, no. 3 (1945): 85–89. In the same issue Sun also published his observations on American petroleum prospecting methods.

114. Huang Jiqing et al., "Xinjiang youtian dizhi diaocha baogao" 新疆油田地质调查报告 (Geological report of Xinjiang Oilfields), MSS; T. K. Huang et al., "Report on Geological Investigation of Some Oil-Fields in Sinkiang," *Memoirs of the Geological Survey of China*, ser. A, no. 21 (February 1947).

CONCLUSION

1. Chu Anping editorial in *Guancha*, as quoted in Suzanne Pepper, *Civil War in China: The Political Struggle, 1945–1949* (Lanham, MD: Rowman and Littlefield, 1999): 194–95.

2. In Guomindang histories this is known as the Battle of Xu-Bang.

3. Besides Li Siguang, Zhu Kezhen of the Institute of Meteorology and Tao Menghe of the Institute of Social Sciences were the leading opponents of the move to Taiwan at the Academia Sinica. See Chen Shiwei, "Government and Academy in Republican China," for more details.

4. Li Yang 李揚, "Nanjing jiefang qianxi kejijie de fan banqian" 南京解放前夕科技界的反搬遷 (The scientific community's resistance to relocation on the eve of Nanjing's liberation) *Dizhixue shi luncong* 4 (2002): 208. Li immediately gave these geologists his support and placed his personal savings at the disposal of the Research Institute so that they could make adequate preparations in his absence.
5. Tang Yaoqing 汤耀庆, Pan Yuntang 潘云唐, and Liu Qing 刘庆, "Li Chunyu" 李春昱, in *Zhongguo xiandai dizhixuejia zhuan*, ed. Huang Jiqing and He Shaoxun, vol. 1, p. 267. As it happens, Communist advances occurred much more quickly than anticipated and there was little time between the GMD evacuation and the CCP takeover.
6. Similar scenes occurred at provincial and branch surveys across the country, each of which was taken over at a different time in 1949, as Communist forces rolled down from the north. For examples, see Yang Chaoqun 样超群, "Liangguang dizhi diaochasuo jianshi" 两卦广地质调查所简史 (Brief history of the Liangguang Geological Survey) and Meng Zongzhi 孟宗智, "Qian zhongyang dizhi diaochasuo xibei fensuo" 前中央地质调查所西北分所 (Northwest Branch of the former Geological Survey) in *Dizhixue shi luncong* 4 (2002): 200–202, 203–5.
7. See Li Yang, "Nanjing jiefang qianxi kejijie" (p. 210), and compare entries *Qian dizhi diaochasuo* (pp. 209–10, 221–23, etc.).
8. See *Dizhi yanjiusuo shidi xiuye ji* for new teaching material culled from student fieldwork.
9. There are even references to fieldwork giving geological undergraduates an "edge" in university basketball competitions against math and physics students. "Dizhijie xiaoxi" 地質界消息 (Geological community news), *Dizhi lunping* 2, no. 3 (1937): 310.
10. Examples of these ongoing pedagogical traditions are easily found in memoirs and reminiscences. See, for example, Chi Jishang 池際尚, "Mianhuai Yuan laoshi dui so de guanhuai he peiyang" 緬懷袁老師對我的關懷和培養 (Fond recollections of teacher Yuan's regard and nurturing), in *Taohua man tianxia* 桃花满天下 (Peach blossoms everywhere, students all over the world), ed. Yuan Zunyi 袁遵仪 (Wuhan, China: Dizhi daxue chubanshe, 1993): 38–39; Zhou Mulin 周慕林, "Qian dizhi diaochasuo de huiyi yu fansi" 前地質調查所的回憶與反思 (Recollections and Reflections on the former Geological Survey), in *Qian dizhi diaochasuo*, ed. Cheng Yuqi and Chen Mengxiong, 198–200.
11. The lyrics were composed over two years from 1938 to 1940, and then submitted to the Council of the Geological Society of China for comments. Some slight revisions were made to the original text, but the anthem was accepted with music by Li Jinhui, and it became official in 1941. "Dizhijie xiaoxi: Zhongguo dizhi xuehui huige" 地質界消息—中國地質學會會歌 (Geological community news: The song of the Geological Society of China), *Dizhi lunping* 5, no. 6 (1940): 573–74, plus inset sheet music.
12. "Dizhijie xiaoxi: Zhongguo dizhi xuehui huige," 574.
13. De Margerie, "Geological Map of the World."
14. Roy Macleod and Philip F Rehbock, "Developing a Sense of the Pacific: The 1923 Pan-Pacific Science Congress in Australia," *Pacific Science* 54, no. 3 (2000): 209–25; N. Yamasaki, "Geological Mapping in Japan, Korea and China," in *Proceedings of the First Pan-Pacific Scientific Conference* (Honolulu: Honolulu Star-Bulletin, Ltd., 1921): 677–79.
15. Huang Jiqing, "Sanshi nianlai zhi Zhongguo dizhixue" 三十年來之中國地質學 (Thirty years of Chinese geology), *Kexue* 28, no. 6 (February 1946): 264. This article was written on July 16, 1945.

16. Second Historical Archives, Nanjing 395 (87).
17. Andrew Huang, "The Inflation in China," *Quarterly Journal of Economics* 62, no. 4 (August 1948): 562–75. Huang's article was based on UN data, but precise statistics were difficult to establish because changes were so rapid and varied so greatly across the country. The causes of inflation were vastly expanded military outlays on the part of the government, costs of demobilization, disrupted agricultural and industrial production, and introduction of paper currency with insufficient metal reserves. During the War of Resistance, prices had already risen drastically, and according to Jonathan Spence, a forty-nine-pound bag of flour was only twenty-two yuan in the summer of 1937. *The Search for Modern China* (New York: W. W. Norton Company, 1990), 502.
18. Yang Zhongjian 楊鍾健, "You jige guoji xueshu huiyi tanqi" 由幾個國際學術會議談起 (Taking a few international conferences as a starting point), *Da gongbao (Shanghai ban)* 大公報 (上海版) (L'Impartial [Shanghai edition]), May 25, 1948, p. 3.
19. Zhang Jiuchen, *Dizhixue yu minguo shehui*, 165.
20. Yang Zhongjian, *Xin Yanjie* 新眼界 (New perspectives) (Changsha: Hunan renmin chubanshe, 1986), 236.
21. Yang Zhongjian, *Yang Zhongjian huiyilu*, 135–36, 154, 173.
22. Xia Xiaohe 夏曉和, "Wo suo zhidao de qian dizhi diaochasuo" 我所知道的前地質調查所 (The former Geological Survey that I know), in *Qian dizhi diaochasuo (1916–1950) de lishi huigu*, ed. Cheng Yuqi and Chen Mengxiong, 219–20.
23. Taiwan was an exception for several reasons: Japanese had left significant data and equipment on the island, the island was not directly involved in the civil war, and by early 1948 the central government was considering Taiwan as a possible retreat. All of these factors made fieldwork more practicable and funding more reliable for geology, and Chinese geologists made several very promising discoveries in paleontology, structural geology, petroleum geology, and soil science. See, for example, Chen Bingfan 陳秉範, "Taiwan lao youtian de xin kanfa" 臺灣老油田之心看法 (A new look at Taiwan's old oilfields), *Dizhi pinglun* 13, nos. 1–2 (1948): 141–42; Zhang Lixu 張麗旭, "Taiwan diceng zhi jiantao" 臺灣地層之檢討 (An examination of Taiwan's strata), *Dizhi pinglun* 13, nos. 3–4 (1948): 185–98; and 13, nos. 5–6 (1948): 291–310.
24. Liu Lifan et al., *Zhongguo shiyou tongshi*, 262–66, 269.
25. See Qin Nai 秦鼐, "Zhongguo baiwan fen zhiyi he sanbaiwan fen zhiyi dizhitu de bianji he chuban" 中國百萬分之一和三百萬分之一地質圖的編輯和出版 (Compilation and publication of the 1:1,000,000 and 1:3,000,000 geological maps of China) in *Qian dizhi diaochasuo (1916–1950) de lishi huigu*, ed. Cheng Yuqi and Chen Mengxiong, 59–60. Huang Jiqing also used the postwar lull to compile and edit the unpublished works of Ding Wenjiang, a project that had been delayed for over a decade. The result appeared as *Geological Reports of Dr. V. K. Ting* (Nanjing: National Geological Survey of China, 1947).
26. Zhongguo dizhi diaochasuo 中央地質調查所 (Geological survey of China), *Zhongguo dizhitu* 中國地質圖 (Geological map of China) (Nanjing: Zhongyang dizhi diaocha suo, 1948).
27. Huang Jiqing, "Zhongguo dizhituzhi bianzhi" 中國地質圖之編制 (Compilation of the geological map of China), *Dizhi pinglun* 13, nos. 3–4 (1948): 268–69. For more information on the map project, see Qin Nai, "Zhongguo baiwan fen zhiyi he sanbaiwan fen zhiyi dizhitu de bianji he chuban and Zhang Jiuchen," *Dizhixue yu minguo shehui: 1916–1950*, 191–93.
28. *Zhongyang dizhi diaochasuo xibei fensuo gaikuang*, 2–3. The Northwest Branch Survey

in Lanzhou, Gansu, was not taken over by the Communists until the end of September 1949. See Liu Nailong 刘乃隆, "Wo zai Lanzhou xibei fensuo de banian" 我在兰州西北分所的八年我在蘭州西北分所的八年 (My eight years at the Northwest Branch Survey at Lanzhou), in *Qian dizhi diaochasuo (1916–1950) de lishi huigu*, ed. Cheng Yuqi and Chen Mengxiong, 190–92.

29. Taiwan had been under Japanese possession since the end of the Sino-Japanese War in 1895, and none of China's geologists had ever set foot on the island before. Attendance at the conference was lower than in other years, but the island was not wracked by civil war, and many geologists stayed for field trips. See "Proceedings of the Twenty-Third Annual Meeting Held at Taipei, November 18–20, 1947," *BGSC* 28, nos. 1–2 (1948): 105–6. Chinese geologists, like Ma Tingying, who were assigned to the Taiwan branch remained on the island and later became the core of the Taiwan University Department of Geology.
30. Yin Zanxun, "Zhongguo dizhi xuehui," 275. See Li Siguang, "Di shiba jie guoji dixuehui zai Lundun kaihui" 第十八屆國際地學會在倫敦開會 (The 18th International Geological Congress, held in London), *Dizhi lunping* 13, nos. 5–6 (1948): 350. For an earlier Chinese view on the importance of hosting a major international scientific congress, see Weng Wenhao, "Disici taipingyang kexue huiyi jilue" 第四次太平洋科學會議紀略 (Brief notes on the Fourth Pan Pacific Science Congress), *Kexue* 14, no. 5 (January 1, 1930): 633.
31. Yang Zhongjian, "You jige guoji xueshu huiyi tanqi," 3.
32. Li Chunyu, "Xu" 序 (Introduction), in *Zhongyang dizhi diaochasuo xibei fensuo gaikuang*, 1.
33. "Cooperation Between the National Geological Survey of China and the Peking Union Medical College for Research on Tertiary and Quaternary Deposits in North China," February 14, 1927, IVPP Archives, Davidson Black.
34. "Plan of the Fuel Laboratory," in Academia Sinica, Institute of Modern History Archives 17–24 2(3).
35. This upturned the usual flow of information established in colonial science, which often featured native collectors with little or no understanding of the scientific meaning of their work and foreign scientists who processed the raw materials they received into "science" at the scientific metropoles. An example can be found in chap. 2 in Raphael Pumpelly's early work. For a broader overview, see Jane Camerini, "Remains of the Day: Early Victorians in the Field," in *Victorian Science in Context*, ed. Bernard Lightman (Chicago: University of Chicago Press, 1997): 354–77.
36. Weng Wenhao, "Weng xu," *Dizhi yanjiusuo shidi xiuye ji*, 2.
37. "Dizhixue de xin fangxiang he xin renwu" 地質學的新方向和新任務 (New directions and responsibilities of geology), *Dizhi lunping* 15, nos. 1–3 (1950): 3–4.
38. Cheng Yuqi, "Jiefang shinianlai Zhongguo dizhixue de chengjiu" 解放十年來中國地質科學的成就 (Achievements of Chinese geology in the ten years since liberation), *Dizhi lunping* 19, no. 9 (1959): 387.
39. For an excellent volume tackling the thorny problem of how to approach the early years of the PRC, see Jeremy Brown and Paul G. Pickowicz, eds., *Dilemmas of Victory: The Early Years of the People's Republic of China* (Cambridge, MA: Harvard University Press, 2007).
40. Joseph Needham "Chinese Science Revisited (I)" *Nature* 4345 (February 7, 1953): 239.
41. For an illuminating example of the struggle between "expert" opinions and the "red" mass line, see chaps. 4 and 5 of Schmalzer, *People's Peking Man*.

42. "Dizhixue de xin fangxiang he xin renwu," 地質學的新方向和新任務 (New directions and responsibilities of geology), *Dizhi lunping* 15, nos. 1–3 (1950): 3.
43. This idea can be found in other fields, as is suggested by the panoramic view presented in Frank Dikötter, *The Age of Openness: China Before Mao* (Los Angeles: University of California Press, 2008).

SELECTED BIBLIOGRAPHY

ARCHIVES

Academia Sinica, Institute of Modern History Archives

Columbiana Archives, Columbia University

Davidson Black Collection, Institute of Vertebrate Paleontology and Paleoanthropology (IVPP) Archives

Granger-de Chardin papers, Georgetown University Archives

HF Osborn Papers, American Museum of Natural History (AMNH) Special Collection

J. Marvin Weller Papers, 1898–1976, University of Illinois at Urbana-Champaign Archives

Peking University Archives

Rockefeller Foundation Archives

Roy Chapman Andrews Papers, American Museum of Natural History, Department of Vertebrate Paleontology Special Collection

Second Historical Archives, Nanjing

Shanghai Academy of Social Sciences

EAST ASIAN LANGUAGE MATERIALS

"Baogao" 報告 (Report). *Guoli Beiping yanjiuyuan yuanwu huibao* 國立北平研究院院務彙報 (National Academy of Beiping Institutional Affairs Report) 1, no. 2 (July 1930): 1–10.

"Beida dizhi yanjiuhui xiaji lüxing tuan de tiyi" 北大地質研究會夏季旅行團的提議 (Proposals for the Beida Geological Society's summer tour). *Beijing daxue dizhi yanjiuhui niankan* 國立北京大學地質研究會年刊 (The annual of the geological society of the university, Peking) 1 (October 10, 1921): 1–3.

"Benhui chengli hui yanshuo ci: He Jie jiaoshou yanshuo ci" 本會成立會演説詞: 何杰教授演説詞 (Opening lecture of our Society's inaugural meeting: Professor He Jie's text). *Guoli Beijing daxue dizhi yanjiuhui niankan* 國立北京大學地質研究會年刊 (The annual of the geological society of the university, Peking) 1 (October 10, 1921): 5–6.

"Benhui chuang kan dizhi zazhi jihua shu" 本會創刊地質雜誌計劃書 (Plan for the founding of our Society's geological journal). *Guoli Beijing daxue dizhi yanjiuhui niankan* 國立北京大學地質研究會年刊 (The annual of the geological society of the university, Peking) 1 (October 10, 1921): 12–14.

"Benhui jishi" 本會紀事 (Records of our Society). *Dixue zazhi* 地學雜誌 (Earth studies journal) 1, nos. 3–4 (1910): 4.

"Benhui shisheng chahuahui jishi" 本會師生茶話會記事 (Notes from our Society's teacher-student tea party). *Guoli Beijing daxue dizhi yanjiuhui niankan* 國立北京大學地質研究會年刊 (The annual of the geological society of the university, Peking) 1 (October 10, 1921): 7–9.

"Benyuan ge yanjiusuo tushu yiqi shebei zhi jinkuang ji yaowen" 本院個研究所圖書儀器設備之近況及要聞 (Recent status and important news regarding the books, instruments, and facilities of our research institutes). *Guoli zhongyang yanjiuyuan yuanwu yuebao* 國立中央研究院院務月報 (Monthly bulletin of the Academia Sinica) 1, no. 9 (March 1930): 22.

"Benyuan yu nongkuangbu jianxia Beiping dizhi diaochasuo zuijin gongzuo baogao" 本院與農礦部兼轄北平地質調查所最近工作報告 (Report on recent activities of the joint Academia Sinica-Ministry of Agriculture and Mines Geological Survey at Beiping). *Guoli zhongyang yanjiuyuan yuanwu yuebao* 國立中央研究院院務月報 (Monthly bulletin of the Academia Sinica) 1, no. 3 (September 1929): 11–12.

"Bianji buzhang Luanzhou xunnan Bai lieshi xiaozhuan" 編輯部長灤州殉難白烈士小傳 (Profile of editorial chief Bai who gave his life at Luanzhou). *Dixue zazhi* 地學雜誌 (Earth studies journal) 3, no. 1 (1912): 85.

Cai Yuanpei 蔡元培. "Zai Nankai xuexiao quanxiao huanyinghui yanshuo ci" 在南開學校全校歡迎會演説辭" (Speech at the welcome meeting and Nankai School), May 23, 1917. *Cai Yuanpei jiaoyu wenxuan* 蔡元培教育文選 (Selected writings on education by Cai Yuanpei). Taibei: Fuxing Shuju, 1956.

Chen Bingfan 陳秉範. "Taiwan lao youtian de xin kanfa" 臺灣老油田之心看法 (A new look at Taiwan's old oilfields). *Dizhi lunping* 地質論評 (Geological review) 13, nos. 1–2 (1948): 141–42.

Chen Duxiu 陳獨秀. "Jinri zhi jiaoyu fangzhen" 今日之教育方針 (Today's educational policy). *Qingnian zazhi* 青年雜誌 (La jeunesse) 1, no. 2 (October 15, 1915).

Cheng Yuqi 程裕淇. "Jiefang shinianlai Zhongguo dizhixue de chengjiu" 解放十年來中國地質科學的成就 (Achievements of Chinese geology in the ten years since liberation). *Dizhi lunping* 19, no. 9 (1959): 387–91.

Cheng Yuqi, and Chen Mengxiong 陈梦熊, eds. *Qian dizhi diaochasuo (1916–1950) de lishi huigu* 前地质调查所 (1916–1950) 的历史回顾 (Historical reminiscences of the former Geological Survey [1916–1950]). Beijing: Dizhi chubanshe, 1996.

Chen Mengxiong 陈梦熊. "Kangri zhanzheng shiqi xibei diqu de dizhi kuangchan diaocha gongzuo" 抗日战争时期西北地区的地质矿产调查工作 (Geological work in northwestern China during the anti-Japanese war period). *Hebei dizhi xueyuan xuebao* 河北地质学院学报 (Journal of Hebei College of Geology) 17, no. 1 (1994): 105–6.

Chen Qun 陳群 et al. *Li Siguang zhuan* 李四光傳(Biography of Li Siguang). Beijing: Renmin chubanshe, 1984.

Chen Shuyu 陳淑渝 and Tao Xin 陶忻, eds. *Zhongguo minquan baozhang tongmeng* 中國民權保障同盟 (Chinese League for the Guarantee of Civil Liberties). Beijing: Shehui kexue chubanshe, 1979.

Chi Jishang 池際尚."Mianhuai Yuan laoshi dui so de guanhuai he peiyang" 緬懷袁老師對我的關懷和培養 (Fond recollections of teacher Yuan's regard and nurturing). In *Taohua man tianxia* 桃花满天下 (Peach blossoms everywhere, students all over the world), edited by Yuan Zunyi 袁遵仪. Wuhan, China: Zhongguo dizhi daxue chubanshe, 1993.

Dang Wei 党为. Jin sanshinian lai de meiguo qingshi yanjiu: yi xinqingshi wei xiansuo

近三十年來的美国清史研究: 以新清史为线索 (Thirty years of American Qing history research: Taking the new Qing history as a thread). PhD diss., Peking University, 2010.

Deng Lilan 邓丽兰. "Nanjing zhengfu shiqi de zhuanjia zhengzhi lun: sichao yu shijian" 南京政府时期的专家政治论: 思潮与实践 (The rhetoric of expert government during the Nanjing decade: Theoretical currents and realities). *Tianjin shehui kexue* 天津社会科学 (Tianjin social sciences) 2 (2002): 115–21.

Ding Wenjiang 丁文江. "Gongshangbu shiban dizhi diaocha shuomingshu" 工商部试办地质调查说明书 (Synopsis of pilot geological investigations of the Ministry of Industry and Commerce). Beijing: Gongshangbu, 1913.

———. "Manyou sanji" 漫遊散記 (Random notes from my wanderings). *Duli pinglun* 獨力評論 (Independent review) 1 (July 10, 1932): 13–18.

———. "Wo suo zhidao de Weng Yongni" 我所知道的翁詠霓 (The Weng Yongni that I know). *Duli pinglun* 97 (April 22, 1934): 2–21.

———. "Xu"序 (Introduction). *Dizhi huibao* 地質彙報 (Memoirs of the Geological Survey of China) 1 (1919): 2.

———. "Xuanxue yu renshengguan" 玄學與人生觀 (Metaphysics and the philosophy of life). In *Kexue yu renshengguan* 科學與人生觀 (Science and the philosophy of life), edited by Hu Shi 胡适, 15–44. Shanghai: Yadong tushuguan 1925.

———. "Zhongguo dizhi xuezhe de zeren" 中國地質學者的責任 (The responsibility of Chinese geologists). *Guoli Beijing daxue dizhi xuehui huikan* 國立北京大學地質學會會刊 (Bulletin of the Geological Society of the National University, Peking), no. 5 (1931): 1–13.

———. "Zhongyang yanjiuyuan de shiming" 中央研究院的使命 (Mission of the Academia Sinica). *Dongfang zazhi* 東方雜誌 (Eastern miscellany) 32, no. 2 (January 1935): 5–8.

Ding Wenjiang et al. *Zhonghua Minguo xin di tu* 中華民國新地圖 (New atlas of the Republic of China). Shanghai: Shen bao guan, 1934.

Dizhi diaochasuo 地質調查所. *Dizhi diaochasuo yange shilue* 地質調查所沿革事略 (Short account of the development of the Geological Survey) n.p., 1922.

———. *Dizhi yanjiusuo shidi xiuye ji* 地質研究所師弟修業記 (Student-teacher studies of the Geological School). Beijing: Jinghua yinshu ju, 1916.

———. *Nongshangbu dizhi yanjiusuo yilan* 農商部地質研究所一覽 (Overview of the Geological School of the Ministry of Agriculture and Commerce). Beijing: Jinghua yinshuju, 1916.

"Dizhijie xiaoxi" 地質界消息 (Geological community news). *Dizhi lunping* 2, no. 1 (1937): 89–126.

"Dizhijie xiaoxi" 地質界消息 (Geological community news). *Dizhi lunping* 2, no. 3 (1937): 305–16.

"Dizhijie xiaoxi" 地質界消息 (Geological community news). *Dizhi lunping* 2, no. 6 (1937): 585–98.

"Dizhijie xiaoxi" 地質界消息 (Geological community news). *Dizhi lunping* 3, no. 1 (1938): 95–104.

"Dizhijie xiaoxi" 地質界消息 (Geological community news). *Dizhi lunping* 3, no. 2 (1938): 195–212.

"Dizhijie xiaoxi" 地質界消息 (Geological community news). *Dizhi lunping* 3, no. 4 (1938): 465–68.

"Dizhijie xiaoxi" 地質界消息 (Geological community news). *Dizhi lunping* 5, no. 5 (1940): 459–66.

"Dizhijie xiaoxi" 地質界消息 (Geological community news). *Dizhi lunping* 5, no. 6 (1940): 573–78, plus inset sheet music.

"Dizhijie xiaoxi" 地質界消息 (Geological community news). *Dizhi lunping* 8, nos. 1–6 (1943): 208–51.

"*Dizhi lunping* shegao jianzhang" 地質論評投稿簡章 (General regulations for draft submissions to the *Dizhi lunping*). *Dizhi lunping* 2, no. 2 (1937): 222.

"Dizhixue de xin fangxiang he xin renwu" 地質學的新方向和新任務 (New directions and responsibilities of geology). *Dizhi lunping* 15, nos. 1–3 (1950): 1–8.

"Dizhi yanjiuhui tushu zhanji biao" 地質研究會圖書暫記表 (Provisional inventory of Geological Society book and map collection). *Guoli Beijing daxue dizhi yanjiuhui niankan* 國立北京大學地質研究會年刊 (The annual of the geological society of the university, Peking) 1 (October 10, 1921): 15.

"Dizhi yanjiusuo yaowen" 地質研究所要聞 (Important news from the Research Institute of Geology). *Guoli zhongyang yanjiuyuan yuanwu yuebao* 國立中央研究院院務月報 (Monthly bulletin of the Academia Sinica) 1, nos. 5–6 (1929): 47.

Du Rulin 杜汝霖 and Liu Yamin 刘亚民. "Shenqie huainian laoyibei dezhi xuejia Zhang Renjian xiansheng" 深切怀念老一辈地质学家张人鉴先生 (Heartfelt reminiscenses of elder geologist Zhang Renjian). *Shijiazhuang jingji xueyuan xuebao* 石家庄经济学院学报 (Journal of Shijiazhuang University of Economics) 24, no. 2 (2001): 194–208.

Fan Bozhang 范柏樟. "Li Siguang chuangli de Guilin kexue shiyanguan" 李四光创立的桂林科学实验馆 (The Guilin Scientific Laboratory established by Li Siguang). *Zhongguo keji shiliao* 中国科技史料 (Chinese historical materials of science and technology) 11, no. 1 (1990): 75–78.

Fu Zhenlun 傅振伦. *Zhongguo fangzhi xue* 中国方志学 (The study of Chinese local gazetteers). Fuzhou, China: Fujiansheng difangzhi bianzuan weiyuanhui bangongshi, 1984.

Gao Zhenxi 高振西. "'Sinian' zhi yiyi zai Zhongguo dizhixue shang zhi bianqian" 'Sinian' 之意義在中國地質學上之變遷 ("Variations of the term 'Sinian' as used in Chinese geology"). *Guoli Beijing daxue dizhi xuehui huikan* 國立北京大學地質學會會刊 (Bulletin of the Geological Society of the National University, Peking) 4 (1930): 59–67

———. "Zuo jiaoshi de Ding Wenjiang xiansheng" 做教師的丁文江先生 (Mr. Ding Wenjiang as a teacher). *Ding Wenjiang zhuanji ziliao* 丁文江傳記資料 (Biographical materials of Ding Wenjiang), vol. 1, edited by 朱傳譽 Zhu Chuanyu. Taibei: Tianyi chubanshe, 1979.

Gao Zhenxi and Liu Yi 刘毅. "Zhongguo dizhi bowuguan shiye de fazhan gaikuang" 中国地质博物馆事业的发展概况 (Summary of the development of the Geological Survey of China's museum project). *Dizhixue shi luncong* 地质学史论丛 (Compendium of the history of geology) 3 (1995): 69–75.

Gong Zhaomin 宫兆敏, Sun Jing 孙晶, and Chen Baoguo 陈宝国. "Cong 'juejiao yizai' kan xinan lianda de rencai peiyang" 从'绝徼移栽'看西南联大的人才培养 (Considering the cultivation of talent from the perspective of "transplants to the far periphery"). *Zhongguo dizhi jiaoyu* 中国地质教育 (Chinese geological education), no. 1 (2007): 105–9.

"Gongshangbu luqu dizhi yanjiusuo shisheng chuangao" 工商部錄取地質研究所師生傳告 (Announcement of faculty and students selected for the Geological School of the Ministry of Industry and Commerce). *Zhengfu gongbao* 政府公報 (Government bulletin) (August 5, 1913).

Guan Guoxuan 関国宣. "Hu Shi yu Zhongguo minquan baozhang tongmeng" 胡适與中國民權保障同盟 (Hu Shi and the Chinese League for the Guarantee of Civil Liberties). *Zhuanji wenxuan* 傳記文選 (Selected biographies) 52, no. 6: 36–42.

Gu Lang 顾琅. *Zhongguo shi da kuangchang diaocha ji 中国十大矿厂调查记* (Record of investigation of China's ten largest mines). Shanghai: Shangwu Shuju, 1916.

"Guoli zhongyang yanjiuyuan diyici gongzuo baogao" 國立中央研究院第一次工作報告 (First annual research report of the Academia Sinica). *Guoli zhongyang yanjiuyuan zong baogao* 國立中央研究院総報告 (Annual report) 1 (1928).

"Guoli zhongyang yanjiuyuan dizhi yanjiusuo zhangcheng" 國立中央研究院地質研究所章程 (Constitution of the Research Institute of Geology, National Academia Sinica). *Guoli zhongyang yanjiuyuan zong baogao* 國立中央研究院総報告 (Annual report) 2 (1929).

Guo Shuanglin 郭双林 and Dong Xi 董习. "Lixihuofen yu *Lixihuofen nanjue shuxin ji*" 李希霍芬与《李希霍芬男爵书信集》(Richthofen and *Baron von Richthofen's Letters*). *Shixue yuekan* 史学月刊 (Journal of historical science) 11 (2009): 52–60.

Gu Xiaoshui 谷小水. *"Shaoshu ren" de zeren*『少數人』的責任 (Responsibility of "the few"). Tianjin, China: Tianjin guji chubanshe, 2005.

Gu Yuxiu 顧毓琇. "Xueshu yu jiuguo" 學術與救國 (Scholarship and national salvation). *Duli pinglun* 獨立評論 (Independet review) 134 (January 6, 1935): 6–8.

He Bingdi 何炳棣. *Zhongguo huiguan shilun* 中國會館史論 (Historical essays on Chinese native place associations). Taibei: Taiwan xuesheng shuju, 1966.

Hou Jiang 侯江. "Kangzhan neiqian Beibei de zhongyang dizhi diaochasuo yu Zhongguo xibu kexueyuan" 抗战内迁北碚的中央地质调查所与中国西部科学院 (Geological Survey of China and Science Institute of West China as relocated to Beibei during the War of Resistance). *Dizhixue kan* 地质学刊 (Geology magazine) 32, no. 4 (2008): 312–23.

Hua Hengfang 華衡芳. *Jinshi shibie* 金石識別 (Classification of metals and rocks). Shanghai: Jiangnan zhizaoju, 1872.

———. "Zi xu" 自序 (Author's preface), *Dixue qianshi* 地學淺釋 (Elements of geology). Shanghai: Jiangnan zhizaoju, 1873.

Huang Jinlin 黄金麟. *Lishi, shenti, guojia: Jindai Zhongguo de shenti xingcheng, 1895–1937* 歷史，身體，國家: 近代中國的身體形成 1895–1937 (History, the body, and nation: The formation of the body in modern China). Taibei: Lianjing chuban shiye gongsi, 2001.

Huang Jiqing 黃汲青. "Gansu Yaoshui yidai dizhi jianbao" 甘肅永昌窯水一帶地質簡報 (Draft report of the Gansu Yaoshui region) (1942), MSS, National Geological Library Special Collections.

———. "Sanshi nianlai zhi Zhongguo dizhixue" 三十年來之中國地質學 (Thirty years of Chinese geology) *Kexue* 28, no. 6 (February 1945): 249–64

———. *Tianshan zhi lu* 天山之麓 (Foothills of the Tianshan range). Chongqing, China: Duli chubanshe, 1945.

———. "Xinhai geming qian dizhi kexue de zhongguo xianqu" 辛亥革命前地質科學的中國先驅 (China's Pre-Xinhai Revolution geological pioneers). *Zhongguo keji shiliao* 中國科技史料 (China historical materials of science and technology) 1 (1982): 2–13.

———. "Zhongguo dizhi kexue de zhuyao chengjiu (a)" 中國地質科學的主要成就 (a) (Main achievements of Chinese geology [a]). *Zhongguo keji shiliao* 中國科技史料 (China historical materials of science and technology) 4, no. 3 (1983): 1–11.

———. "Zhongguo dizhituzhi bianzhi" 中國地質圖之編制 (Compilation of the geological map of China). *Dizhi pinglun* 13, nos. 3–4 (1948): 268–69.

Huang Jiqing and He Shaoxun 何绍勋, ed. *Zhongguo xiandai dizhixuejia zhuan* 中国现代地质学家传 (Biographies of contemporary Chinese geologists). Changsha, China: Hunan kexue jishu chubanshe, 1990.

Huang Jiqing et al. "Wusu dushanzi youtian" 乌苏独山子油田 (Wusu Duzhanzi Oilfield) (1943), MSS, National Geological Library Special Collections.

———. "Xinjiang youtian dizhi diaocha baogao" 新疆油田地质调查报告 (Geological report of Xinjiang oilfields), MSS, Huang Haosheng personal collection.

Hu Bosu 胡伯素. "Dizhixi biyesheng ji zaixiao xuesheng renshu zhi tongji" 地質系畢業生及在校學生人數之統計 (Statistics of Department of Geology graduates and enrolled students). *Guoli Beijing daxue dizhi xuehui huikan* 國立北京大學地質學會會刊 (Bulletin of the Geological Society of the National University, Peking) 4 (April 1930): 275–80, plus inset.

Hunan dizhi diaochasuo gaikuang 湖南地質調查所概況 (Overview of the Hunan Geological Survey), MS n.d.

Huo Youguang 霍有光. "Waiguo shili jinru Zhongguo jindai dizhi kuangchan lingyu ji yingxiang" 外国势力进入中国近代地质矿产领域及影响 (The reach and influence of foreign involvement in modern Chinese geology and mining). *Zhongguo keji shiliao* 中国科技史料 (Chinese historical materials of science and technology) 15, no. 4 (1994): 3–20.

Hu Shi 胡适. *Ding Wenjiang zhuan* 丁文江传 (Biography of Ding Wenjiang). Haikou, China: Haikou chubanshe, 1993.

Ishii Yamajiro (石井八万次郎). "清國浙江杭州附近地質調查概報" (General report of the geological investigation of the Hangzhou, Zhejiang area of the Qing nation). *Tokyo chishitsugaku zasshi* 東京地質學雜誌 (Journal of the Geological Society of Japan) 16, no. 185 (1909): 53–65.

Jiangsusheng zhengxie wenshi ziliao weiyuanhui 江苏省政协文史资料委员会, ed. *Yidai zongshi—Sun Yunzhu jiaoshou jinian zhuanji* 一代宗师—孙云铸教授纪念专辑 (Master of a generation—Prof. Sun Yunzhu commemorative collection). Nanjing: Huadong shiyou dizhiju yinshuachang, 1995.

Jiangxisheng dizhi diaochasuo gaikuang 江西省地質調查所概況 (Overview of the Jiangxi Provincial Geological Survey) Nanchang: Jiangxi zhengfu jianshe ting, 1948.

"Jingjibu gongkuang tiaozhengchu zhengji jishu renyuan zhanxing banfa" 經濟部工礦調整処徵集技術人員暫行辦法 (Ministry of Economic Affairs Industrial and Mining Adjustment Administration Provisional Personnel Recruitment Measures) *Dizhi lunping* 4, no. 2 (1939): 151–52.

Ji Rongsen 計榮森, transcriber. *Jiangxuehui jishi* 講學會記事 (Minutes of lecture meetings). Vol. 1 (handwritten) (8/26/1930), meeting notes.

———. *Zhongguo dizhi xuehui gaikuang* 中國地質學會概況 (Overview of the Geological Society of China). Chongqing, China: Zhongguo dizhi xuehui, 1942.

"Kangzhan shiqi gongchang neiqian shiliao xuanji (1) 抗战时期工厂内迁史料选辑(一) (Selected historical documents on War of Resistance relocation of factories to the interior [1]). *Minguo dang'an* 民國檔案 (Republican archives) 2 (1987): 35–53+2.

Kuang Rongguang. "Zhili dizhi tu" 直隸地質圖 (Geological map of Zhili). *Dixue zazhi* 地學雜誌 (Earth studies journal) 1, no. 1 (1910).

———. "Zhili kuangchan tu" 直隸礦產圖 (Mining map of Zhili). *Dixue zazhi* 地學雜誌 (Earth studies journal) 1, no. 2 (1910).

———. "Zhili shiceng guji" 直隸石層古跡 (Paleontological relics of Zhili strata). *Dixue zazhi* 地學雜誌 (Earth studies journal) 1, nos. 3–4 (1910).

"Kuangshi shi tushu shi yuelan guize" 礦石室圖書室閱覽規則 (Regulation for use of the specimen room and library). *Guoli Beijing daxue dizhi yanjiuhui niankan* 國立北京大學地質研究會年刊 (The annual of the geological society of the university, Peking) 1 (October 10, 1921): 12.

Liang Bo (梁波) and Feng Hui (冯炜). "Mantie dizhi diaochasuo" 满铁地质调查所 (Geological Survey Institute of South Manchuria Railway Company). *Kexue yanjiu* 科学研究 (Scientific research) 20, no. 3 (2002): 251–55.

Liang Qichao 梁啟超. *Yinbingshi quanji* 飲冰室全集 (Complete works from the ice drinker's studio). Kunming, China: Yunnan daxue chubanshe, 1941.

Liang Zhu 梁柱. "Cai Yuanpei de xueshu guan ji qi daxue kecheng jianshe sixiang tanxi" 蔡元培的学术观及其大学课程建设思想探析 (Analysis of Cai Yuanpei's academic views and his thoughts on building a university curriculum). *Beijing daxue xuebao (Zhexue shehui kexue ban)* 北京大学学报(哲学社会科学版) (Journal of Peking University [Humanities and social sciences]) 3 (2004): 132–40.

Li Chuanyong 李传永. "Gexin Zhongguo dilixue de xianqu—Zhang Xiangwen" 革新中国地理学的先驱—张相文 (Pioneer who broke new ground in Chinese geography—Zhang Xiangwen). *Keji renwu* 科技人物 (Sci-tech people) 5 (2003): 93–95.

Li Chunyu 李春昱. "Guanyu cezhi ershiwan fenzhiyi dizhitu zhi shangtao (fu taolun san)" 關於測制二十萬分之一地質圖之商討 (附討論三) (Discussion concerning the surveying and production of a 1:200,000 geological map). *Dizhi lunping* 5, no. 6 (1940): 549–63.

Li Erkang 李爾康. "Woguo huaxue gongye gaikuang ji fazhan tujing" 我國化學工業概況及發展途徑 (Overview and development of our country's chemical industry). *Gongye zhongxin* 工業中心 (Industrial center) 6, nos. 7–8 (1937): 268–80; and 6, nos. 9–12 (1937): 339–57.

Li Erong 李鄂荣. "'Dizhi' yici heshi chuxian yu woguo wenxian" '地质'一词何时出现于我国文献 (When did the term "dizhi' first appear in Chinese documents). *Zhongguo keji shiliao* 中国科技史料 (Chinese historical materials of science and technology) 5, no. 3 (1984): 53–57.

Li Siguang 李四光. *Chuanguo dipingxian* 穿过地平线 (Crossing the horizon). Beijing: Baihuawen chubanshe, 1998.

———. "Di shiba jie guoji dixuehui zai Lundun kaihui" 第十八屆國際地學會在倫敦開會 (The 18th International Geological Congress, held in London). *Dizhi lunping* 13, nos. 5–6 (1948): 347–50.

———. "Dizhi gongzuozhe zai kexue zhanxian shang zuo le yixie shenme?" 地質工作者在科學戰綫上做了一些什麼? (What have geological workers done on the frontlines of science?). *Xinhua yuebao* 新華月報 (New China monthly) 29 (March 1952): 167–72.

———. "Dongya konghuang zhong Zhongguo mei tie gongji wenti" 東亞恐慌中中國煤鉄供給問題 (The problem of China's coal and iron supplies during the East Asian crisis). *Wuhan daxue like jikan* 武漢大學理科季刊 (Wuhan University science and engineering quarterly) 5, no. 2 (1934): 173–78.

———. "Guofang yu beifang" 國防與北防 (National defense and the North). *Xiandai pinglun* 8, no. 191 (1928): 183–86.

———. *Li Siguang quanji* 李四光全集 (Complete works of Li Siguang). Wuhan, China: Hubei renmin chubanshe, 1996.

———. "Ranliao de wenti" 燃料的問題 (The fuel problem). *Xiandai pinglun* 現代評論 (Contemporary commentary) 7 (1926): 157–81.

———. "Tingqiao fenlei biaozhun ji erdieji zhi qi xinshu" 蜓殼分類標準及二叠紀之七新屬 (Classification standards for fusulinida shells and seven new Permian genuses). *Guoli zhongyang yanjiuyuan dizhi yanjiusuo xiwen jikan* 國立中央研究院地質研究所西文集刊 (Western language journal of the Research Institute of Geology, National Academia Sinica) no. 14 (November 1933): 1–21.

———. "Zhanguo hou Zhongguo neizhan de tongji he zhiluan de zhouqi" 戰國後中國內戰的統計和治亂的周期 (Statistics of China's post-warring states internal wars and the period of pacification). *Qingzu Cai Yuanpei xiansheng liushiwu sui lunwenji* 慶祝蔡元培先生六十五歲論文集 (Essay collection in honor of Cai Yuanpei's sixty-fifth birthday), vol. 1 (Beiping: Zhongyang yanjiuyuan lishi yuyan suo, 1933): 157–66.

———. *Zhongguo disiji bingchuan* 中國第四紀冰川 (Quaternary glaciation in China). Beijing: Kexue chubanshe, 1975.

Liu Lifan 刘立范, Zhong Yuanqiang 张元强, and Zhang Shuyan 张叔岩. *Zhongguo shiyou tongshi* 中国石油通史 (General history of Chinese petroleum). Vol. 2. Beijing: Zhongguo shihua chubanshe, 2003.

"Liu Ou guanfeisheng guiyue" 留歐官費生規約 (Stipulations for government-sponsored students studying in Europe). *Jiaoyu zazhi* 教育雜誌 (Education journal) 5, no. 11 (1913).

Liu Shaomin 劉昭民. *Zhonghua dizhixue shi* 中華地質學史 (History of Chinese geology). Taibei: Taiwan shangwu yinshu guan, 1985.

Li Xuetong 李学通. "Ding Wenjiang liuying xuefei kao" 丁文江留英学费考 (Study of Ding Wenjiang's overseas study costs). *Lishi yanjiu* 历史研究 (Historical research) 6 (2009): 181–86.

———. "Dizhi diaochasuo yange zhu wenti kao" (Study of several problems concerning the development of the Geological Survey). *Zhongguo keji shiliao* 中国科技史料 (Chinese historical materials of science and technology), no. 4 (2003): 351–58.

———. *Huanmie de meng: Weng Wenhao yu Zhongguo zaoqi gongyehua* 幻灭的梦—翁文灏与中国早期工业化 (Vanishing dream: Weng Wenhao and China's early industrialization). Tianjin, China: Tianjin guji chubanshe, 2005.

———. "Nongshangbu dizhi yanjiusuo shimo kao" 农商部地质研究所始末考 (Investigation of the ins and outs of the Ministry of Agriculture and Commerce Geological School). *Zhongguo keji shiliao* 中国科技史料 (Chinese historical materials of science and technology) 22, no. 2 (2001): 22–23.

———. *Shusheng congzheng: Weng Wenhao* 书生从政: 翁文灏 (A scholar in politics: Weng Wenhao). Lanzhou: Lanzhou daxue chubanshe, 1996.

———. *Weng Wenhao nianpu* 翁文灏年谱 (Chronicle of Dr. Weng Wenhao). Jinan, China: Shandong jiaoyu chubanshe, 2005.

———. "Zhongguo dizhi shiye chuqi ruogan shishi kao" 中国地质事业初期若干史实考 (Investigation of certain historical facts concerning the early period of China's geological endeavors). *Zhongguo keji shi zazhi* 中国科技史杂志 (The Chinese journal for the history of science and technology) 27, no. 1 (2006): 61–74.

———. "Zhongrui xibei kexue kaochatuan zujian de zhengyi" 中瑞西北科学考察团组建的争议 (The controversy over the establishment of the Sino-Swedish scientific expedition to northwestern China). *Zhongguo keji shiliao* 中国科技史料 (Chinese historical materials of science and technology) 25, no. 2 (2004): 95–105.

Li Yang 李揚. "Nanjing jiefang qianxi kejijie de fan banqian" 南京解放前夕科技界的反搬遷 (The scientific community's resistance to relocation on the eve of Nanjing's liberation). *Dizhixue shi luncong* 地質學史論叢 (Compendium of the history of geology) 4 (2002): 206–11.

Li Yuanqing 李元卿, "Yumen youtian de kaifa yu guogong hezuo" 玉门油矿的开发与国共合作 (The opening of Yumen Oilfield and Nationalist-Communist cooperation). *Shiyou daxue xuebao (shehui kexue ban)* 石油大学学报 (社会科学版) (Journal of China University of Petroleum (Social Sciences edition)) 16, no. 2. 文化与石油文化 (February 1, 2000): 37–39.

Lu Kaiyu 卢开宇. "Zhongguo dixue hui zai Xinhai geming qianhou de huodong" 中国地学会在辛亥革命前后的活动 (Activities of the Chinese Earth Studies Society before and after the Xinhai Revolution). *Shi lin* 史林 (Historical circles) 1 (2003): 65–72.

Lu Xun 鲁迅. *Lu Xun yiwen quanji* 鲁迅佚文全集(Complete collection of Lu Xun's lost works). Edited by Liu Yunfeng 刘云峰. Beijing: Qunyan chubanshe, 2001.

Lu Yiyi 卢宜宜, "Luokefeile jijinhui de Zhongguo xiangmu (1913–1941)" 洛克菲勒基金会的中国项目 (1913 1941) (The Rockefeller Foundation's China projects, 1913–1941). *Zhongguo keji shiliao* 中国科技史料 (Chinese historical materials of science and technology) 19, no. 2 (1998): 24–28.

Mai Meide 麥美德 (Luella Miner). *Dizhixue* 地質學 (Geology). Beijing: Xiehe nü shuyuan, 1911.

Mao Shizhen 茆诗珍 and Xu Fei 徐飞. "Liu Mei youtong yu Zhongguo zaoqi dizhixue" 留美幼童与中国早期地质学 (American-educated youth and China's early geology). *Kexue jishu yu bianzhengfa* 科学技术与辩证法 (Science, technology and dialectics) 22, no. 6 (December 2005): 88–91.

Ma Shengyun 马胜云. *Li Siguang nianpu* 李四光年谱 (Chronicle of Li Siguang). Beijing: Dizhi chubanshe, 1999.

Meng Zongzhi 孟宗智. "Qian zhongyang dizhi diaochasuo xibei fensuo" 前中央地质调查所西北分所 (Northwest branch of the former Geological Survey). In *Dizhixue shi luncong* 地质学史论丛 (Compendium of the history of geology) 4 (2002): 203–5.

Neige guanbao 内閣官報 (Inner cabinet government news). 宣統三年九月初七 (October 10, 1918), no. 48, n.p.

Nongshangbu gongbao 農商部公報 (Ministry of Agriculture and Commerce bulletin) 2, no. 7 (February 15, 1916): 49.

"Nongshangbu ling" 農商部令 (Ministry of Agriculture and Commerce decrees). *Nongshang gongbao* 農商公報 (Ministry of Agriculture and Commerce bulletin) 3, no. 5 (1916).

Peng Guangqin 彭光欽. "Kexue de yingyong" 科學的應用 (The application of science). *Duli pinglun* 199 (May 3, 1996): 11–13.

Shen Lisheng 申力生. *Zhongguo shiyou gongye fazhan shi* 中國石油工業發展史 (History of the development of China's petroleum industry). Beijing: Shiyou gongye chubanshe, 1988.

Shi Baoheng 石宝珩, Wang Yangzhi, and Liu Bingyi 刘炳义. *Zhongguo shiyou shi yanjiu* 中国石油史研究 (Researches on China's petroleum history). Beijing: Shiyou gongye chubanshe, 1992.

Song Guangbo 宋广波. "Dizhi yanjiusuo ruogan shishi bukao" 地质研究所若干史实补考 (Reexamination of certain historical facts about the Geological School). *Zhongguo kejishi zazhi* 中国科技史杂志 (The Chinese journal for the history of science and technology) 27, no. 2 (2006): 162–69.

Sun Benwen 孫本文. "Kangzhan qinianlai zhi kexuejie" 抗戰七年來之科學界 (The scientific community of the past seven years of the War of Resistance). In *Zhongguo zhanshi xueshu* 中國戰時學術 (China's wartime academics). Beijing: Tiandi chubanshe, 1945.

Sun Guanlong 孙关龙. "*Zhonghua mingguo xin ditu* ji qi bianzhizhe zhiyi Zeng Shiying xiansheng" *中华民国新地图及其编制者之一曾世英先生* (Zeng Shiying, compiler of the new atlas of the Republic of China). Zhongguo keji shiliao 中国科技史料 (Chinese historical materials of science and technology) 11, no. 2 (1990): 45–52.

Sun Jianchu 孫健初. "Cong Zhongguo dili dizhi shuodao shiyou zhi fenbu" 從中国地理地質說到石油之分佈 (Discussion of petroleum distribution based on China's geography and geology). *Ziyuan weiyuanhui jikan* 资源委员会季刊 (National Resources Commission quarterly) 5, no. 3 (1945): 85–89.

———. "Fazhan Zhongguo youkuang jihua gangyao" 發展中國油礦計劃綱要 (Outline of a plan to develop China's petroleum resources). In *Sun Jianchu di zhi lun wen xuan ji* 孙健初地质论文选集 (Selected geological papers of Sun Jianchu). Beijing: Shiyou gongye chubanshe, 1998.

———. "Gansu yongchang jiaoshui yidai dizhi jianbao" 甘肃永昌窖水一带地质简报 (Brief geological report of the Gansu Yongchang-Jiaoshui region) (1942), MSS, National Geological Library Special Collections.

Sun Ronggui 孙荣圭. "Guanyu Zhongguo dizhijie changqi zhenglun de wenti" 关于中国地质界长期争论的问题 (Regarding issues long-debated by Chinese geological circles). *Zhongguo keji bao* 中国科技报 (Chinese science and technology newspaper), July 23, 1986.

Sun Yutang 孫毓棠. *Zhongguo jindai gongye shiliao, 1840–1895* 中國近代工業史料 *1840–1895* (Historical materials of modern Chinese industry). Beijing: Kexue chubanshe, 1957.

Sun Zhongshan 孫中山. *Sanmin zhuyi* 三民主義 (Three principles of the people). Changsha, China: Yuelu shushe, 2000.

Tao Fangxuan 陶方宣. *Hu Shi de quanzi* 胡适的圈子 (Hu Shi's circle). Jinan, China: Shandong huabao chubanshe, 2010.

Tao Menghe 陶孟和. "Zhuiyi Zaijun" 追憶在君 (Remembering Zaijun). *Duli Pinglun* 獨立評論 (Independent review) 188 (February 16, 1936): 33–34.

Tao Shilong 陶世龙. "Cong Pangpeile dao Weilishi" 从庞培勒到维里士 (From Pumpelly to Willis). *Dizhixue shi luncong* 地质学史论丛 (Compendium of the history of geology) 3 (1995): 15–22.

———. "Guanyu 'Zhongguo pinyu lun'" 关于 '中国贫油论' (On the "China is petroleum-poor" theory). *Renmin ribao* 人民日报 (People's daily), September 25, 1986.

Tian Qirui 田奇瑞 et al. "Ni zuzhi 'Beijing daxue dizhi yanjiuhui' gongqi" 擬組織『北京大學地質研究會』公啓 (Public announcement of the plan to organize a Beijing University Geological Research Society). *Guoli Beijing daxue dizhi yanjiuhui niankan* 國立北京大學地質研究會年刊 (The annual of the geological society of the university, Peking) 1 (October 10, 1921): 2–4.

Wang Ermin 王爾敏. "Zhongguo jindai xuehui yaolun" 中國近代學會約論 (On the study societies in modern China). *Shihuo yuekan* (食貨月刊 (Social and economic conditions monthly) 1, no. 5 (1971): 287–95.

Wang Genyuan 王根元. "Zhongguo dizhi jiaoyu shilue" 中国地质教育史略 (Brief history of geological education in China). *Diqiu kexue—Zhongguo dizhi daxue xuebao* 地球科学-中国地质大学学报 (Earth science journal of China University of Geosciences) 11, no. 2 (March 1986): 207–15.

Wang Gongmu 王恭睦. "Beida dizhi yanjiuhui xiaji lüxingtuan de tiyi " 北大地質研究會夏季旅行團的提議 (Proposal for a Beida Geology Society summer field trip group). *Guoli Beijing daxue dizhi yanjiuhui niankan* 國立北京大學地質研究會年刊 (The annual of the geological society of the university, Peking) 1 (October 10, 1921): 1–4.

Wang Lixin 王立新. "Meiguo chuanjiaoshi yu yapian zhanzheng hou de 'kaiyan kan shijie' sichao" 美国传教士与鸦片战争后的'开眼看世界'思潮 (American missionaries and post-Opium War "Kaiyan kan shijie" thought). *Meiguo yanjiu* 美国研究 (American studies quarterly) 11, no. 2 (1997): 27–51.

Wang Qisheng 王奇生. *Liuxue yu jiuguo* 留学与救国 (Study abroad and saving the nation). Guilin, China: Guangxi shifan daxue chubanshe, 1995.

Wang Shanshan 王珊珊. "Wanqing zhengfu tielu sixiang bianqian yanjiu" 晚清政府铁路思想变迁研究 (Research on the development of official late Qing railway thinking). MA thesis, Shandong University, 2008.

Wang Yangzhi 王仰之. "Xie Jiarong—Kancha Zhongguo shiyou dizhi de xianqu" 謝家榮—勘查中國石油地質的先驅 (Xie Jiarong: Pioneer in the exploration of China's petroleum geology), *Zhongguo keji shiliao* 中国科技史料 (Chinese historical materials of science and technology) 12, no. 3 (1991): 54–55, 59.

———. "Yang Zhongjian nianpu" 杨钟健年谱 (Chronicle of Yang Zhongjian). *Xibei daxue xuebao (ziran kexue ban)* 西北大学学报(自然科学版) (Journal of Northwest University [Natural science edition]) 39, no. 2 (1983): 117–28.

———. *Zhongguo dizhi diaochasuo shi* 中国地质调查所史 (History of the Geological Survey of China). Beijing: Shiyou gongye chubanshe, 1996.

———. *Zhongguo shiyou biannianshi* 中国石油编年史 (Annals of Chinese petroleum). Beijing: Shiyou gongye chubanshe, 1996.

Wang Zancheng 王贊承. "Xinhai shouyi yangxia guangfu jishi" 辛亥首義陽夏光復紀實 (Eyewitness Account of the Recovery of Xinhai Uprising at Hanyang and Hankou). In *Xinhai geming huiyilu* 辛亥革命回憶錄 (Recollections of the Xinhai Revolution), vol. 2, pp. 45–46. Beijing: Zhonghua shuju, 1961.

Wei Yuan 魏源. *Haiguo tuzhi* 海国圖志 (Illustrated gazetteer of the maritime states) bk. 4, leaf 43.

———. *Wei Yuan quan ji* 魏源全集 (Complete works of Wei Yuan). Changsha, China: Yuelu shushe, 2004.

Weng Wenhao 翁文灝. "Disici taipingyang kexue huiyi jilue" 第四次太平洋科學會議紀略 (Brief notes on the Fourth Pan Pacific Science Congress). *Kexue* 科學 (Science) 14, no. 5 (January 1, 1930): 615–39.

———. *Dizhen* 地震 (Earthquakes). Shanghai: Shangwu yinshuguan kexue xiao congshu, 1929.

———. "Duiyu Ding Zaijun xiansheng de zhuiyi" 對於丁在君先生的追憶 (Recollections of Mr. Ding Zaijun). In *Ding Wenjiang zhuanji ziliao* 丁文江傳記資料 (Biographical materials of Ding Wenjiang), edited by Zhu Chuanyu 朱傳譽, 110–14. Taibei: Tianyi chubanshe, 1979.

———. "Lixihuofen yu Zhongguo zhi dizhi gongzuo" 李希霍芬與中國之地質工作 (Richthofen and Chinese geological work). *Fangzhi yuekan* 方志月刊 (Local records monthly) 6, no. 12 (1933): 37–38.

———. *Weng Wenhao xiansheng yanlun ji* 翁文灝先生言論集 (Collection of Mr. Weng Wenhao's speeches). Shanghai: Bingzi xuehui yinxing, 1935.

———. "Weng Wenhao xiansheng zai zhi dizhi diaochasuo tongren shu" 翁文灝先生再致地質調查所同人書 (Second letter from Weng Wenhao to his colleagues at the Geological Survey). *Dizhi lunping* 3, no. 1 (1938): 97–100.

———. "Zhongguo de kexue gongzuo" 中國的科學工作 (China's scientific work). *Duli pinglun* 獨立評論 (Independent review) 34 (January 8, 1933): 5–9.

———. "Zhongguo shitan zhi fenlei 中國石炭之分類 (Classification of Chinese coal). *Bulletin of the Geological Survey of China* 8 (1926): 59–82.

Weng Wenhao 翁文灝 and Xie Jiarong 謝家榮. "Gansu yumenkuang baogao" 甘肅玉門油礦報告 (Gansu Province Yumen Min report). *Hunan shiye zazhi* 湖南實業雜誌 (Hunan industrial journal) 54 (1922).

Wu Fengming 吴凤鸣. "Guanyu Gu Lang ji qi dizhi kuangchan zhuzuo de pingshu" 關於顧琅及其地質礦產著作的評述 (Commentary on Gu Lang's geological and mining works). *Zhongguo keji shiliao* 中国科技史料 (Chinese historical materials of science and technology) 5, no. 3 (1984).

Wu Jing 武静. "Qingmo liuxuesheng guiguo kaoshi zhidu ji sikao" 清末留学生归国考试制度及思考 (Late Qing system and rationale for testing returned students). *Tianjinshi jiaokeyuan xuebao* 天津市教科院学报 (Journal of Tianjin Academy of Educational Science). 1 (2011): 25–28.

Wu Yuexing 武月星. *Zhongguo xiandaishi dituji, 1911–1949* 中國現代史地圖集, 1911–1949 (Atlas of contemporary Chinese history). Beijing: Zhongguo ditu chubanshe, 1999.

Wu Zhaoqian 吴昭谦. "Qingchao monian Wu Rulun dui chuanbo xifang dixue zhi gongxian" 清朝末年吴汝纶对传播西方地学之贡献 (The contribution of Wu Rulun to the dissemination of Western geoscience in the late Qing dynasty). *Anhui dizhi* 安徽地质 (Anhui geology) 8, no. 2 (1998): 72–75.

Xia Xiangrong 夏湘蓉 and Wang Genyuan 王根元. *Zhongguo dizhi xuehui shi* 中国地质学会史 (History of the Geological Society of China). Beijing: Dizhi chubanshe, 1982.

Xie Jiarong 謝家榮. "Dizhi diaochasuo hezuo banfa" 地質調查合作辦法 (Geological Survey collaboration policies). *Dizhi lunping* 1, no. 5 (1936): 521–22.

———. "Fakan ci" 發刊辭 (Statement on the founding of the journal). *Dizhi lunping* 1, no. 1 (1936): 1–2.

Xiong Yi 熊毅. "Turang gongzuo shiwu nian" 土壤工作十五年 (Fifteen years of soil research). *Turang jikan* 土壤季刊 (Soils quarterly) 5, no. 3 (1946): 144.

Xu Hongzu 徐宏祖. *Xu Xiake youji* 徐霞客遊記 (Travel diaries of Xu Xiake). Edited by Ding Wenjiang. Shanghai: Shangwu yinshuju, 1928.

Xu Jiyu 徐繼畬. "Xu Jiyu shi" 徐繼畬識 (Xu Jiyu). *Yinghuan zhilue* 瀛環志略 (A brief account of the maritime circuit). Shan yun lou, 1866.

"Xu yi" 序一 (Preface one). *Guoli Beijing daxue dizhi yanjiuhui niankan* 國立北京大學地質研究會年刊 (The annual of the geological society of the university, Peking) 1 (October 10, 1921): 1.

Yang Chaoqun 祥超群. "Liangguang dizhi diaochasuo jianshi" 两卦广地质调查所简史 (Brief history of the Liangguang Geological Survey). In *Dizhixue shi luncong* 地质学史论丛 (Compendium of the history of geology) 4 (2002): 200–202.

Yang Chuan 楊銓. "Xuehui yu kexue" 學會與科學 (Learned societies and science). *Kexue* (Science) 1, no. 7 (1915): 707–11.

Yang Cuihua 楊翠華. *Zhongjihui dui kexue de zannzhu* 中基會對科學的贊助 (The China Foundation's sponsorship of science). Taibei: Zhongyang yangjiuyuan jindaishi yanjiusuo, 1991.

Yang Jingyi 杨静一. "Pangpeili yu jindai dizhixue zai Zhongguo de chuanru" 庞佩利与近代地质学在中国的传入 (Pumpelly and the introduction of modern geology to China). *Zhongguo keji shiliao* 中国科技史料 (Chinese historical materials of science and technology) 17, no. 3 (1996), 18–27.

Yang Xiaoming 杨小明 and Li Qiang 李强. "1912–1937 nian Zhongguo dizhi kexue zhishi zengzhang de jianzhihua yanjiu" 1912–1937 年中国地质科学知识增长的建制化研究 (Study of the structural organization of the proliferation of Chinese geological knowledge). *Zhongguo kejishi zazhi* 中国科技史杂志 (The Chinese journal for the history of science and technology) 31, no. 2 (2010): 165–75.

Yang Xinxiao 杨新孝. "Ji Beida dixhi yanjiuhui" 记北大地质研究会 (Remembering Beida's Geological Society (author's photocopy).

Yang Zhongjian 杨钟健. "Dui rexin caiji guhuashi tongzhi jin yiyan" 對熱心採集骨化石同志進一言 (A word to those comrades who are eager to collect fossil bones). *Dizhi lunping* 5, no. 6 (1940): 543–48.

———. "Feichang shiqi zhi dizhijie" 非常時期之地質界 (The geological community in extraordinary times). *Dizhi lunping* 2, no. 6 (1937): 509–14.

———. Letter to wife, December 14, 1941 (personal collection).

———. "Li Siguang laoshi huiyilu" 李四光老师回忆录 (Recollections of teacher Li Siguang). In *Li Siguang jinian wenji* 李四光纪念文集 (Collection of papers commemorating Li Siguang), edited by Li Siguang yanjiuhui choubeizu, Dizhi xuehui dizhi lixue zhuanye weiyuanhui 李四光研究会筹备组, 地质学会地质力学专业委员会. Beijing: Dizhi chubanshe, 1981.

———. *Xin Yanjie* 新眼界 (New perspectives). Changsha, China: Hunan renmin chubanshe, 1986.

———. *Yang Zhongjian huiyilu* 杨钟健回忆录 (Memoir of Yang Zhongjian). Beijing: Dizhi chubanshe, 1983.

———. "You jige guoji xueshu huiyi tanqi" 由幾個國際學術會議談起 (Taking a few international conferences as a starting point). *Da gongbao (Shanghai ban)* 大公報 (上海版) (L'Impartial [Shanghai edition]) May 25, 1948.

Yang Zunyi 杨遵仪. *Tao li man tianxia: Jinian Yuan Fuli jiaoshou bainian danchen* 桃李满天下: 纪念袁复礼教授百年诞辰 (Students all over the world: Commemorating the hundredth anniversary of Professor Yuan Fuli's birth). Wuhan, China: Zhongguo dizhi daxue chubanshe, 1993.

Yao Zhenli 姚振黎. "*Yinghuan zhilue* yanjiu"《瀛環志略》研究 (Research on *Yinghuan zhilue*). In *Diqijie qingdai xueshu yantaohui lunwenji* 第七屆清代學術研討會論文集 (Collected papers of the seventh Qing dynasty academic symposium), 71–100. Gaoxiong, Taiwan: National Zhongshan University Chinese Literature Department, 2002.

Yin Zanxun 尹贊勛. "Zhongguo dizhi xuehui" 中國地質學會 (Geological Society of China). *Kexue dazhong* 科學大衆 (Popular science) 4, no. 6 (1948): 271–75.

Yu Guang 于洸. "Li Siguang jiaoshou zai Beijing daxue 李四光教授在北京大学 (Professor Li Siguang and Peking University). *Hebei dizhi xueyuan xuebao* 河北地質學院學報 (Academic journal of the Hebei Geological Institute) 15, no. 5 (1992): 534–40.

Yu Guang 于洸 et al. "Hongyang chuantong, bawo jiyu, zaichuang huihuang: Qingzhu Beijing daxue jianli 100 zhounian, Beida dizhixue xi jianli 89 zhounian" 杨传统、把握机遇、再创辉煌—庆祝北京大学建立100周年、北大地质学系建立89周年 (Promoting tradition, grasping opportunities, recreating splendor: Celebrating 100 years of Peking University and 89 years since the founding of the Peking University Department of Geology). In *Beijing daxue guoji dizhi kexue xueshu yantaohui lunwenji* 北京大学国际地质科学学术研讨会论文集 (Collected papers of the Peking University international academic symposium of geological science). Edited by Beijing daxue dizhixue xi 北京大学地质系. Beijing: Dizhen chubanshe, 1998.

Zeng Shiying 曾世英. "Bianzuan liyan" 编纂例言 (Introductory remarks on compilation). *Zhonghua minguo xin ditu* 中華民國新地圖 (*New Atlas of the Republic of China*). Shanghai: Shanghai shenbao guan, 1934.

Zhang Hongzhao 章鸿钊. "Dixue hui ying xing shiwu zhi shangque" 地學會應行事務之商榷 (Discussions on the proper duties of a geosciences society). *Dixue zazhi* 地學雜誌 (Earth studies journal) 3, nos. 3–4 (1912): 2–3.

———. *Liuliu zishu* 六六自述 (Self-narrative at sixty-six). Wuhan, China: Wuhan dizhi xueyuan chubanshe, 1987.

———. "The Lure of Lime Fertilization in the Hangzhou Region" (Hangshu yi shihui daifei zhi yinyou) 杭屬以石灰代肥之引誘. *Dixue zazhi* 地學雜誌 (Earth studies journal) 3, no. 1 (1911): 1–8.

———. "Shijie geguo zhi dizhi diaocha shiye" 世界各國之地質調查事業 (The workings of the geological surveys of the countries of the world), pts. 1–4, *Dixue zazhi* 地學雜誌 (Earth studies journal) 1, nos. 3–4 (1910), 1–4; 2, no. 11 (1911), 1–6; no. 12 (1911), 1–5; no. 13 (1911): 1–4.

———. *Zhongguo dizhixue fazhan xiaoshi* 中國地質學發展小史 (Short history of the development of geology in China). Shanghai: Shangwu yinshuguan, 1940.

———. "Zhonghua dizhi diaocha siyi" 中華地質調查私意 (My suggestions for China's geological investigations), pts. 1 and 2. *Dixue zazhi* 地學雜誌 (Earth studies journal) 3, no. 1 (1911): 1–15; and 3, nos. 3–4 (1912): 14–20.

Zhang Jian 张剑. *Kexue shetuan zai jindai Zhongguo de mingyun: Yi Zhongguo kexueshe wei zhongxin* 科学社团在近代中国的命运: 以中国科学社为中心 (The science association and the change of society in modern China: A study on the science society of China [sic]). Jinan, China: Shandong jiaoyu chubanshe, 2005.

Zhang Jian and Huang Ting 黄婷. "Zhu Jiahua de kexue guannian yu guomin zhengfu shiqi kexue jishu de fazhan" 朱家骅的科学观念与国民政府时期科学技术的发展 (Zhu Jiahua's scientific views and Republican-era development of science and technology). *Jindai Zhongguo* 近代中国 (Contemporary China), no. 14 (2004): 291–321.

Zhang Jiangyi 張江一 et al. *Sun Jianchu zhuan* 孫健初傳 (Biography of Sun Jianchu). Beijing: Shiyou gongye chubanshe, 1989.

Zhang Jiuchen 张九辰. *Dizhixue yu minguo shehui: 1916–1950* 地质学与民国社会: 1916–1950 (Geology and Republican society: 1916–1950). Jinan, China: Shandong jiaoyu chubanshe, 2005.

———. "Zhongguo jindai dixue zhuyao xueke mingcheng de xingcheng yu yanhua chutan" 中国近代地学主要学科名称的形成与演化初探 (Preliminary investigation of the formation and evolution of common terms for geo-scientific subjects in modern China). *Zhongguo keji shiliao* 中国科技史料 (Chinese historical materials of science and technology) 22, no. 1 (January 1, 2001): 26–36.

Zhang Jiuchen, and Fengxian Xu 徐凤先. *Zhongguo xibei kexue kaochatuan zhuanlun* 中国西北科学考查团专论 (A focused study of the Chinese expedition to the Northwest). Beijing: Zhongguo ke xue ji shu chu ban she, 2009.

Zhang Lixu 張麗旭. "Taiwan diceng zhi jiantao" 臺灣地層之檢討 (An examination of Taiwan's strata). *Dizhi pinglun* 13, nos. 3–4 (1948): 185–98; and 13, nos. 5–6 (1948): 291–310.

Zhang Peifu 张培富 and Xia Wenhua 夏文华. "Guojia keyan jigou yu Zhongguo xiandai kexue wenhua: Yi 80 nianlai zhongyang yanjiuyuan yanjiu zhe zhuangkuang yu jinlu wei kaocha neirong" 国家科研机构与中国现代科学文化——以80年来对中央研究院研究之状况与进路为考察内容 (National research institutions and modern scientific culture in China: Taking a review of research on the conditions and development of the Academia Sinica as a study object). *Shanxi daxue xuebao (Zhexue shehui kexue ban)* 山西大学学报(哲学社会科学版) (Journal of Shanxi University [Philosophy and social science edition]) 33, no. 6 (2010): 11–17.

Zhang Shuyan 张叔岩. "Woguo zaoqi shiyou gongye de fazhan yu yinjin xishou guowai xianjin jishu" 我国早期石油工业的发展与引进吸收国外先进技术 (Early development of the petroleum industry in China and the introduction and assimilation of advanced foreign technologies). In *Zhongguo shiyou shi yanjiu* 中国石油史研究 (Researches in Chinese petroleum history), edited by Shi Baoheng 石宝珩, Wang Yangzhi, and Liu Bingyi 刘炳义, 188–94. Beijing: Shiyou gongye chubanshe, 1992.

Zhang Xiangwen 張相文. "Zhongguo dixue hui qi" 中國地學會啓 (Opening of the Chinese Earth Studies Society). *Dixue zazhi* 地學雜誌 (Earth studies journal) 1, no. 1 (1910): 1–2.

Zhao Guobin 趙國賓. "Benhui choubei shidai jiyao" 本會籌備時代紀要 (Essential notes from the preparatory phase of our society). *Guoli Beijing daxue dizhi yanjiuhui niankan*

國立北京大學地質研究會年刊 (The annual of the geological society of the university, Peking) 1 (October 10, 1921): 4.

———. "Benhui yinianlai de huigu he nianlai niban de shixiang" 本會一年來的回顧和年來擬辦的事項 (Reflections on the past year of our society and planned items for the coming year). *Guoli Beijing daxue dizhi yanjiuhui niankan* 國立北京大學地質研究會年刊 (The annual of the geological society of the university, Peking) 1 (October 10, 1921): 1–3.

Zhongguo dizhi diaochasuo 中央地質調查所 (Geological Survey of China). *Zhongguo dizhi tu* 中國地質圖 (Geological map of China). Nanjing: Zhongyang dizhi diaocha suo, 1948.

"Zhongguo dizhi xuehui chubanpin" 中國地質學會出版品 (Publications of the Geological Society of China). *Dizhi lunping* 2, no. 3 (1937): 304.

Zhongguo renmin daxue Qingshi yanjiusuo 中国人民大学清史研究所 (People's University Institute of Qing history). *Qingdai de kuangye* 清代的矿业 (Qing Period Mining). Beijing: Zhonghua shuju, 1983.

Zhongyang dizhi diaochasuo xibei fensuo 中央地質調查所西北分所 (Northwest Branch of the National Geological Survey). *Zhongyang dizhi diaochasuo xibei fensuo gaikuang* 中央地質調查所西北分所概況 (Overview of the Northwest Branch of the National Geological Survey). Nanjing: Zhongyang dizhi diaochasuo, 1948.

"Zhongyang yanjiuyuan minguo shiqi niandu linshifei yusuan biao" 中央研究院民國十七年度臨時費預算表 (Year 17 [of the republic] interim expenditure calculations chart for the Academia Sinica). *Guoli zhongyang yanjiuyuan shiqi niandu zong baogao* 國立中央研究院十七年度総報告 (Year 17 annual report of the Academia Sinica) (1928): 56–60.

Zhou Jianbo 周建波. *Yangwu yundong yu Zhongguo zaoqi xiandaihua sixiang* 洋務運動與中國早期現代化思想 (The Foreign Affairs movement and early modernization thought in China). Jinan, China: Shandong Renmin Chubanshe, 2001.

Zhou Shuren 周树人. "Zhongguo dishi luelun" 中國地質略論 (Brief outline of Chinese geology). *Zhejiang Chao* 浙江胡 (Zhejiang tide) 8 (1903): 60.

Zhou Tiandu 周天度. *Cai Yuanpei zhuan* 蔡元培傳 (Biography of Cai Yuanpei). Taibei: Xinchao she, 1994.

Zhou Yan 周彥 and Li Dan 李丹. "Qingmo xinzheng yu Zhongguo jindaihua" 清末新政与中国近代化 (New policies of the late Qing and Chinese modernization). *Lishi dang'an* 历史档案 (Historical archives) 2 (2010): 62–67.

Zhou Yaohua 周耀华. "Lushan disiji bingchuan de san ci da lunzheng" 庐山第四纪冰川的三次大论争 (Three great debates on Quaternary glaciation on Lushan), in *Tiandi zongheng* 天地纵横 (Heaven and earth, warp and weft), ed. Yin Peiji 尹培基. Beijing: Dizhi chubanshe, 1992.

Zhou Zhenfu 周振甫. *Yan Fu shiwen xuan* 嚴復詩文選 (Selections from Yan Fu's poems and prose). Taibei: Taida chubanshe, 1964.

Zhu Chuanyu 朱傳譽, ed. *Ding Wenjiang zhuanji ziliao* 丁文江傳記資料 (Biographical materials of Ding Wenjiang). Taibei: Tianyi chubanshe, 1979.

Zhu Hanguo 朱汉国. *Nanjing guomin zhengfu jishi* 南京国民政府纪实 (Records of the nationalist government at Nanjing). Hefei, China: Anhui renmin chubanshe, 1993.

WESTERN-LANGUAGE MATERIALS

Aalto, Kenneth R. "American Contributions to the Geological Mapping of Hokkaido, Late Nineteenth Century." *Earth Sciences History* 30, no. 1 (2011): 1–19.

Academia Sinica. *Academia Sinica with Its Research Institutes.* Shanghai: Science Press, 1929.

Ahnert, E .E. "The Geological Society and Science in Asia." *BGSC* 1, nos. 1–4 (1922): 10.

Anderson, Benedict. *Imagined Communities: Reflections on the Origins and Spread of Nationalism.* London: Verso, 1991.

Andersson, J. G. "The National Geological Survey of China," *Geografiska Annaler* 3 (January 1, 1921): 305–10.

"Andrews Is Hit in Statement by Group in Peking." *Peking Leader*, April 30, 1929.

Andrews, Roy Chapman. "Andrews Tells About Deadlock." *Peking Leader*, April 20, 1929.

———. "China as a Field for Scientific Research." *BGSC* 1, nos. 1–4 (1922): 8.

"Annual Meeting of Geological Society Holds First Day's Session." *Peking Leader*, Sunday, January 6, 1924.

Argand, Émile. "La Tectonique de l'Asie." *International Geological Congress XIII*, vol. 1, no. 5 (1922): 171–372.

Barbour, George. "Peiping Scientists Tell at Meeting Here of New Approach to 'First Man.'" *Peking Leader*, December 29, 1929.

Barrell, Joseph. "Review of *Research in China.*" *Science* 29, no. 737 (February 12, 1909): 257–60.

Basalla, George. "The Spread of Western Science." *Science* 156 (May 5, 1987): 611–22.

Bateson, Gregory. Steps to an Ecology of Mind: Collected Essays in Anthropology, Psychiatry, Evolution, and Epistemology. Chicago: University of Chicago Press, 1972.

Belsky, Richard. *Localities at the Center: Native Place, Space, and Power in Late Imperial Beijing.* Cambridge, MA: Harvard University Asia Center, 2005.

Bennett, Adrian Arthur. *John Fryer: The Introduction of Western Science and Technology into Nineteenth-Century China.* Cambridge, MA: Harvard University Press, 1967.

Bieler, Stacey. *"Patriots" or "Traitors"? A History of American-Educated Chinese Students.* Armonk, NY: M. E. Sharpe, 2004.

Black, Davidson. *Asia and the Dispersal of Primates: A Study in Ancient Geography of Asia and Its Bearing on the Ancestry of Man.* Peking: Geological Survey of China, 1925.

———. "The Geological Society and Science in China." *BGSC* 1, nos. 1–4 (1922): 9.

Blackwelder, Eliot. "Moving Underground Water in the Accumulation of Oil and Gas." *Economic Geology* 17, no. 3 (1922): 217.

———. "Review of *Geology of the Yang-tze Valley (China).*" *Science* 40, no. 1026 (August 28, 1914): 312–13.

Bowler, Peter. *The Norton History of the Environmental Sciences.* New York: W. W. Norton & Co., 1993.

Brook, Timothy. *Geographical Sources of Ming-Qing History.* Ann Arbor: Center for Chinese Studies, University of Michigan, 2002.

Brown, Jeremy, and Paul G. Pickowicz, eds. *Dilemmas of Victory: The Early Years of the People's Republic of China.* Cambridge, MA: Harvard University Press, 2007.

Brown, Mary Bullock. *An American Transplant: The Rockefeller Foundation and Peking Union Medical College.* Berkeley: University of California Press, 1980.

Brownell, Susan. *Training the Body for China: Sports in the Moral Order of the People's Republic.* Chicago: University of Chicago Press, 1995.

Buck, Peter. *American Science and Modern China, 1876–1936.* Cambridge: Cambridge University Press, 1980.

Callahan, William. "National Insecurities: Humiliation, Salvation, and Chinese Nationalism." *Alternatives* 29 (2004): 199–218.

Callon, Michel. "Some Elements of a Sociology of Translation: Domestication of the Scal-

lops and the Fishermen of St Brieuc Bay." *Power, Action and Belief. A New Sociology of Knowledge?* Edited by J. Law. London: Routledge & Kegan Paul, 1986.
Camerini, Jane. "Remains of the Day: Early Victorians in the Field." In *Victorian Science in Context.* Edited by Bernard Lightman. Chicago: University of Chicago Press, 1997.
———. "Wallace in the Field." *Osiris* 11 (1996): 44–65.
Carlson, Ellsworth. *The Kaiping Mines. 1877–1912.* Cambridge, MA: Harvard University Press, 1957.
Cavendish, Patrick. "The 'New China' of the Kuomintang." In *Modern China's Search for a Political Form,* edited by Jack Gray, 138–69. London: Oxford University Press, 1969.
Champlin, Peggy. *Raphael Pumpelly: Gentleman Geologist of the Gilded Age.* Tuscaloosa: University of Alabama Press, 1994.
Chang, H. T. (Zhang Hongzhao). "On the History of the Geological Science in China." *BGSC* 1, nos. 1–4 (1922): 4–7.
Chen, Shiwei. "Government and Academy in Republican China: History of Academia Sinica, 1927–1949." PhD diss., Harvard University, 1998.
Chi, Y. S.(計榮森). *Bibliography of Chinese Geology for the Years 1936–1940* (中國地質文獻目錄民国二十五年至二十九年). Pehpe, China: National Geological Survey of China, 1942.
———. *Science Bibliography of China, V. Section of Geology* (中國科學著作目錄, 第五組: 地質學). Nanking, China: National Research Council, Academia Sinica, 1936.
Chinese Ministry of Information. *China Handbook, 1937–1943.* New York: Macmillan Co., 1943.
Chishitsu Ch sajo (Japan). *Imperial Geological Survey of Japan: With a Catalogue of Articles Exhibited at the Panama-Pacific International Exposition Held at San Francisco, United States of America in 1915.* Tokyo: Imperial Geological Survey of Japan, 1915.
Chow, Kai-Wing. *Publishing, Culture, and Power in Early Modern China.* Stanford, CA: Stanford University Press, 2004.
Chow Tse-tsung. *The May 4th Movement: Intellectual Revolution in Modern China.* Cambridge, MA: Harvard University Press, 1960.
Chung, Juliette. "Transnational Science: The Japanese Establishment of Shanghai Natural Science Institute and the Knowledge of Taxonomy in China, 1923–1945." Paper presented at the annual meeting of the History of Science Society, Vancouver, British Columbia, November 2–5, 2000.
Chung Yu Wang (Wang Chongyou). *Bibliography of the Mineral Wealth and Geology of China.* London: C. Griffin & Co., 1912.
Clark, Anthony. *Beating Devils and Burning Their Books: Views of China, Japan, and the West.* Ann Arbor, MI: Association for Asian Studies, 2010.
Clifford, Nicholas. *"A Truthful Impression of the Country": British and American Travel Writing in China 1880–1949.* Ann Arbor: University of Michigan Press, 2001.
Cohen, Paul A. *Discovering History in China: American Historical Writing on the Recent Chinese Past.* New York: Columbia University Press, 1984.
"Constitution of the Geological Society of China." *BGSC* 8, no. 1 (1929): 11–13 (Chinese version unpaginated facing pages 10 and 11).
Cooper, William S., and Tang Pei-sung. "Science in China." *Science* 91, no. 2358 (March 8, 1940): 238–39.
Crossley, Pamela Kyle. *The Wobbling Pivot, China Since 1800: An Interpretive History.* Malden, MA: John Wiley & Sons, 2010.
Da Mota, Teresa Salomé Alves. "A Mere Shadow of an Institution: The Unhappy Story of

the Portuguese Geological Survey (PGS) in the Period Between the Two World Wars." *Annals of Science* 64, no. 1 (2007): 19–40.

Davis, Clarence B. "Railway Imperialism in China, 1895–1939." In *Railway Imperialism*, ed. Clarence B. Davis, Kenneth E. Wilburn, and Ronald E. Robinson. New York: Greenwood, 1991.

de Certeau, Michel. *The Practice of Everyday Life.* Berkeley: University of California Press, 1988.

de Margerie, E. "The Geological Map of the World." *Compte-rendu de la XIIe session du Congrès géologique international, Canada, 1913:* 173–87.

De Weerdt, Hilde. "Maps and Memory: Readings of Cartography in Twelfth- and Thirteenth-Century Song China." *Imago Mundi* 61, no. 2 (June 2009): 145–67.

Dikötter, Frank. *The Age of Openness: China before Mao.* Los Angeles: University of California Press, 2008.

Ding Yizhuang. "Reflections on the 'New Qing History' School in the United States." *Chinese Studies in History* 43, no. 2 (Winter 2009/2010): 92–96.

"Distinguished Scientists Speak at Final Session Monday of Geological Society." *Peking Leader*, Wednesday, January 9, 1924: 5.

Drake, Fred. "A Mid-Nineteenth-Century Discovery of the Non-Chinese World." *Modern Asian Studies* 6, no. 2 (1972): 205–24.

Drake, Fred. *China Charts the World: Hsü Chi-yü and his Geography of 1848.* Cambridge, MA: Harvard University Press, 1975.

Duara, Prasenjit. "De-Constructing the Chinese Nation." In *Chinese Nationalism*, edited by Jonathan Unger, 31–55. Armonk, NY: M. E. Sharpe, 1996.

———. "Historicizing National Identity, or Who Imagines What and When." In *Becoming National*, edited by Geoff Eley and Ronald Grigor Suny, 151–70. New York: Oxford University Press, 1996.

———. *Rescuing History from the Nation: Questioning Narratives of Modern China.* Chicago: University of Chicago Press, 1995.

Dugdale, E. T. S., trans. *German Diplomatic Documents, 1871–1914. Vol. 3, The Growing Antagonism, 1898–1910.* New York: Harper & Brothers, 1930.

Eagan, William. "The Debate over the Canadian Shield, 1880–1905." *Isis* 80, no. 2 (June 1989): 232–53.

Edney, J. "The Bituminous Coal Mines West of Peking." *Journal of the North China Branch of the Royal Asiatic Society* (Shanghai) 5 (1867): 243–50.

Elman, Benjamin. "China and the World History of Science, 1450–1770," *Education about Asia* 12, no. 1 (Spring 2007): 40–44.

———. *A Cultural History of Civil Examinations in Late Imperial China.* Berkeley: University of California Press, 2000.

———. "Naval Warfare and the Refraction of China's Self-Strengthening Reforms into Scientific and Technological Failure." *Modern Asian Studies* 38, no. 2 (2003): 283–326.

Fairbank, John King. *Trade and Diplomacy on the China Coast: The Opening of the Treaty Ports, 1842–1854.* Cambridge, MA: Harvard University Press, 1953.

Fan, Fa-ti. *British Naturalists in Qing China: Science, Empire, and Cultural Encounter.* Cambridge, MA: Harvard University Press, 2004.

———. "Hybrid Discourse and Textual Practice: Sinology and Natural History in the Nineteenth Century." *Historia Scientiarum* 38 (2000): 26–56.

"A Feast of Wit at Annual Dinner of Geological Society Saturday." *Peking Leader*, Wednesday, January 7, 1925: 3.

Fiskesjö, Magnus, and Chen Xingcan. *China Before China: The Discovery of China's Prehistory.* Stockholm: Museum of Far Eastern Antiquities, 2004.

Fitzgerald, John. "The Nationless State: The Search for a Nation in Modern Chinese Nationalism." In *Chinese Nationalism*, edited by Jonathan Unger, 56–85. Armonk, NY: M. E. Sharpe, 1996.

Fogel, Joshua. Introduction to *The Teleology of the Modern Nation-State.* Philadelphia: University of Pennsylvania Press, 2005.

Fritzsche, Sybille. "Narrating China: Western Travelers in the Middle Kingdom after the Opium War." PhD diss., University of Chicago, 1995.

Fuller, M. L. "Explorations in China." *Bulletin of the American Association of Petroleum Geologists* 3 (1919): 99–116.

———. "Oil Prospects in Northeastern China." *Bulletin of the American Association of Petroleum Geologists* 10, no. 11 (1926): 1073–17.

Furth, Charlotte. *Ting Wen-chiang: Science and China's New Culture.* Cambridge, MA: Harvard University Press, 1970.

Galison, Peter, and Bruce Hevly, eds., *Big Science: The Growth of Large Scale Research.* Stanford, CA: Stanford University Press, 1992.

Gallenkamp, Charles. *Dragon Hunter: Roy Chapman Andrews and the Central Asiatic Expeditions.* New York: Viking, 2001.

Geertz, Clifford. "The Integrative Revolution: Primordial Sentiments and Civil Politics in the New States." *Old Societies and New States.* New York: Free Press, 1963.

Gellner, Ernest. *Nations and Nationalism.* Ithaca, NY: Cornell University Press, 1983.

Gentzler, Jennings Mason. *Changing China: Readings in the History of China from the Opium War to the Present.* New York: Praeger Publishers, 1977.

Geological Reports of Dr. V. K. Ting. Nanjing: National Geological Survey of China, 1947.

"Geological Society Hedin Presents Complete Published Collection of His Tibet Exploration to Survey." *Peking Leader*, November 10, 1923.

Gershon, Jack. *Horatio Nelson Lay and Sino-British Relations, 1854–1864.* Cambridge, MA: Harvard University Press, 1972.

Glasgow University Calendar. Glasgow: n.p., 1910–11.

Glasgow University Calendar. Glasgow: n.p., 1911–1912.

Glosser, Susan. *Chinese Visions of Family and State, 1915–1953.* Berkeley: University of California Press, 2003.

Goodman, Bryna. *Native Place, City, and Nation: Regional Networks and Identities in Shanghai, 1853–1937.* Berkeley: University of California Press, 1995.

Grabau, A. W. "The Annual Meeting of the Geological Society of China" *Peking Leader*, Sunday, January 14, 1923.

———. "The First Year of the Third Asiatic Expedition's Activities: Report of Last Friday's Meeting." *Peking Leader*, Tuesday, October 3, 1922.

Grabau, Amadeus. "Contributions to Geologic Science by Graduates of the National University." *Guoli Beijing daxue ziran kexue jikan* 國立北京大學自然科學季刊 (Science quarterly of the National University of Peking) 1, no. 3 (April 1, 1930): 235–49.

———. "A Decade of Research in Chinese Geology." 北京大學地質研究會年刊 *Guoli Beijing daxue dizhi xuehui huikan* 國立北京大學地質學會會刊 (Bulletin of the Geological Society of the National University, Peking) 4 (April 1930): 1–8.

———. "Ordovician Fossils from North China" *Paleontologica Sinica*, ser. B, vol. 1, 1922.

———. "The Sinian System." *BGSC* 1, nos. 1–4 (1922): 48–88.

Greene, Mott. *Geology in the Nineteenth Century.* Ithaca, NY: Cornell University Press, 1982.

Haas, William Joseph. *China Voyager: Gist Gee's Life in Science.* Armonk, NY: M. E. Sharpe, 1996.

Hallam, Anthony. *Great Geological Controversies.* Oxford: Oxford University Press, 1989.

Harland, W. B. "On the Physiographical History of Western Szechwan, with Special Reference to the Ice Age in the Red Basin." *Journal of the West China Border Research Society,* ser. B, vol. 15 (1945): 1–19.

"Hedin Thrills Audiences Here." *Peking Leader,* April 26, 1929: 1, 7–8.

Heinrich, Larissa. *The Afterlife of Images: Translating the Pathological Body between China and the West.* Durham, NC: Duke University Press, 2008.

Herman, Theodore. "Group Values Towards the National Space: The Case of China." *Geographical Review* 49, no. 2 (April 1959): 164–82.

Hevly, Bruce. "The Heroic Science of Glacier Motion. "*Osiris* 11 (1996): 66–86.

Ho, Kim Kwan. *Japanese Perspectives on China's Early Modernization: the Self-Strengthening Movement, 1860–1895.* Ann Arbor: Center for Chinese Studies, University of Michigan, 1974.

Honig, Emily. *Creating Chinese Ethnicity: Subei People in Shanghai, 1850–1980.* New Haven, CT: Yale University Press, 1992.

Hood, Dora. *Davidson Black: A Biography.* Toronto: University of Toronto, 1964.

Hoover, Herbert. *The Memoirs of Herbert Hoover: Years of Adventure, 1874–1920.* Vol. 1. New York: Macmillan, 1951.

Hornibrook, John. "Local Elites and Mechanized Mining in China: The Case of the Wen Lineage in Pingxiang County, Jiangxi." *Modern China* 27, no. 2 (April 2001): 202–28.

Hostetler, Laura. *Qing Colonial Enterprise: Ethnography and Cartography in Early Modern China.* Chicago: University of Chicago Press, 2001.

Hou, Chi-ming. *Foreign Investment and Economic Development in China, 1840–1937.* Cambridge, MA: Harvard University Press, 1965.

Hsieh, C. Y. "The Petroleum Resources of China." *Bulletin of the Geological Survey of China,* no. 30 (1937): English 53–67, Chinese 17–24.

———. "Petroleum Resources of China."*Report of the XVIII International Geological Congress, USSR* 4 (1940): 283–93.

———. "Proceedings of the Society." *BGSC* 1, nos. 1–4 (1922): 32–39.

———. "Proceedings of the First Annual Meeting of the Geological Society of China." *BGSC* 2, nos. 1–2 (1923): 1–16.

Huang, Andrew. "The Inflation in China." *Quarterly Journal of Economics* 62, no. 4 (August 1948): 562–75.

Huang, Martin. *Negotiating Masculinities in Late Imperial China.* Honolulu: University of Hawaii Press, 2006.

Huang, T. K., et al. "Report on Geological Investigation of Some Oil-Fields in Sinkiang." *Memoirs of the Geological Survey of China,* ser. A, no. 21 (February 1947).

Huc, Evariste Régis. *Souvenirs d'un voyage dans la Tartarie, le Thibet, et la Chine pendant les années 1844, 1845 et 1846.* Paris: Librairie d'Adrien le Clerc, 1850.

Israel, John. *Lianda: A Chinese University in War and Revolution.* Stanford, CA: Stanford University Press, 1998.

Jia Lanpo, and Huang Weiwen. *The Story of Peking Man: From Archaeology to Mystery.* New York: Oxford University Press, 1990.

Jones, David. *The Image of China in Western Social and Political Thought.* New York Palgrave, 2001.

Karl, Rebecca. *Staging the World: Chinese Nationalism at the Turn of the Twentieth Century.* Durham, NC: Duke University Press, 2002.

Kim Kwan Ho. *Japanese Perspectives on China's Early Modernization: the Self-Strengthening Movement, 1860–1895.* Ann Arbor: Center for Chinese Studies, University of Michigan, 1974.

King, Thomas M. S. J., and Mary Wood Gilbert, eds. *The Letters of Teilhard de Chardin and Lucile Swan.* Washington, DC: Georgetown University Press, 1993.

Kirby, William. "The Chinese War Economy." In *China's Bitter Victory: The War with Japan, 1937–1945,* edited by James C. Hsiung and Steven I. Levine, 185–212. Armonk, NY: M. E. Sharpe, 1992.

———. "Engineering China: Birth of the Developmental State, 1928–1937." In *Becoming Chinese, Passages to Modernity and Beyond,* edited by Wen-hsin Yeh, 137–60. Berkeley: University of California Press, 2000.

———. *Germany and Republican China.* Stanford, CA: Stanford University Press, 1984.

———. "The Internationalization of China." *China Quarterly* 150 (June 1997): 443–58.

Kolatch, Jonathan. *Sports, Politics, and Ideology in China.* New York: Jonathan David Publishers, 1972.

Kuklick, Henrika, and Robert Kohler. "Science in the Field." *Osiris,* 2nd ser., vol. 11 (1996).

Kwok, Daniel. *Scientism in Chinese Thought, 1900–1950.* New York: Biblo, 1971.

Lach, Donald. *Asia in the Making of Europe.* Chicago: University of Chicago, 1965.

Lai Chi-kong. *Recent Historiography on the Self-Strengthening Movement.* Davis, CA: Institute of Governmental Affairs, UC Davis, 1989.

Lam, Tong. *A Passion for Facts: Social Surveys and the Construction of the Chinese Nation-State, 1900–1949.* Berkeley: University of California Press, 2011.

Lattimore, Owen. *Inner Asian Frontiers of China.* New York: Capitol Pub. Co. American Geographical Society, 1951.

Lee, En-Han. "China's Response to Foreign Investment in Her Mining Industry (1902–1911)." *Journal of Asian Studies* 28, no. 1 (November 1, 1968): 55–76.

Lee, J. S. "Canon of Marine Transgression in Post-Paleozoic Times." *BGSC* 7, no. 1 (1928): 81–128.

———. "Classification and Correlation of the Paleozoic Coal-Bearing Formation in North China." *BGSC* 5, no. 2 (1925): 113–34.

———. "Confirmatory Evidence of Pleistocene Glaciation from the Huangshan, Southern Anhui." *BGSC* 15, no. 3 (1936): 279–84.

———. "Data Relating to the Study of the Problem of Glaciation in the Lower Yangtze Valley." *BGSC* 13, no. 3 (1934): 395–432.

———. "The Fundamental Cause of Evolution of the Earth's Surface Features" with appended Discussion. *BGSC* 5, nos. 3–4 (December 1926): 209–62.

———. "Fusulina from the Pingting Basin, Northeastern Shansi." *BGSC* 3, no. 1 (1924): 15–16.

———. " Fusulinidae of North China. "*Paleontologia Sinica* 4, ser. B., no. fasc. 1 (1927).

———. "The Geology of China." MS thesis, University of Birmingham, 1918.

———. "Grabauina, A. Transitional Form between Fusulinella and Fusulina." *BGSC* 3, no. 1 (1924): 51–54.

———. "A Graphical Method to Aid Specific Determination of Fusulinoids and Some Results of Its Application to the Fusulinae from N. China." *BGSC* 2, nos. 3–4 (1923): 51–97.

———. "New Terms in the Description of the Fusulinidae." *BGSC* 3, no. 1 (1924): 13–14.

———. "Note on Traces of Recent Ice-Action in N. China." *Geological Magazine* 59, no. 1 (January 1922): 14–21.

———. "An Outline of Chinese Geology." *Geological Magazine* 58, no. 6 (May 1921): 259–65, 324–29, 370–77, 409–20.

———. "Reflections on Twenty Years' Experience." *BGSC* 22, nos. 1–2 (1942): 21–47.

———. "Some Characteristic Structural Types in Eastern Asia and Their Bearing upon the Problem of Continental Movement."*Geological Magazine* 66, no. 782 (August 1929): 358–75; 66, no. 783 (September 1929): 413–31; 66, no. 784 (October 1929): 457–522.

———. "The Stratigraphy of the Anthracolithic Formation in the Liuhokou Coalfield, N. China." *國立北京大學地質研究會年刊* 1 (October 10, 1921): 1–18.

Lee, Wah Seyle. "Bibliography of the Geology of China." MA thesis, Stanford University, 1916.

Leonard, Jane Kate. *Wei Yuan and China's Rediscovery of the Maritime World.* Cambridge, MA: Harvard University Press, 1984.

Levine, Steven and James Chieh Hsiung, ed. *China's Bitter Victory: The War With Japan, 1937–1945.* Armonk, NY: M. E. Sharpe, 1992.

Lin, Cheng. *The Chinese Railways, Past and Present.* Shanghai: United China Press, 1937.

Lin, Xiaoqing Diana. *Peking University: Chinese Scholarship and Intellectuals, 1898–1937.* Albany, NY: State University of New York Press, 2006.

Liu, Lydia H. *The Clash of Empires: The Invention of China in Modern World Making.* Cambridge, MA: Harvard University Press, 2004.

Louie, Kam. *Theorising Chinese Masculinity: Society and Gender in China.* Cambridge: Cambridge University Press, 2002.

Macleod, Roy, and Philip F Rehbock. "Developing a Sense of the Pacific: The 1923 Pan-Pacific Science Congress in Australia." *Pacific Science* 54, no. 3 (2000): 209–25.

Martin, W. A. P. "Notes on China, Geological and Geographical." *American Journal of Science* 47 (1869): 98–103.

Mason, Mary Gertrude.*Western Concepts of China and the Chinese, 1840–1876.* New York, (s.n.) 1939.

Matthews, William Dillard. *Climate and Evolution.* New York: New York Academy of Sciences, 1915.

Mazur, Allan. *A Romance in Natural History: The Lives and Works of Amadeus Grabau and Mary Antin.* Syracuse, NY: Garret, 2004.

"Membership, 1922." *BGSC* 1, nos. 1–4 (1922): 97–99.

"Mongolian Explorers Meet and Tell of Work Accomplished and Expedition Plans." *Peking Leader*, Wednesday, April 15, 1925: 3.

Morris, Andrew. *Marrow of the Nation: A History of Sport and Physical Culture in Republican China.* Los Angeles: University of California Press, 2004.

"Much Enthusiasm When Third Asiatic Expedition Reports to Geological Society." *Peking Leader*, Friday, September 25, 1925.

Nakano Takamasa. "Some Prevailing Trends of the Historical Development of Geosciences in the Far East." *Geoforum* 1, no. 3 (1970): 63–80.

Nan, Y. T., and L. P. Wu. "Note on Some Uranium Minerals from Eastern Kuangsi." *BGSC* 23, nos. 3–4 (1943), 169–72.

National Geological Survey of China. *The National Geological Survey of China 1916–1931: A Summary of Its Work during the First Fifteen Years of Its Establishment.* Peiping, March 1931.

Needham, Joseph. *Chinese Science.* London: Pilot Press, 1945.

———. "Chinese Science Revisited (I)." *Nature* 4345 (February 7, 1953): 237–39.

———. *Science and Civilisation in China*. Vol. 3. Cambridge: Cambridge University Press, 1959.

Needham, Joseph, and Dorothy Needham, eds. *Science Outpost: Papers, 1942–1946*. London: Pilot Press, 1948.

Nichols, Francis H. "Notes from the Diary of the Late Francis H. Nichols in China." *Bulletin of the American Geographical Society* 37, no. 6 (1905): 339–56.

Nordlund, Christer. "'On Going Up in the World': Nation, Region and the Land Elevation Debate in Sweden." *Annals of Science* 58 (2001): 17–50.

Obruchev, V. A. "Expedition to the Barlyk and Tarbagatai in 1905." MSS 1907.

Oldroyd, David. *The Highlands Controversy: Constructing Geological Knowledge through Fieldwork in Nineteenth-Century Britain*. Chicago: University of Chicago Press, 1990.

Oldroyd, David, and Jing Yi Yang. "Bailey Willis (1857–1949): Geological Theorizing and Chinese Geology." *Annals of Science* 60 (2003): 1–37.

———. "On Being the First Western Geologist in China: The Work of Raphael Pumpelly." *Annals of Science* 53 (1996): 107–36.

Oreskes, Naomi. *The Rejection of Continental Drift*. New York: Oxford University Press, 1999.

Osterhammel, Jürgen. "Forschungsreise und Kolonialprogramm: Ferdinand von Richthofen und die Erschießung Chinas im 19. Jahrhundert." *Archiv für Kulturgeschichte* 69 (1987): 150–95.

Osterhammel, Jürgen, and Ute Wardenga. "Ferdinand von Richthofen als Erforscher Chinas: Ein Beitrag zur Entstehung und Verarbeitung von Reisebeobachtungen im Zeitalter des Imperialismus." *Berichte zur Wissenschaftsgeschichte* 13, no. 3 (1990): 141–55.

Pan, C. H. "Non-Marine Origin of Petroleum in North Shensi, and the Cretaceous of Szechuan, China." *Bulletin of the American Association of Petroleum Geologists* 25, no. 11 (November 1941): 2049–68.

Pepper, Suzanne. *Civil War in China: The Political Struggle, 1945–1949*. Lanham, MD: Rowman and Littlefield, 1999.

Perdue, Peter. "Where Do Incorrect Political Ideas Come From? Writing the History of the Qing Empire and the Chinese Nation." Edited by Joshua Fogel. *The Teleology of the Modern Nation-State*. Philadelphia: University of Pennsylvania Press, 2005.

Peterson, Willard J. "The Life of Ku Yen-wu (1613–1682)." Pts. 1 and 2. *Harvard Journal of Asiatic Studies* 28 (1968): 114–56; 29 (1969): 201–47.

Phimister, Ian. "Foreign Devils, Finance and Informal Empire: Britain and China, c. 1900–1912." *Modern Asian Studies* 40, no. 3 (July 1, 2006): 737–59.

Plating, John D. *The Hump: America's Strategy for Keeping China in World War II*. College Station, TX: Texas A&M University Press, 2011.

Polo, Marco, Henry Yule, and Henri Cordier. *The Travels of Marco Polo: The Complete Yule-Cordier Edition: Including the Unabridged Third Edition (1903) of Henry Yule's Annotated Translation, as Revised by Henri Cordier, Together with Cordier's Later Volume of Notes and Addenda (1920)*. New York: Dover Publications, 1993.

Porter, Roy. "Gentlemen and Geology: The Emergence of a Scientific Career, 1660–1920." *Historical Journal* 21 (1978): 809–36.

Price, Derek de Solla. *Little Science, Big Science . . . and Beyond*. New York: Columbia University Press, 1986.

"Proceedings of the 6th Annual Meeting of the Society." *BGSC* 8, no. 1 (1929): 1–10.

"Proceedings of the 9th Annual Meeting of the Geological Society." *BGSC* 12, no. 1 (1932):1–18.

"Proceedings of the 10th Annual Meeting of the Geological Society." *BGSC* 13, no. 1 (1933)1–14.

"Proceedings of the 11th Annual Meeting of the Geological Society." *BGSC* 14, no. 1 (1935): 1–22.

"Proceedings of the 12th Annual Meeting of the Geological Society." *BGSC* 15, no. 1 (1936): 1–16.

"Proceedings of the 13th Annual Meeting." *BGSC* 17, no. 1 (March 1937): i–xiv.

"Proceedings of the 14th Annual Meeting of the Geological Society." *BGSC* 18, no. 1 (1938): 1–8.

"Proceedings of the 17th Annual Meeting of the Geological Society." *BGSC* 21, nos. 2–4 (1941): 111–29.

"Proceedings of the Twenty-Third Annual Meeting Held at Taipei, November 18–20, 1947." *BGSC* 28, nos. 1–2 (1948): 105–6.

Pugach, Noel."Standard Oil and Petroleum Development in Early Republican China." *Business History Review* 45, no. 4 (Winter 1971): 452–73.

Pumpelly, Raphael. *Across America and Asia.* New York: Leopold & Holt, 1870.

———. *Explorations in Turkestan: Expedition of 1904.* Washington, DC: Carnegie Institution of Washington, 1908.

———. *Geological Researches in China, Mongolia, and Japan.* Washington, DC: Smithsonian Institution, 1866.

———. *My Reminiscences.* Vol. 1. New York: H. Holt, 1918.

———. "The Relation of Secular Rock Disintegration to Loess, Glacial Drift and Rock Basins." *American Journal of Science* 17 (1879): 133–44.

———. "Report of the Chinese Government on a Preliminary Examination of the Coal Districts of the Si Shan, in the Province of Chili." *Papers Relating to Foreign Affairs, Accompanying the Annual Message of the President of the Second Session Thirty-Eighth Congress. 1864 Part III.* Washington, 1865.

———. "Richthofen's 'China.'" *Nation,* April 4, 1878: 231–32.

Pusey, James Reeve. *China and Charles Darwin.* Cambridge, MA: Harvard University Press, 1983.

Pye, Lucian. *The Spirit of Chinese Politics.* Cambridge, MA: MIT Press, 1968.

Qian Zhao. "A Review of Studies of the May Fourth Movement in China over the Past Decade." *Chinese Studies in History* 43, no. 4 (Summer 2010): 73–89.

Raj, Kapil. "When Human Travellers Become Instruments: The Indo-British Exploration of Central Asia" in the Nineteenth Century." In *Relocating Modern Science: Circulation and the Construction of Knowledge in South Asia and Europe, 1650–1900,* 181–222. New York: Palgrave Macmillan, 2007.

Rankin, Mary Backus. "Nationalistic Contestation and Mobilization Politics: Practice and Rhetoric of Railway-Rights Recovery at the End of the Qing." *Modern China* 28, no. 3 (July 1, 2002): 315–61.

Ravenstein, E. G. "Obituary: Ferdinand Freiherr von Richthofen." *The Geographical Journal* 26, no. 6 (December 1905): 679–80.

Reardon-Anderson, James. "Science in Wartime China." *China's Bitter Victory: The War with Japan, 1937–1945.* Edited by James C. Hsiung and Steven I. Levine. Armonk, NY: M. E. Sharpe, 1992.

———. *The Study of Change: Chemistry in China, 1840–1949.* Cambridge: Cambridge University Press, 1991.

Regal, Brian. *Henry Fairfield Osborn: Race, and the Search for the Origins of Man.* Burlington, VT: Ashgate, 2002.

Rhoads, Edward J. M. *Stepping Forth into the World: The Chinese Educational Mission to the United States, 1872–81*. Hong Kong: Hong Kong University Press, 2011.

Richthofen, Ferdinand von. *Baron Richthofen's Letters: 1870–1872*. Shanghai: North China Herald, n.d.

———. *Ferdinand Freiherr von Richthofen: Tagebücher aus China* Vols. 1 and 2. Berlin: Dietrich Reimer (Ernst Vohsen), 1907.

———. "First Preliminary Notes of Geological Explorations in China (Letter Addressed to Prof. Whitney, Shanghai, March 1, 1969)." *Proceedings of the Academy of Arts and Sciences* 8 (1873).

———. "The Natural System of Volcanic Rocks." *Memoir California Academy Sciences* 1 (pt. 2) (1868): 1–94.

———. "On the Existence of the Nummulitic Formation in China." *American Journal of Science*, 3rd ser. 1 (1871): 110–13.

———. "On the Porcelain Rocks of China." *American Journal of Science*, 3d ser. 1 (1871): 179–81.

Richthofen, Ferdinand von, et al. *China: Ergebnisse eigener Reisen und darauf gegründeter Studien*. Berlin: D. Reimer, 1877.

Roberts, Frances Markley. *Western Travelers to China*. Hong Kong: Kelly & Walsh Ltd, 1932.

Rogaski, Ruth. *Hygienic Modernity: Meanings of Health and Disease in Treaty-Port China*. Berkeley: University of California Press, 2004.

Rudwick, Martin. *Bursting the Limits of Time: The Reconstruction of Geohistory in the Age of Revolution*. Chicago: University of Chicago Press, 2005.

———. *The Great Devonian Controversy: The Shaping of Scientific Knowledge among Gentlemanly Specialists*. Chicago: University of Chicago Press, 1985.

———. *Meaning of Fossils: Episodes in the History of Palaeontology*. New York: American Elsevier, 1972.

———. *Worlds before Adam: The Reconstruction of Geohistory in the Age of Reform*. Chicago: University of Chicago Press, 2008.

Sacca, John Wands. "Like Strangers in a Foreign Land: Chinese Officers Prepared at American Military Colleges, 1904–37." *Journal of Military History* 70, no. 3 (July 2006): 703–42.

Schmalzer, Sigrid. *The People's Peking Man: Popular Science and Human Identity in Twentieth-Century China*. Chicago: University of Chicago Press, 2008.

Schmidt, Benjamin. "Mapping an Exotic World: The Global Project of Dutch Geography, circa 1700." *The Global Eighteenth Century*. Edited by Felicity Nussbaum. Baltimore, MD: Johns Hopkins University Press, 2003.

Schneider, Laurence A. "The Rockefeller Foundation, the China Foundation, and the Development of Modern Science in China." *Social Science and Medicine* 16, no. 12 (1982): 1217–21.

Schram, Stuart. *The Political Thought of Mao Tse-tung*. Harmondsworth, UK: Penguin Books, 1969.

Schrecker, John. *Imperialism and Chinese Nationalism: Germany in Shantung*. Cambridge, MA: Harvard University Press, 1971.

Schwartz, Benjamin. *In Search of Wealth and Power: Yen Fu and the West*. Cambridge, MA: Harvard University Press, 1964.

"Scientific Conditions in China" *Science* 100, no. 2590 (August 18, 1944): 141–42.

Secord, James A. *Controversy in Victorian Geology: The Cambrian-Silurian Debate*. Princeton NJ: Princeton University Press, 1986.

Seuss, Eduard. *Das Antlitz der Erde.* Prague: F. Tempsky, 1885–1909.
Shaw, Charles F. *The Soils of China.* Shanghai: The Geological Survey of China, 1931.
Shen, Grace Yen. "Taking to the Field: Geological Fieldwork and National Identity in Republican China." *Osiris* 24, no. 1 (January 1, 2009): 231–52.
Singh, Arvinder. "Understanding China's 'Go-West' Campaign," *China Report* 38, no. 1 (February 1, 2002): 121–27.
Smith, Anthony D. *Nationalism and Modernism: A Critical Survey of Recent Theoroies in Nations and Nationalism.* New York: Routledge, 1998.
Smith, Joanna F. Handlin. "Benevolent Societies: The Reshaping of Charity During the Late Ming and Early Ch'ing." *Journal of Asian Studies* 46, no. 2 (May 1, 1987): 309–37.
Spence, Jonathan. *The Search for Modern China.* New York: W. W. Norton Company, 1990.
Steinmetz, George. "'The Devil's Handwriting: Precolonial Discourse, Ethnographic Acuity, and Cross-Identification in German Colonialism." *Comparative Studies in Society and History* 45, no. 1 (May 2003): 41–95.
Stoecker, Helmuth. *Deutschland und China um 19. Jahrhundert.* Berlin: Rutten & Loening, 1958.
Strassberg, Richard. *Inscribed Landscapes: Travel Writing from Imperial China.* Berkeley: University of California Press, 1994.
Strauss, Julia. *Strong Institutions in Weak Polities: State Building in Republican China, 1927–1940.* Oxford: Clarendon Press, 1998.
Struve, Lynn. "Huang Zongxi in Context: A Reappraisal of His Major Writings." *Journal of Asian Studies* 47, no. 3 (August 1988): 474–502.
Suess, Eduard. *Das Antlitz der Erde.* Prague: F. Tempsky, 1885–1909.
Sun, Y. C. "Proceedings of the Second Annual Meeting." *BGSC* 3, no. 1 (1924): 3–4.
Sun Youli. *China and the Origins of the Pacific War, 1931–1941.* London: St. Martin's Press, 1993.
Swinhoe, Robert. "Notes on the Island of Formosa." *Journal of the Royal Geographical Society of London* 34 (1864): 6–18.
T'an, H. C., and C. Y. Lee. "Oil Fields in Szechuan Province." *Bulletin of the Geological Survey of China* 22 (1933): English 1–31, Chinese 1–38.
Tang Pei-sung. "Aspirations, Reality, and Circumstances: The Devious Trail of a Roaming Plant Physiologist." *Annual Review of Plant Physiology* 34, no. 1 (1983): 1–20.
———. "The Third General Meeting." *BGSC* 1, nos. 1–4 (1922): 37.
Thompson, Roger. "If Shanxi's Coal Is Lost, Then Shanxi Is Lost!: Shanxi's Coal and an Emerging National Movement in Provincial China, 1898–1908." *Modern Asian Studies* 45, no. 5 (2011): 1261–1288.
Thongchai Winichakul, *Siam Mapped: A History of the Geo-body of a Nation.* Honolulu: University of Hawaii Press, 1994.
Thorp, James. *Geography of the Soils of China.* Nanjing: National Geological Survey of China, 1936.
———. *Instructions to Chinese Soil Survey Field Parties.* Peiping: National Geological Survey of China, 1935.
Ting, V. K. "The Aims of the Geological Society of China." *BGSC* 1, nos. 1–4 (1922): 8.
———. "Chinese Students." *Westminster Review* 169 (January 1908): 48–55.
———. "Foreword." *Bulletin of the Geological Survey of China* 1 (July 1919), i.
———. "Mineral Resources of China." *Mining Magazine* 17 (1917): 188–90.
———. "Mining Legislation and Development in China" (Appendix II). In *Mineral Enterprise in China* by William W. Colling, 212–21. New York: Macmillan, 1918.

———. "Modern Science in China." *Asia* 36, no. 2 (February 1936): 131–34.

———. "Presidential Address." *BGSC* 3, nos. 1–4 (1924): 9–11.

Tirpitz, Alfred von. *My Memoirs.* Vol. 2. New York: Dodd, Mead & Co., 1919.

Townsend, James. "Chinese Nationalism." *Chinese Nationalism.* Edited by Jonathan Unger. Armonk, NY: M. E. Sharpe, 1996.

Tsien, Tsuen-hsuin. "Western Impact on China through Translation." *Far Eastern Quarterly* 13, no. 305 (1997): 305–27.

Tuan Yi-fu. *Topophilia: A Study of Environmental Perception, Attitudes, and Values.* New York: Columbia University Press, 1974.

Tyrrell, G. W. "Dr. V. K. Ting." *BGSC* 20 (1940): 369–70.

Van Dyke, Paul. *The Canton Trade: Life and Enterprise on the China Coast, 1700–1845.* Hong Kong: Hong Kong University Press, 2007.

Vaughan, T. Wayland. "Status of Areal Geological Mapping in the Pacific Region." *Proceedings of the First Pan-Pacific Scientific Conference, Honolulu, Hawaii, August 2 to 20, 1920.* Honolulu: Honolulu Star-Bulletin, 1921.

Walcott, Charles. "Cambrian Geology and Paleontology." *Smithsonian Miscellaneous Collections* 57 (1914):14.

Waley-Cohen, Joanna. *The Culture of War in China: Empire and the Military under the Qing Dynasty.* London: I. B.Tauris, 2006.

———. "The New Qing History." *Radical History Review* 88 (Winter 2004): 193–206.

———. *The Sextants of Beijing: Global Currents in Chinese History.* New York: W. W. Norton, 1999.

Wang, C. C. "On the Oil Geology of N. Shensi." *Bulletin of the Geological Survey of China* 12, no. 20 (1933): English 65–82, Chinese 45–58.

Wang, Chung Yu. "Science in China." *Science* 98, no. 2548, new ser. (October 29, 1943): 376–78.

Wang Hongzhen et al. "Development of Geological Science in China in Twentieth Century." *Journal of China University of Geosciences* 11, no. 3 (September 2000): 197–203.

Wang Hui (Howard Y. F. Choy, trans.). "The Fate of 'Mr. Science' in China: The Concept of Science and Its Application in Modern Chinese Thought." *positions* 3, no. 1 (Spring 1995): 1–68.

Wang, W. H. (翁文灝). "Geology of China, The." *Encyclopaedia Sinica.* Edited by Samuel Couling. London: Oxford University Press, 1917.

Wang, Y. C. *Chinese Intellectuals and the West, 1872–1949.* Chapel Hill, NC: University of North Carolina Press, 1966.

Wang, Zuoyue. "Saving China Through Science: The Science Society of China, Scientific Nationalism, and Civil Society in Republican China." *Osiris* 17 (2002): 291–322.

Ward, Julian. *Xu Xiake. 1587–1641: The Art of Travel Writing.* Richmond, UK: Curzon, 2001.

Warwick, Andrew. "Exercising the Student Body: Mathematics and Athleticism in Victorian Cambridge." *Science Incarnate: Historical Embodiments of Natural Knowledge,* edited by Christopher Lawrence and Steven Shapin, 288–36. Chicago: University of Chicago Press, 1998.

Wataru Masuda. *Japan and China: Mutual Representations in the Modern Era.* Translated by Joshua Fogel. New York: St. Martin's Press, 2000.

Weller, J. M., F. A. Sutton, and C. C. Sun. "Geological Reconnaissance in Kansu and Ching-Hai Provinces, Northwest China." MSS, 1938. National Geological Library.

Weston, Timothy. *The Power of Position: Beijing University, Intellectuals, and Chinese Political Culture, 1898–1929.* Berkeley: University of California Press, 2004.

Wilkinson, Endymion. *Chinese History: A Manual.* Cambridge, MA: Harvard University Press, 2000.

Willis, Bailey. *Friendly China: Two Thousand Miles Afoot among the Chinese.* Stanford, CA: Stanford University Press, 1949.

———. "Memorial of Pumpelly, Raphael." *Bulletin of the Geological Society of America* 36 (1925): 45–84

Willis, Bailey, et al. *Research in China.* Washington, DC: Carnegie Institution of Washington, 1907–1913.

Wissmann, Hermann von. "Diequartäre Vertgletscherung in China." *Zeitschrift der Gesellschaft für Erdkunde zu Berlin,* nos. 7/8: 241–62.

———. "The Pleistocene Glaciation in China." *BGSC* 17, no. 1 (March 1937): 160.

Wong, Parkin. "The Mineral Resources of China." MA thesis, Cornell University, 1914.

Wong, R. Bin. *China Transformed: Historical Change and the Limits of European Experience.* Ithaca, NY: Cornell University Press, 1997.

———. "Two Kinds of Nation, What Kind of State?" *Nation Work: Asian Elites and National Identities,* edited by Timothy Brook and Andre Schmid. Ann Arbor: University of Michigan Press, 2000.

Wong Wen-hao. "Biographical Note of V. K. Ting." *BGSC* 16, no. 1 (1937): i–xi.

———. "Contribution à l'étude de la porphyrite quartzifère de Lessines." *Mémoires de l'Institut géologique de l'Université de Louvain,* t. 1 (1913): 298–325.

———. "Mineral Resources of China. "*Memoirs of the Geological Survey of China,* ser. B, vol. 1 (1919).

———. "Richthofen and Geological Work in China." *BGSC* 12, nos. 1–2 (1933): 311–14.

Wong, W. H. "L'influence seismogenique de certaine structures geologiques en Chine." In the Congres geol. Int. 13me Session, Belgique, *Comptes Rendus,* fasc. 2: 1161–97.

———. "Proceedings of the 5th Annual Meeting of the Geological Society of China." *BGSC* 6, no. 1 (1927): 1–7.

Working Group of INHIGEO Japan Members. "Special Issue: An Introduction to the History of Geological Sciences in Japan." *JAHIGEO Newsletter* 13 (May 2011): 2–26.

Wright, David. *Translating Science: The Transmission of Western Chemistry into Late Imperial China, 1840–1900.* Boston: Brill, 2000.

———. "The Translation of Modern Western Science in Nineteenth-Century China, 1840–1895." *Isis* 89, no. 653 (1998): 653–73.

Wright, Tim. *Coal Mining in China's Economy and Society, 1895–1937.* Cambridge: Cambridge University Press, 1984.

Wu, Shellen Xiao. "Underground Empires: German Imperialism and the Introduction of Geology in China, 1860–1919." PhD diss, Princeton University, 2010.

Wu, T'ien-wei. "Contending Political Forces." In *China's Bitter Victory: The War with Japan, 1937–1945,* edited by James C. Hsiung and Steven I. Levine, 51–78. Armonk, NY: M. E. Sharpe, 1992.

Xu Xiaoqun. *Chinese Professionals and the Republican State: The Rise of Professional Associations in Shanghai, 1912–1937.* Cambridge: Cambridge University Press, 2001.

Yamasaki, N. "Geological Mapping in Japan, Korea and China." *Proceedings of the First Pan-Pacific Scientific Conference.* Honolulu: Honolulu Star-Bulletin, Ltd., 1921.

Yang, Chen Ning, and Laurence A Schneider. "Oral History Interview with Chen Ning Yang." Niels Bohr Library & Archives, American Institute of Physics, College Park, MD, 1980. http://www.aip.org/history/ohilist/4984.html.

Yang, Jing Yi, and David Oldroyd. "The Introduction and Development of Continental

Drift Theory and Plate Tectonics in China: A Case Study in the Transference of Scientific Ideas from West to East." *Annals of Science* 46 (1989): 21–43.

Yang Tsui-hua. "The Development of Geology in Republican China, 1912–1937." In *Science and Medicine in Twentieth-Century China: Research and Education*, edited by John Z. Bowers, J. William Hess, and Nathan Sivin, 65–89. Ann Arbor: University of Michigan Press, 1989.

———. "Geological Sciences in Republican China, 1912–1937." PhD diss., State University of New York at Buffalo, 1985.

Yeh, Wen-Hsin. *The Alienated Academy: Culture and Politics in Republican China, 1919–1937*. Cambridge, MA: Harvard University Press, 1990.

Yochelson, Ellis. *Charles Doolittle Walcott, Paleontologist*. Kent, OH: Kent State University Press, 1998.

Young, T. I. (楊遵儀). *Bibliography of Chinese Geology up to 1934* (中國地質文獻目錄). Peiping: National Academy of Peiping, 1935.

Zhang Yufa. "Returned Chinese Students from America and the Chinese Leadership (1846–1949)." *Chinese Studies in History* 35, no. 3 (April 2002): 52–86.

Zheng Yangwen. "The 'Peaceful Rise of China' after the 'Century of Unequal Treaties': Will History Matter?" *Negotiating Asymmetry: China's Place in Asia*. Edited by Anthony Reid and Zheng Yangwen. Honolulu: University of Hawaii Press, 2009.

Zhongyang yanjiuyuan chengli wushi zhounian jinian lunwenji 中央研究院成立五十周年紀念論文集 (Collected Papers Commemorating the Fifteeth Anniversary of the Founding of the Academia Sinica). Taibei: Zhongyang yanjiuyuan, 1978.

Zhu, Dajian, and Homer Le Grand. "Plates, Politics and Localism: Geological Theory in China." *History and Anthropology* 11, nos. 2–3 (1999): 291–327.

Zimmer, Oliver. "In Search of Natural Identity: Alpine Landscape and the Reconstruction of the Swiss Nation." *Comparative Studies in Society and History* 40, no. 4 (October 1998): 637–65.

Zi Zhongyun. "The Rockefeller Foundation and China" *American Studies in China* 2 (1995). http://www.mgyj.com/american_studies/1995e/9505.txt.

INDEX